# Internal Migration in the Developed World

The frequency with which people move home has important implications for national economic performance and the well-being of individuals and families. Much contemporary social and migration theory posits that the world is becoming more mobile, leading to the recent 'mobilities turn' within the social sciences. Yet, there is mounting evidence to suggest that this may not be true of all types of mobility, nor apply equally to all geographical contexts. For example, it is now clear that internal migration rates have been falling in the USA since at least the 1980s. To what extent might this trend be true of other developed countries?

Drawing on detailed empirical literature, *Internal Migration in the Developed World* examines the long-term trends in internal migration in a variety of more advanced countries to explore the factors that underpin these changes. Using case studies of the USA, UK, Australia, Japan, Sweden, Germany and Italy, this pioneering book presents a critical assessment of the extent to which global structural forces, as opposed to national context, influence internal migration in the Global North.

*Internal Migration in the Developed World* fills the void in this neglected aspect of migration studies and will appeal to a wide disciplinary audience of researchers and students working in Geography, Migration Studies, Population Studies and Development Studies.

**Tony Champion** is Emeritus Professor of Population Geography at Newcastle University UK. His research interests include migration and its impact on population distribution in the Developed World, with particular reference to counter-urbanisation and city resurgence. He was President of the British Society for Population Studies in 2013–2015.

**Thomas Cooke** is a population and urban geographer and Professor in the Department of Geography at the University of Connecticut, USA. His research focuses on internal migration decline in the United States, the spatial distribution of metropolitan poverty and the family dimension of migration behaviour. He is currently an Editor of *Urban Geography*.

**Ian Shuttleworth** is Senior Lecturer in Human Geography at Queen's University Belfast, UK. His research interests include migration, labour market mobility and social segregation. He also has an interest in divided societies with a special focus on Northern Ireland. He is currently director of the Northern Ireland Longitudinal Study Research Support Unit.

# International Population Studies
Philip Rees

This book series provides an outlet for integrated and in-depth coverage of innovative research on population themes and techniques. International in scope, the books in the series cover topics such as migration and mobility, advanced population projection techniques, microsimulation modeling, life course analysis, demographic estimation methods and relationship statistics. The series includes research monographs, edited collections, advanced level textbooks and reference works on both methods and substantive topics. Key to the series is the presentation of knowledge founded on social science analysis of hard demographic facts based on censuses, surveys, vital and migration statistics. All books in the series are subject to review.

Full details of books published in the series can be found at:
https://www.routledge.com/International-Population-Studies/book-series/ASHSER-1353

**Minority Internal Migration in Europe**
*Edited by Nissa Finney and Gemma Catney*

**Demography at the Edge**
Remote Human Populations in Developed Nations
*Edited by Dean Carson, Rasmus Rasmussen, Prescott Ensign, Lee Huskey and Andrew Taylor*

**Geographies of Ageing**
Social Processes and the Spatial Unevenness of Population Ageing
*Amanda Davies and Amity James*

**The Routledge Handbook of Census Resources, Methods and Applications**
Unlocking the 2011 Census
*Edited by John Stillwell*

**Internal Migration in the Developed World**
Are We Becoming Less Mobile?
*Edited by Tony Champion, Thomas Cooke and Ian Shuttleworth*

# Internal Migration in the Developed World

## Are We Becoming Less Mobile?

**Edited by Tony Champion,
Thomas Cooke and Ian Shuttleworth**

Routledge
Taylor & Francis Group

LONDON AND NEW YORK

First published 2018
by Routledge

2 Park Square, Milton Park, Abingdon, Oxfordshire OX14 4RN
52 Vanderbilt Avenue, New York, NY 10017

*Routledge is an imprint of the Taylor & Francis Group, an informa business*

First issued in paperback 2019

*British Library Cataloguing-in-Publication Data*
A catalogue record for this book is available from the British Library

*Library of Congress Cataloging-in-Publication Data*
A catalog record for this book has been requested

ISBN: 978-1-4724-7806-1 (hbk)
ISBN: 978-0-367-24526-9 (pbk)

Typeset in Times New Roman
by codeMantra

# Contents

vi  *Contents*

# List of figures

# List of tables

# List of contributors

**Martin Bell** is Emeritus Professor and a former Director of the Centre for Population Research at the University of Queensland. He has written extensively on trends and patterns of migration in Australia and on methods and techniques of measuring internal migration, particularly in the context of making cross-national comparisons.

**Corrado Bonifazi** is Director of the Institute for Research on Population and Social Policies of the Italian National Research Council. His research interests are the dynamics of migration at the international level, migration policies, foreign immigration in Italy, Italian emigration, internal migration and dynamics of urbanization.

**Tony Champion** is Emeritus Professor of Population Geography at Newcastle University. His research interests include migration and its impact on population distribution in the Developed World, with particular reference to counter-urbanisation and city resurgence. He was President of the British Society for Population Studies in 2013–2015.

**Elin Charles-Edwards** is Lecturer in the School of Geography, Planning and Environmental Management at the University of Queensland. Her research interest lies in the study of human spatial mobility at a variety of temporal and spatial scales, ranging from diurnal travel in cities to permanent internal and international migration.

**Thomas Cooke** is a population and urban geographer and Professor in the Department of Geography at the University of Connecticut. His research focuses on internal migration decline in the United States, the spatial distribution of metropolitan poverty and the family dimension of migration behaviour. He is currently an Editor of *Urban Geography*.

**Tony Fielding** is Research Professor of Human Geography in the School of Global Studies at the University of Sussex. He lived in Japan for one year followed by three shorter stays of about six months each over the period 1994–2007, during which time he taught courses in the Faculties of Economics, Letters and International Relations at Ritsumeikan and Kyoto Universities.

**William H. Frey** is Senior Fellow with the Metropolitan Policy Program at the Brookings Institution in Washington, DC, and Research Professor with the Population Studies Center and Institute for Social Research at the University of Michigan. He is a demographer and sociologist specializing in U.S. demographics and has written widely on urban populations, migration, immigration, race, aging, political demographics and the U.S. Census.

**Anne Green** is Professor of Regional Economic Development at City REDI, Birmingham Business School, University of Birmingham. A geographer by background, her research interests span employment, non-employment, regional and local labour market issues, migration and commuting and associated policy issues.

**Keith Halfacree** is Reader in Human Geography at Swansea University with specific interests in Cultural, Population and Rural Geography. Particular interests include the causes, scope, dimensions and consequences of migration into more rural areas of diverse types of people originating in more urban locations, as well as the cultural representations and embodied experiences of rurality and nature.

**Frank Heins** is Senior Researcher at the Institute for Research on Population and Social Policies of the Italian National Research Council. His research interests are in the territorial aspects of demographic and socio-economic structures and dynamics, and in migration, both internal and international.

**Nik Lomax** is University Academic Fellow in Spatial Data Analytics at the School of Geography, University of Leeds. His research interests are in the measurement and analysis of spatial mobility with a focus on internal and international migration patterns and their integration into local area population estimates and projections.

**Thomas Niedomysl** is Associate Professor and Senior Lecturer at the Department of Human Geography at Lund University and also affiliated to the Centre for Innovation, Research and Competence in the Learning Economy (CIRCLE). His research interests are in the field of population and regional development, especially internal migration in Sweden but also international migration, place marketing and economic development.

**John Östh** is Senior Lecturer in Geography and GIS at the Department of Social and Economic Geography at Uppsala University. His research is oriented towards methods and tools development, population geography, accessibility and mobility and he administers the PLACE research database held by Uppsala University.

**Nikola Sander** is Research Director at the German Federal Institute for Population Research (BIB). Prior to joining the BIB in 2017, she was Assistant Professor of Population Geography at the University of Groningen, where most of the work for this volume was carried out. Her research interests are in internal and international migration, mobility and population projections. She has also been active in creating data visualisations for exploratory data analysis and communicating research to educators, policy makers and the general public.

**Ian Shuttleworth** is Senior Lecturer in Human Geography at Queen's University Belfast. His research interests include migration, labour market mobility and social segregation. He also has an interest in divided societies with a special focus on Northern Ireland. He is currently director of the Northern Ireland Longitudinal Research Support Unit.

**John Stillwell** is Professor of Migration and Regional Development in the School of Geography, University of Leeds, and is currently Director of UK Data Service Census Support. He has longstanding research interests in internal migration and was co-investigator on the recently completed IMAGE (Internal Migration Around the GlobE) project.

**Enrico Tucci** is Senior Researcher and Demographer at the Italian National Institute of Statistics. Since 2005 he has been working on international cooperation projects and programmes as an expert on migration data management and analysis.

**Philipp Ueffing** is a demographer at the United Nations Population Division. He was formerly Research Associate in the Queensland Centre for Population Research at the University of Queensland. He has specialist expertise in migration data analysis, geographical information systems and population projections.

**Tom Wilson** is Principal Research Fellow in the Northern Institute at Charles Darwin University, Australia. He received his PhD from the University of Leeds and has subsequently worked in the academic, private and government sectors, focusing particularly on demographic forecasting models, Indigenous demography, migration and local and regional population change.

# The book and the series

This book is the second in the International Population Studies series to be published by Routledge and provides an excellent analysis of internal migration across a set of countries, adding to insights available in previous volumes in the series that have dealt with other aspects of internal migration (Finney and Catney, 2012, and Smith *et al.*, 2015). The book has been edited by three academics who are experts in internal migration analysis. Tony Champion is renowned for his work on the turnaround from urbanisation to counter-urbanisation in the 1970s and his analyses of the subsequent urban population resurgence. Thomas Cooke has been influential in identifying for the United States a recent, systematic decline in internal migration at many spatial scales and testing alternative explanations. Ian Shuttleworth is Northern Ireland's "go-to" expert on migration and demography and an expert in use of longitudinal data sets, playing a leading role in development and use of the Northern Ireland Longitudinal Study (NILS).

The book seeks to learn whether the phenomenon of the decline in internal migration, first manifest in the USA and studied by Cooke, is now prevalent across the developed world. To do this, the editors persuaded a team of authors—experts in internal migration trends and patterns in their own countries—to join them in testing for the universality of internal migration decline. Each country study addresses the same general research questions but reveals a rich and varied set of answers. Importantly, the book also discusses how to address the challenges of producing comparable internal migration measures by controlling for particularities of national geographies and data sets, as pioneered by Martin Bell and John Stillwell, investigators who with colleagues have put together the IMAGE repository of migration data and comparable measures for countries across the world that are home to 80 per cent of humanity. I won't steal the book's thunder by revealing the conclusions they came to, except to say that I hope you will find the methodological and case study chapters as fascinating as I did when reading the drafts as Series Editor.

Philip Rees
University of Leeds, UK
March 2017

# References

Finney, N. and Catney, C. (ed) 2012. *Minority Internal Migration in Europe.* Farnham: Ashgate.

Smith, D.P., Finney, N., Halfacree, K. and Walford, N. (eds) 2015. *Internal Migration: Geographical Perspectives and Processes.* Farnham: Ashgate.

# Book editors' preface

Every so often something happens that surprises or even shocks. In the arena of politics, the year 2016 saw the referendum vote in favour of the UK leaving the European Union (EU) and the emergence of a populist as the next US President. Being taken by surprise is perhaps less common in social science and especially demography. Nevertheless, fertility has wrong-footed the experts on several occasions, notably with the post-war baby boom confounding the population projections that assumed a continuation of the low birth rates of the 1930s and then with the higher projections of the 1960s being undermined by the onset of the second demographic transition.

Migration, too, has the capacity to surprise. This is not just in terms of international movements like the unexpectedly large volume of labour migration from the new EU member states to the UK from 2004 and the recent increase in numbers of refugees fleeing civil strife and natural disaster in Africa and the Middle East. It is also the case for internal migration. Witness the discovery of counterurbanisation in the 1970s—initially met with disbelief but then recognised as an obvious development (Champion, 1989), only to be upstaged by the recognition of a back-to-the-city movement as London and other major cities rebounded after several decades of population decline.

The topic of this book is another example of a largely unanticipated turn of events in our understanding of internal migration. As noted in Chapter 1 of this book, the notion of society being increasingly mobile is a firmly entrenched one—so much so that the indications of a sustained decline in migration rates, first observed in the USA as for counter-urbanisation, were originally met with considerable scepticism, if not downright incredulity. After all, Castles and Miller (1994) had depicted the current era as 'the age of migration' which, although referring primarily to international movements linked to globalisation, might also be expected to increase within-country migration. Moreover, one of Lee's (1966) 'laws of migration', extending Ravenstein's seminal work of the 1880s, was that migration is a cumulative process that builds on itself. So were the data from the US Bureau of the Census inaccurate or, if not, was the downturn in frequency of address-changing the result of short-term 'period' factors like those that were eventually shown to have made the 1970s into (just) the 'decade of counter-urbanisation' or a result of longer-term structural changes that extended over several decades? And was the decrease seen in the USA only?

By early in the new millennium, the evidence had mounted sufficiently to confirm the reality of the decline and to show that it extended over more than one decade and was thus unlikely to be a consequence of short-term period effects. Wolf and Longino (2005) were able to lament that the USA was still in denial about a very real long-term societal development which threatened the future wellbeing of a nation built on a tradition of high mobility that allowed economic opportunities to be realised wherever they occurred across the country. Perhaps the biggest wake-up call for the academic community came with Cooke's (2011) analysis of the decline in US inter-county migration rates over the decade 1999–2009 and his follow-up study of 2013. Even after adjusting for methodological changes in the CPS time series and allowing for the effect of recession after the Global Financial Crisis, Cooke noted a significant rise in 'secular rootedness' that continued a trend dating from the 1980s. Then the Census Bureau pronounced 2014 as the year with the lowest annual migration rate recorded since the start of annual recording in 1947.

This is the context from which this book has derived. The immediate stimulus came from a project that aimed to see whether the UK was experiencing any similar tendency towards reduced levels of internal migration (Champion and Shuttleworth, 2016a, 2016b). Its review of the worldwide literature on migration intensity trends discovered a relative dearth of studies taking a long-term perspective that could allow for fluctuations caused by period effects such as the business cycle, while Wright and Ellis (2016) lament the lack of recent attention to internal migration in its entirety. This led to the idea of commissioning a set of in-depth national case studies that would each do their best to piece together the address-changing record for their individual countries, drawing on the available statistical material for as long a period as the data were considered reliable. The sample of countries was selected so as to cover a range of geographical and historical contexts, but was limited to those parts of the world which had passed through their (first) demographic transition by the middle of the twentieth century and focused on the countries deemed most likely to be following the US experience.

Inevitably, the country choice was also strongly influenced by the pool of internal migration researchers willing to participate in this endeavour. It certainly was not a task for the faint-hearted, as had already become evident in the project on the UK, which in the end had to settle for dealing with just England and Wales and even then encountered multiple difficulties in assembling robust time series that covered several decades. The scale of the challenge was also becoming clear at that time through the experience of a project aiming to compile a global inventory of national datasets on internal migration known as IMAGE (Internal Migration Around the GlobE, see www.imageproject.com.au). This project was successful in obtaining datasets from the vast majority of the 193 UN member states, this achievement serving to underline just how much the nature of the datasets and the types of migration covered by them varied from country to country.

Against the background, this book is an exercise in the art of the possible. The authors of the seven country case-study chapters in Part 2 of the book were instructed to confront head-on the challenges of assembling the requisite data and to be open about how much confidence to place in the conclusions that they drew from these. Underlining such cautionary notes, the book includes as part of the background material a chapter devoted to the methodological challenges—which should be compulsory reading for anyone thinking of embarking on a similar study of their own country (Chapter 3). Part 1 of the book also contains a chapter that, alongside presenting the latest results from the IMAGE project's analysis of temporal trends, also provides examples of how such challenges are best addressed (Chapter 4). Before these, Chapter 2 also demonstrates how difficult it can be to pin down the exact causes of any trends in overall migration intensity, citing the multiplicity of changing conditions that can be expected to exert either an upward or a downward effect on migration rates.

Nor does the concluding part of the book shirk from the main message that this is still very much work in progress. The contributors of the two commentary chapters were encouraged to take a critical look at the evidence presented in the case study chapters and to list what they consider to be the main items of unfinished business. Similarly, in the final chapter we attempt to synthesise the central findings of the case studies. Whilst there are enough commonalities between countries to suggest that there are common forces leading to falling internal migration across the developed world, there is also sufficient divergence to indicate that not all countries are following the same path and that there may be different migration drivers/responses operating in different national contexts. Uncertainty remains about what has caused migration to fall in most countries (but not all), but this offers considerable scope for the development of a research agenda.

As just indicated, a huge amount of hard work has gone into getting this far towards improving our understanding of long-term trends in migration rates in the developed world. We are particularly indebted to the contributing authors for being prepared to engage in this endeavour and put so much effort into helping to improve our understanding of this relatively neglected aspect of migration studies. We were very glad when Katy Crossan and the Board of Ashgate accepted our book proposal, and—following its transfer to Routledge—wish to thank Faye Leerink and Priscilla Corbett for taking the book through its production stages. Finally, special thanks to Series Editor Philip Rees for commending the book proposal in the first place and for all his subsequent support, encouragement and constructive criticism. There were moments when we felt that we had bitten off more than we could chew, but Phil kept us on track. Needless to say, any remaining deficiencies in the book are entirely the editors' responsibility.

It is a fervent hope that our initiative in compiling this book will lead to further efforts not only in the seven countries featured in it but also in undertaking similar research on other countries, including those which have emerged from their demographic transition in more recent decades but are now subject to a similar set of demographic, economic, social and technological changes—in other words, period effects. A larger sample of countries would help to solve the key question of whether those sorts of changes are leading to a general downturn in the frequency of address-changing along the lines of the US experience not only in high-income countries but also in lower-income countries or whether there are uneven impacts as mediated through the distinctive geographical, social, cultural and regulatory environments of individual countries and regions. This will be a major task but it is one that is worthwhile pursuing because of the importance of internal migration in shaping national economic, social and cultural life, and in the light of the evidence presented in the book that suggests that migration rate increases are neither inevitable nor universal.

Tony Champion, Newcastle University, UK
Thomas Cooke, University of Connecticut, USA
Ian Shuttleworth, Queen's University Belfast, UK
March 2017

## References

Castles, S. and Miller, M.J. 1994. *The Age of Migration*. Basingstoke: Palgrave Macmillan.

Champion, A.G. 1989. *Counterurbanization: The Changing Pace and Nature of Population Deconcentration*. London: Edward Arnold.

Champion, T. and Shuttleworth, I. 2016a. Is longer-distance migration slowing? An analysis of the annual record for England and Wales. *Population, Space and Place*, published online in Wiley Online Library. doi:10.1002/psp.2024

Champion, T. and Shuttleworth, I. 2016b. Are people moving address less? An analysis of migration within England and Wales, 1971–2011, by distance of move. *Population, Space and Place*, published online in Wiley Online Library. doi:10.1002/psp.2026

Cooke, T.J. 2011. It is not just the economy: Declining migration and the rise of secular rootedness. *Population, Space and Place*, 17(3), 193–203.

Cooke, T.J. 2013. Internal migration in decline. *The Professional Geographer*, 65(4), 664–675.

Lee, E.S. 1966. A theory of migration. *Demography*, 3, 47–57.

Wolf, D.A. and Longino, C.F. 2005. Our 'increasingly mobile society'? The curious persistence of a false belief. *The Gerontologist*, 45, 5–11.

Wright, R. and Ellis, M. 2016. Perspectives on migration theory: Geography. In White, M.J. (ed) *International Handbook of Migration and Population Distribution*. Dordrecht: Springer, 11–30.

# Part I
# Setting the scene

# 1 Introduction

## A more mobile world, or not?

*Tony Champion, Thomas Cooke
and Ian Shuttleworth*

Much of social and migration theory seems to be wedded to the assumption that the world is becoming more mobile, with migration increasing over the long term as an adjunct of economic and social development. As recounted in more detail below, the literature contains frequent mentions of our now living in 'the age of migration', 'a world on the move' and witnessing the emergence of a 'hypermobility' that is eroding attachment to place and creating a 'liquid modernity' that is leading to the end of territory and place. Thus, there is a need to refocus from territorial understandings of space to an understanding based on networks and spaces of flows. The notion that modernisation is associated with an increasingly footloose society has a long currency in migration theory and related social science that can be traced back to the 'laws of migration' set out in the latter half of the nineteenth century and that were placed on a more formal footing nearly a century later. Such ideas have underpinned the rise of the New Mobilities Paradigm (NMP) and the associated 'mobilities turn' that has recently influenced the social sciences in the current era of globalisation.

Yet there is mounting evidence that this belief in increasing mobility does not apply to all types of mobility nor to all geographical contexts. In particular, it is now clear that migration rates have been falling in the USA, and not just due to the so-called Great Recession that was sparked by that country's sub-prime mortgage crisis of 2007, but instead dating back to at least the 1980s. This observation has stimulated researchers in other parts of the world to examine the experience of their own countries in order to discover whether they have followed the US trend. Simultaneously, new insights have been emerging from a newly established worldwide inventory of internal migration data known as the IMAGE project (see below), indicating that several other UN members besides the USA have in recent years seen some reduction in people's propensity to move homes. Additionally, a closer examination of the theoretical literature reveals a rather more nuanced perspective than a deterministic relationship between migration and development.

It is this divergence between what appears to be conventional theoretical expectation and what is now emerging from the empirical migration literature that provides the main stimulus for this book. As outlined in the preface, the editors' interest was prompted by their finding that the UK, another member of the Anglo-American world, did not appear to be experiencing any long-term decline in between-area migration rates similar to the USA. Why should this be the case? Have the many similarities between the USA and the UK in terms of demographic, economic, social and technological changes not been reflected in similar patterns of change in human behaviour? In other words, have what appear to be general social and economic developments across advanced economies associated with globalisation not—so far at least—been able to erode the role of national differences in governance, regulation, institutional arrangements and cultural attitudes? Hence the central purpose of this book to document long-term trends in internal migration in a variety of more advanced countries and explore the factors behind these changes.

The remainder of this chapter sets out this context and rationale in more detail. It begins by reviewing the theoretical context, describing what the current literature suggests about whether migration rates should be expected to be rising or falling. Next, the latest statistical evidence is presented on national rates of address-changing, looking cross-sectionally at the differences between countries before tracing trends over time and distinguishing between types of migration in terms of distance of movement. The rather complex and confusing picture that results from these accounts then forms the basis for selecting the research questions addressed by the book and the methodology adopted for it, especially in terms of the use of the case-study approach and of the reasons for choosing the particular countries that are put under the spotlight. The final section describes the structure of the rest of the book and outlines the role of each subsequent chapter.

## Theoretical expectations

As mentioned at the outset, there appears to be a general acceptance across the social sciences that we are living in an era of rapidly increasing mobility. In this section, this statement is substantiated and its underpinnings are examined before looking at the reactions to the observation that US internal migration rates have declined substantially in recent years. In particular, this review reveals that, despite the widely held belief that modernisation in the form of economic and social development leads to progressively higher levels of population movement, some theoretical frameworks do allow the possibility of a reduction over time in the frequency with which people move homes. In the broadest terms, this possibility involves two lines of reasoning. One is the expectation that overall intensities of migration will fall due to a shift in the composition of the population towards less migratory groups resulting from the demographic transition process, most notably through the rising share of older people with their lower than average rates of migration. The other is that a variety of factors are working to reduce the desire or the ability to move homes, with this applying to all types of people but to varying extents according to their differential susceptibility to such influences.

Looking first at mainstream social theory, common discourses nowadays refer to the concept of 'modernity' and its role in generating a 'hypermobility' that leads to 'dislocation', resulting in a reduction of people's attachment to place and a switch of emphasis to the primacy of flows and networks (Urry, 2007; see also Adey, 2010). Theories of a 'liquid modernity' (Bauman, 2000) are linked to 'de-territorialisation' processes that spell the end of nation states as containers for societies and help to convey the impression that we now live in 'the age of migration' (Castles *et al.*, 2014). Globalisation theory identifies mobility, migration and related population flows as being essential to the constitution of the 'global' (Robins, 2000). However, this perhaps exaggerates the extent to which nation states have ceased to matter. For example, whilst their borders are porous to flows of capital and goods, they are far less open to flows of people despite the rhetoric of globalisation (Shuttleworth, 2007). Indeed, international migration, and the political need to restrict it, has become a theme in the second decade of the twenty-first century, which has led pressures to close national borders to migrants. At the same time, there is empirical evidence that international migration rates are decreasing (Abel and Sander, 2014)—perhaps another case where conventional expectations (and social theory) do not match the data.

These ideas about globalisation and de-territorialisation have been embraced by the NMP (Sheller and Urry, 2006), according to whom 'All the world seems to be on the move' (p. 207). In this context, Cresswell (2006, p. 15) emphasises the centrality of mobility within modernity, highlighting Sennett's (1994) quotation that 'the modern individual is, above all else, a mobile being'. Hannam *et al.* (2006, p. 2) make clear the pervasiveness of this development: 'The global order is increasingly criss-crossed by tourists, workers, terrorists, students, migrants, asylum-seekers, scientists/scholars, family members, business people, soldiers, guest workers and so on.' They go on to point to the multiple and intersecting nature of these mobilities and to the way in which they seem to be producing a more networked patterning of economic and social life. As a result, a 'mobilities turn' is spreading into and transforming the social sciences, not only placing new issues on the table, but also transcending disciplinary boundaries. It seems that a new paradigm is being formed within the social sciences. This is the justification that was used for the 2006 launch of the journal *Mobilities*, designed to 'address this emerging attention to many different kinds of mobilities, both by those engaged in practising and regulating diverse mobilities and by those involved in researching and understanding present-day and historical mobilities' (Hannam *et al.*, 2006, pp. 1–2).

These ideas of accelerating mobility are not new. They build on a long tradition in the literature on migration and related social science, extending back to Ravenstein's pair of classic papers on 'The laws of migration' published in the 1880s. In the second of these, he specifically raises the question

'Does migration increase?' and answers it in the affirmative: 'Wherever I was able to make a comparison I found that an increase in the means of locomotion and a development of manufactures and commerce have led to an increase in migration' (Ravenstein, 1889, p. 288). Viewed in the historical context of industrialisation in nineteenth century Europe, this opinion is unsurprising. Three quarters of a century then elapsed before the subject of migration intensity was next addressed in a systematic manner by Lee (1966) in his refinement and extension of Ravenstein's ideas. In his 'theory of migration', Lee set out several hypotheses relating to the volume of migration, two of which relate to changes in migration rates and numbers: 'Unless severe checks are imposed, both volume and rate of migration tend to increase with time' and 'The volume and rate of migration vary with the state of progress in a country' (Lee, 1966, pp. 53–54). His rationale was primarily four-fold: 'Industrialisation and Westernisation ... increase the diversity of areas', people become more diverse occupationally, developments in transport reduce the barriers posed by intervening obstacles and the act of migration itself makes a person more likely to make a further move. He concludes: 'We may argue that a high rate of progress entails a population which is continually in a state of flux, responding quickly to new opportunities and reacting swiftly to diminishing opportunities' (Lee, 1966, p. 54).

This mooted association between migration intensity and development level was further codified by Zelinsky (1971) in his 'hypothesis of the mobility transition'. Drawing primarily on Lee but also 'join[ing] together a number of ideas already immanent in the literature' (p. 221), he summarised his central thesis thus: 'There are definite, patterned regularities in the growth of personal mobility through space-time during recent history, and these regularities comprise an essential component of the modernisation process.' (pp. 251–252). He specifically designed his 'mobility transition' to sit alongside the demographic transition model, which he relabelled as the 'vital transition' with its two components of the epidemiological and fertility transitions, and suggested that 'A high degree of interaction may exist among all the processes in question' (p. 222). As such, it is a stage-based developmentalist model that delineates a shift from the situation of 'little genuine residential migration' characteristic of pre-modern traditional society through the early and late transitional phases to the advanced society. By this latter stage, 'For the individual migrant, ... one can postulate a lifetime cycle of residential shifts, along with an elaborate schedule of circulatory trips' (Zelinsky, 1971, pp. 245–246). This is 'a state in which the term "sedentary" no longer appears apposite' and people 'frequently migrate in the sense of formal change of residence' (p. 247). His final axiom is that: 'Such evidence as we have indicates an irreversible progression of stages' (p. 222), although he does suggest a final stage where address-changing decreases and is substituted by other types of spatial mobility.

Zelinsky's idea of the irreversibility of the mobility transition is a very logical one in the context of the time when he was writing and, as he said, in terms of the available evidence. For one thing, almost all the contemporary evidence was cross-sectional in nature across a range of societies at different levels of development rather than monitoring temporal trends. This allowed Zelinsky to demonstrate that lifetime migration was much higher in the USA than any other country for which he could find statistics, with his Table II showing that just 25.7% of its population were living in the city or county of birth in 1958. This is much lower than his next country, Switzerland, with 44.3% living in their commune of birth in 1941. All his other examples had levels of above 50% for their own smallest reporting areas, with over 90% for the regions of the USSR in 1926 and the prefectures of Taiwan in 1930. Even here, however, Zelinsky (1971, p. 237) urges caution, saying that 'Table II must be taken with a grain of salt', not just because of the wide variation in the dates of the observations but more importantly because of the 'non-equivalence among reporting units'. He is fully aware of Ravenstein's (1885) first law that states that most migration takes place over short distances, meaning that 'the chance that a migration will be noted rises as the size of the areal unit decreases' (p. 237)—a good example of what has since become known as the Modifiable Area Unit Problem (MAUP), which has traditionally bedevilled comparative analysis of migration but no longer does so, as will be shown in Chapter 3 of this book.

With regard to evidence on trends in migration over time, the conclusion that Zelinsky reaches from a review of a variety of studies and data relating to each of the four phases that many countries had by then experienced is a broad one: 'each phase represents a major gain in aggregate territorial mobility' (Zelinsky, 1971, p. 243). At that time, there was clearly a dearth of time-series data that could be used to trace the experience of a country as it passed through all of these phases. Perhaps somewhat worryingly, he did not seem to be aware that internal migration rates had been calculated for the USA on a yearly basis since 1947, though his mention of the current US rate of 20% (tucked away in a footnote without a source cited) was likely based on this data. In any case, the picture depicted by that time series largely corroborated Zelinsky's overall conclusion, showing the US rate lying around this one in five level for the first two decades of its existence, with only minor fluctuations. It was only in the late 1960s that the rate dipped significantly below this, but migration researchers know full well not to read too much into short-term changes that can result from period factors such as the business cycle. However, as shown in Figure 1.1, this dip proved to be start of a downward trend that has continued to the present day, apart from the temporary uplifts recorded after each of the four economic recessions that have occurred since the late 1970s (namely 1980–1981, 1990–1991, 2001 and 2008–2009).

The evidence of Figure 1.1 means that there can be no room left for doubt about Americans moving homes much less frequently now than 30 years ago. Even so, it took a long time for this idea to sink in, as lamented by Wolf and Longino in their 2005 paper 'Our "increasingly mobile society"? The

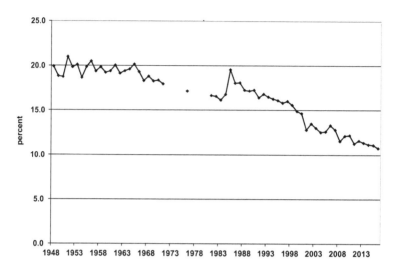

*Figure 1.1* Annual rate of address changing within the USA, 1948–2016.

Source: Plotted by the authors using Current Population Survey data published by the US Census Bureau (see www.census.gov/data/tables/time-series/demo/geographic-mobility/ historic.html).

curious persistence of a false belief', but perhaps it is not so surprising given the prevalence of theories about the close link between modernisation and increasing mobility just outlined. It is almost as if the US experience since the mid-1980s has turned theory on its head. How can theory accommodate this complete reversal in trend? The answer is that it can manage this quite easily, to the extent that there has now been a multiplicity of explanations put forward to account for this slowdown in migration rates.

In the first place, it should be stressed that the earlier theoretical litera-ture did anticipate decline in migration rates, or at least acknowledged the possibility of it. As mentioned above, Lee (1966) added the proviso 'Unless severe checks are imposed', one clear instance of which was the immigra-tion restrictions imposed by the USA after the first World War. Equally important is Zelinsky's incorporation of it into the mobility transition hypothesis, where he makes clear that the different types of movement will vary in their magnitude over time; for instance, with 'frontierward' and rural-urban migration gathering momentum in the early transitional stage, and then falling back in the late transitional one (Zelinsky, 1971, p. 233, Figure 2). Particularly relevant in the present context is his expec-tation of a levelling-off of urban-to-urban and intra-urban migration in his 'phase IV' advanced society and the concurrent rise of 'circulation', defined as 'a great variety of movements, usually short-term, repetitive, or cyclical in nature, but all having in common the lack of any declared intention of a permanent or long-lasting change in residence' (p. 226). Such circulation would likely substitute for some of the potential residential migration during this phase and is a process which he speculates would intensify in his 'phase V' future super-advanced society, despite some of it now being absorbed in its turn by improved (tele)communication (see also Zelinsky, 1971, pp. 230–231, Table 1).

On this basis, a declining intensity of internal migration is not incompatible with a continuing rise in the other forms of spatial mobility embraced by the NMP. Indeed, part of the current misconceptions about internal migration have arisen from the NMP saying so little about residential migration defined in terms of people changing their usual, or 'permanent', address within a country (see, for instance, Caletrio, 2016 and Chapter 13 of this book). This is arguably a shortcoming because, despite the importance of other forms of spatial mobility, where people live and when, where and how often they move remains significant in matching labour supply to demand, in influencing life chances and in shaping patterns of socio-spatial segregation. Despite this, the vast majority of the academic literature linked to the 'age of migration' (for example, Castles *et al.*, 2014) concentrates on between-country movements, as do media coverage and political debates. The international dimension is perhaps most notably evident in the list of types of moves provided by Hannam *et al.* (2006) quoted earlier. To the extent that internal migration has been mentioned, the emphasis has been on the growth of temporary movement rather than the decline of more permanent moves, with little appreciation of the possibility that the former, along with the acceleration of other mobilities, may be causally related to the latter.

Admittedly, the distinction between permanent and more temporary forms of address-changing is a grey area with no hard-and-fast cut-off points, as made clear by Bell and Ward (2000). Even today in highly urbanised societies there are some people who are deemed 'chronically mobile', behaving like the nomads of the pre-industrial era, but also others who at any one time have only one dwelling place but remain there for only months or even weeks. In terms of temporary migration, a whole raft of different types of movement is involved, the variety of which is only partly conveyed by adjectives like 'seasonal' and 'circular', though their key distinguishing feature is the return to the starting point which forms the person's usual residence after at least one night away. Since the mid-twentieth century, temporary migrations for pleasure, or 'consumption-related mobility', have become more numerous than work moves, or 'production-related mobility' (Bell and Ward, 2000), but it is the latter that dominates the literature. In Australia, for instance, alongside the long-established traditions of construction workers moving between building sites and of seasonal labourers in agriculture, travelling for employment has been increasingly common among business executives, professionals and technicians who make periodic, intercity trips to meet clients, suppliers or colleagues in multi-location enterprises (Charles-Edwards, 2004; Swarbrooke and Horner, 2001). Weekly commuting has been well-documented in the UK (Green *et al.*, 1999), while some industries like mining and oil production are now highly dependent on a more extended cycle of travel to remote sites in the form of 'fly-in-fly-out' operations (Gillies, 1997; Houghton, 1993). Harder to classify are other types of temporary absence from home such as by prisoners and hospital patients, as well as students away at university in term-time. All this adds up to a formidable list of types of spatial mobility that, to a greater or lesser extent, would now seem to be substituting for what might previously have involved a permanent change of usual address.

Yet, important as this theorisation of declining internal migration rates may be, it forms only part of a considerably larger number of explanations that has been put forward in the literature on the USA in recent years. Anticipating the fuller discussion of these in the USA in Chapter 5 of this book, Table 1.1 provides a flavour of this range of hypothesised, and in some cases already tested, drivers. Some of them relate to aspects of population composition that are shifting towards the less migratory groups, most notably the elderly, owner-occupiers and dual-earner households. In terms of the space economy, it has been suggested that places are becoming less differentiated in their industrial structures at the same time as certain types of jobs are becoming more concentrated in the major cities. Both are perhaps somewhat paradoxically seen to militate against the need to move for job reasons, as too does the filling of more vacancies by international migrant labour. Changing terms of employment and increasing regulation in labour and housing markets can raise the barriers facing potential movers. Rapidly evolving computing and telecommunications technology can facilitate home-working, while rising car ownership and more fuel-efficient engines means it is less necessary to move homes when changing jobs, as too does the growth of weekly commuting. Meanwhile, the combination of the trend towards an ever more consumer-oriented society and the growing importance of local social capital is seen as making it both more expensive and less attractive to move homes, leading to greater rootedness and to the waning of the 'migration instinct'.

Taken together, the large number of factors listed in Table 1.1 now makes the decline in US internal migration rate since the mid-1980s seem like the most natural and least strange of developments. Certainly, in the decade or more since Wolf and Longino's (2005) lament, aided by a great deal more US research (notably Cooke, 2011, 2013; Molloy *et al.*, 2011; Partridge *et al.*, 2012; see Chapter 5), there has come to be a much fuller acceptance of this state of affairs in the general media and business communities of the USA (see, for instance, Brooks, 2016; Bunker, 2015; Jaffe, 2012; Karahan and Li, 2016; Lowery, 2013; Miller, 2014; Samuelson, 2014). At the same time, however, the identification of this list of drivers does not completely negate all the arguments previously put forward to explain what was theorised in the past as leading to ever-increasing rates of migration. The case can made, as it is more fully in Chapter 2 of this book, that there are forces in play that are still working in that direction, such as the shift of social structures towards

*Table 1.1* A selection of explanations for the decline in internal migration rates in the USA

| | |
|---|---|
| 1 | Ageing of population into less mobile age groups |
| 2 | Rise of the dual-income/dual-career household making labour migration more difficult |
| 3 | Rise of the dual-income household making labour migration less necessary when one partner loses job |
| 4 | Greater possibility of long-distance commuting making a residential move less necessary on changing job/workplace |
| 5 | Fewer gains to be made from moving long distance as places have become more similar in industrial/occupational structure, amenities and house prices |
| 6 | Stronger hold of the largest agglomerations with their greater diversity of jobs and superior opportunities for career progression |
| 7 | Less incentive to move to labour-shortage regions as international labour migration has increasingly filled the vacancies there |
| 8 | Increasing labour-market rigidity, e.g. via employer benefits, state regulation of professionals |
| 9 | Better access of accurate information about potential destinations via new media |
| 10 | New and improved telecommunications making home-working and other forms of remote operation more viable |
| 11 | Easier and cheaper air travel allowing more weekly commuting and short visits to work sites, family, etc. |
| 12 | Growth in owner-occupation making moving home more costly |
| 13 | Greater regulation in the housing market |
| 14 | The rising importance of social capital leading to greater rootedness in the local community |
| 15 | The waning of the 'restless nation', 'culture of migration' and 'roadie' images |

Source: Compiled by the authors from a review of the literature cited in Chapter 5.

those more skilled occupational groups that are traditionally characterised by higher migration rates. One can envisage a form of tug-of-war between this group of drivers and the ones listed in Table 1.1. Evidently, the US situation is one in which such an opposition of forces has moved in favour of the latter over the past three decades.

This conceptual discussion raises a number of questions that shape the rest of the book. The key question that prompted the set of country case studies that form the main part of the book is whether the US pattern is being replicated in other parts of the developed world. If the recent US migration decline is a result of development stage (for example, the transition from Zelinsky's Stage IV to Stage V), then other high-income countries might be expected to have experienced similar trends, given that the economic, social and demographic explanations set out in Table 1.1 bear all the hallmarks of developments that are widely shared by them. However, if countries differ in their migration experiences despite their comparable levels of development, then other interesting questions arise. Perhaps some trends have gone further in the USA than in other developed countries, or the drivers that are pushing internal migration down in the USA are stronger than elsewhere. Alternatively, there might be country-specific factors that modify and mediate structural forces that are acting across all advanced economies and which mean that the purchase of these migration drivers differs according to national context.

This book does not consider how general structural or period effects impinge on all countries because only advanced/developed countries are discussed in any detail. Nevertheless, there may be period effects—such as technological advances and globalisation tendencies—that have an effect on all countries regardless of their developmental stage. This type of global context is often missed in theories such as that of Zelinsky. It would be naïve, for example, to expect countries to proceed from Stage I to Stage V of the Zelinsky model in exactly the same way at different times, given that the experience of a developing country in the early twenty-first century, with growing use of mobile phones and the internet, is very different from that of a developing country in the early twentieth century, with the growing use of cars and trains then. The next section reviews the extant literature on the experience of countries across the world in order to see how far they parallel the US situation or diverge from it.

## Empirical observations

There is now an extensive literature on international differences in internal migration rates in the sense of studies that have explicitly aimed at making comparisons between countries. Its growth over time has been very uneven, considerably paralleling the pattern of theoretical contributions described above as one has fed off the other. This means that the topic generated rather little attention in the three-quarters of a century after Ravenstein, but then the classic papers by Lee (1966) and Zelinsky (1971) rekindled interest for a while, and more recently there has been a surge of activity as the migration rate decline in the USA has become more fully recognised. Even so, it is also the case that cross-sectional comparisons of national *levels* of migration at a single point in time remain far more common than studies that have attempted to examine differences between countries in terms of the *trends* over time in their migration rates. In what follows, a mainly chronological approach is employed and it will show how the challenges of comparing migration intensity on a consistent basis between countries and over time have been tackled.

In the earliest cross-national literature, the key theme was the difference in internal migration rates between the New World and the Old, focusing primarily on the contrasts between the USA and Europe. As concluded by Ferrie (2006) in his review of historical statistics on US internal migration, 'Americans are an unusually peripatetic people' (p. 1). As far back as 1834, de Toqueville had sensed their comparative rootlessness, observing that, 'A man ... settles in a place, which he soon afterwards leaves to carry out his changeable longings elsewhere' (p. 145). In a chapter entitled 'Why the Americans are so restless in the midst of their prosperity', he attributed this primarily to the prevalent values and attitudes, especially an open class structure and a belief in individual opportunity. Ravenstein (1889) provided statistical confirmation by showing from census data that the proportion of the US population living outside their state of birth was higher than that of Europeans living outside their locality of birth. Given that the latter areas were much smaller than the US states, such that their inhabitants had less far to move in order to leave them, he had no doubt that Americans 'are greater wanderers, less tied to home associations, than are the inhabitants of Europe' (p. 280).

While this type of 'life-time migration' data needs to be used with caution for the reasons set out in Chapter 3 of this book, it is valuable where data on change of address between two time points are not available. It is also the measure that Zelinsky (1971) used for testing his hypothesis of the link between migration rate and modernisation, as documented in the previous section, and also by Parish (1973) in his case study of Europe. While both these studies confirmed the general increase in migration rates over time, they also showed that rates varied considerably within Europe. In particular, Parish covered eight countries and in most of these cases spanned a century of change from the 1850s. At the end of his reference period, the proportion of people born outside their administrative area of current residence ranged from 11% for Italy's 'compartments' (*sic*) to 57% for Belgium's communes (Parish, 1973, p. 598). Nevertheless, Parish was as aware as Zelinsky before him that such comparisons are fraught with danger owing to differences in the size of reporting areas, pointing out that the average radius of the Italian compartments was 48 times that of the Belgian commune. Even so, if comparison is limited to the five countries with commune-level data, there remains a considerable margin between Belgium's 57% and Italy's 35%, with Switzerland also high at 56%, the Netherlands relatively low at 39% and Finland's 47% being remarkably high given the much larger average size of its communes. Both authors also recognised other difficulties in interpreting the results of comparing birthplace and current residence, namely not knowing when the moves had occurred, how many intervening moves people had made and whether any of these had returned them to their origin such that they would not be counted as life-time migrants.

It is, however, Long (1988) who is generally credited with the first robust comparisons of internal migration. By this time he was able to take advantage of the fact that an increasing number of national censuses were including a direct migration question on whether people had been living at the same address at a given time prior to the census, normally one year or five years ago though in some cases at the time of the previous census. This provided the opportunity of not only focusing on recent rates of movement as opposed to a life-time change of place, but also of recording all address-changing rather than just those involving between-place moves. His analysis of data from the 1970/1971 round of censuses for 12 countries (Long, 1988, p. 260, Table 8.1) represented a major step forward in our knowledge of international variations in migration, not only confirming the wide range of variation between countries across the world but also showing that the USA was not exceptional in its rootlessness. He calculated that, in Canada and Hong Kong too, the average person would experience 12–13 years in which they would move homes, with Australia and Puerto Rico not far behind at 10–11 and New Zealand at 9–10. At the other end of the scale, for Ireland the average would be just 3–4 moves, with France, Great Britain, Hungary, Japan and Taiwan being intermediate. Additionally, for those countries for which data were available for both one-year and five-year address-changing, he was able to show from the ratio between the two rates that the USA, New Zealand and Australia had higher rates of repeat moving than did Japan and Great Britain at this time (Long, 1988, p. 261, Table 8.2).

Long followed up this seminal work with two papers (1991, 1992) which covered the 1980/1981 round of censuses and looked at change in rates over time by comparing with the earlier round for the seven countries that featured in both datasets. The cross-sectional results for 1980/1981 produced a very similar ranking as for a decade earlier, again drawing on both one-year and five-year change of address data. The highest migration intensities were for New Zealand, Canada, the USA and Australia, followed by Hong Kong, Switzerland, Israel, Puerto Rico, Great Britain, Sweden, Japan, France, the Netherlands and Belgium. Ireland came bottom of the list again, though in a less extreme position than previously. There was still a 3:1 ratio between the extremes of national rates globally and a 2:1 ratio within Europe.

As regards change between the two periods, the picture that Long found was rather varied. New Zealand, Canada and Ireland posted a higher rate in the 1980/1981 round than in the 1970/1971 one, but the reverse was the case for the USA, Great Britain, Japan and Puerto Rico. Further complexity arose from the fact that, in some cases where there was additional data by distance of move, there was variation between the migration types shown; for instance, with the US five-year rate declining over time for within-county address-changing but rising for moves taking place between counties within states. There was also evidence of the one-year data showing a different direction of change from the five-year data, leading Long (1991, p. 144) to issue the following caution: 'Mobility changes between two censuses may not be a valid indicator of underlying trends because the two points in time may not reflect comparable positions on economic cycles'. There is little wonder then at his overall conclusion of there being 'no simple trend' (p. 142)!

One way of allowing for such short-term factors is to track migration rates continuously over time, as undertaken by OECD (1990) for 1970–1987 for a selection of countries, though with some missing data, including for the USA as they used the same source as used earlier in this chapter in Figure 1.1. These national time series relate to just one spatial level for each country and are plotted in Figure 1.2. As noted previously, great caution should be exercised in comparing the countries at any single year, because the regional units vary considerably in area, but it is assumed that the regional geography is fixed over time so that meaningful conclusions can be drawn about whether these inter-regional migration intensities are increasing or declining. For countries with data back to the early 1970s, Italy experienced a continuous reduction that had halved the rate by 1987, with Germany seeing nearly as substantial and uninterrupted reduction. Finland, Canada, Japan and the USA registered falls earlier on in the period, followed by a tendency towards stabilisation after a certain point. Finally, in Australia, Norway, Sweden and the UK, initial falls were followed by slight uplifts in rate in the 1980s. The OECD's interpretation of these patterns cites just economic factors, notably relating the initial falls in rate in European countries to the sharp recession of 1974–1975 and the subsequent cessation or reversal of this trend to economic recovery. Nevertheless, the length of the period covered here helps to allow for such cyclic behaviour, suggesting that for most of these countries inter-regional migration was running at a lower intensity at the end than at the start.

The situation in Europe was updated further by Rees and Kupiszewski (1999), with data for the 1990s for 10 countries being compared (in all but one case) with the 1980s. They got round the problem of some countries being able to supply data just for between-area moves rather than all changes

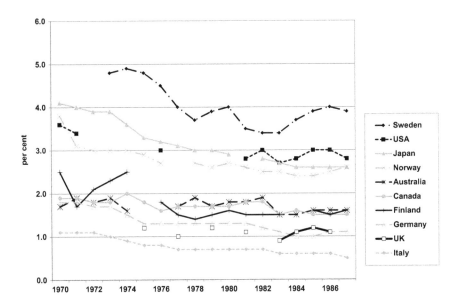

*Figure 1.2* Proportion of the working-age population that changed their region of residence during the year, 1970–1987, for selected OECD countries.
Source: Plotted by the authors using data from OECD (1990, p. 85, Table 3.3).

of address by using Courgeau's *K* Index of internal migration intensity (Courgeau *et al.*, 2012; also see Chapters 3 and 4 of this book). Looking across the values of *K* for the early 1980s (mainly 1984) and the early 1990s (mainly 1994) led them to a four-way classification. High migration intensity was found to be characteristic of Norway, the Netherlands and Great Britain, while the lowest intensities were for the Czech Republic, Estonia, Italy, Poland and Romania. In-between with medium intensity was the single case of Portugal. Germany was also classified separately, in this case because it switched from a low to a high intensity migration regime between the two decades, attributed to the ongoing effect of that country's reunification in 1990 (see Chapter 10 of this book for further details). The changes in *K* value over time for the other countries were all much smaller and rather mixed in direction of trend. As well as Germany, the Netherlands and Romania also experienced an increase in migration intensity, while Norway's value remained unchanged and the five others analysed (there were no 1980s data for Estonia) saw a fall in intensity. Bear in mind that, for all countries except Portugal, the data refer to single years, so these temporal comparisons may be affected by short-term factors operating at either or both dates.

Since the turn of the millennium, there has been a substantial proliferation of studies of international differences in internal migration rates, notably expanding the number of countries compared cross-sectionally but in some cases including trends over time. OECD (2005) managed both in looking at the proportion of the working-age population that changed region of residence over the year, using data for 17 countries in or around 2003 and data for 5 and/or 10 years previously for all but four of these. The former confirmed that the proportions 'tend to be lower in Europe than in the USA or in countries belonging to the Asia/Pacific area' and that 'In Europe, however, the situation is not uniform across countries', with Southern and Eastern European countries generally having very low gross migration rates and with France and the UK having relatively high ones but still lower than the US rate (OECD, 2005, p. 89). As regards the trend over the previous 10 years, the overall verdict was that, except in Japan, the decline in inter-regional migration observed in previous decades, reported by OECD (1990) as mentioned above, had ended. Indeed, 'Some increasing trend in mobility is noticeable in other European countries such as France and the Netherlands and since the late 1990s in Germany' (OECD, 2005, p. 90). Closer inspection of their data, however, also shows varying degrees of decline in rate for the UK, Greece, Canada, Australia and New Zealand, though strangely not for the USA. Wisely, though, they do stress the need for caution in making these cross-national comparisons owing to the lack of consistency between data sets, with variations between countries in the age span of the population analysed and with the small print on definitions in the Annex saying that the data for at least one country relate to five-year rather than one-year migration.

Bonin *et al.* (2008) restrict their attention to the EU but cover all the 27 countries that were members at the time and draw on two surveys. One is the European Labour Force Survey, which they use to compare countries on their rates of one-year moving between their NUTS two regions annually between 1995 and 2006. Their cross-sectional analysis for 2006 confirms the distinction between the very low regional migration rates for most countries of Southern and Eastern Europe and the much higher rates for countries such as France, the Netherlands, Sweden and the UK. Their time-series data are rather patchy, especially for the 12 new member states, but the average for the EU-15 countries rose somewhat in the late 1990s and then subsided more markedly, ending up below the 1995 rate. There was found to be considerable variation in trend between countries, with notable falls in rate for Italy, Greece, Portugal and Sweden and increases for Belgium and Spain. For most countries, however, the rate jumps around from one year to the next, perhaps because of relatively small sample size, so some care is needed in interpreting these results.

The other source used by Bonin *et al.* (2008) is the Eurobarometer survey of EU citizens aged 15 and over, which had a special focus on geographic migration behaviour in its spring waves of 2005 and 2007. One advantage of this source is that it covers all changes of address rather than just inter-regional migration, while it also included questions on lifetime migration and migration intentions. As regards the latter, focused on whether people think that they are likely to make a within-country move during the next five years, the EU-25 average from the 2005 survey came out at 27%, with the cross-country variation being closely correlated with the LFS-based rates of completed migration just outlined. The proportion was over 40% for France and Sweden and over a third for Finland, Denmark and the UK, while at the other end of the scale lay Portugal, Slovakia, the Czech Republic and Austria with rates of 15% or less.

In terms of lifetime moves, of particular interest given that it is very rarely reported in multi-national studies is their examination of the proportion of 15+ year-olds still living at their parental home and of the number of moves made by them after leaving home. As plotted in Figure 1.3, the weighted average for all 25 countries shows 17.5% still at home and another 17.5% who have made five or more moves. The countries with by far the highest shares of the latter are Finland, Denmark and Sweden, while those with the largest proportions still living with their parents are Slovenia, Slovakia, Poland, Malta and Greece, all in excess of 25%. While the precise rankings will be affected somewhat by country variations in mean age, this insight serves to reinforce the other evidence available. In their own words, '... the share of the population still living at home is a good predictor for the overall intensity of geographic mobility in a country' (Bonin *et al.*, 2008, p. 36). Further details of the 2005 Eurobarometer survey results on geographic mobility, including links with job mobility, can be found in Vandenbrande *et al.* (2006).

Two more recent studies that similarly cover all distances of address-changing give more emphasis to the role of housing factors than labour

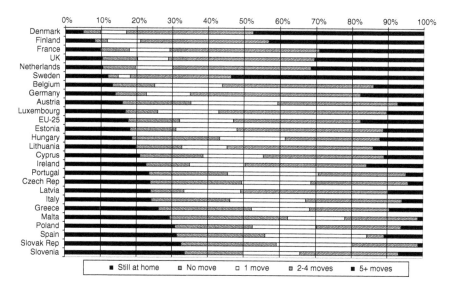

*Figure 1.3*  Native-born adults, 2005, by whether still living at the parental home or by number of moves since leaving it, for 25 EU countries and the EU-25 weighted average.

Source: Plotted by the authors using data from Bonin *et al.* (2008, p. 36, Table 8), with countries ranked on basis of ascending proportion still living at parental home.

market ones, perhaps not surprisingly given the greater importance of the former in prompting short-distance moves than inter-regional ones. Lennartz *et al.* (2016) report on what the EU's Statistics on Income and Living Conditions (SILC) show about the proportion of 18–34 year olds still living in the parental home in 2007 and on how this relates to the structure of housing tenures. Their 'co-residence' (with parents) measure displays the same stark contrast between the Nordic and Southern European arenas as found by Bonin *et al.* (2008), with around three in five being in this situation in the latter compared to one in five or fewer in the former. This pattern corresponds neatly with the much higher share of this age group renting in Scandinavia, with upwards of 40% there, as opposed to under 20% in Greece and Italy and below 10% in Spain and Portugal. On the other hand, the direction of causality behind this association is arguable, given other differences between these two parts of Europe such as in higher education participation, job searching behaviour and family values.

Caldera Sanchez and Andrews (2011) also examine the relationship between geographic mobility and housing markets, again using SILC data but supplementing this with similar household microdata for Switzerland, the USA and Australia. Despite adopting an unusual measure of migration based on place of residence two years before the survey and on whole households rather than individual persons, the pattern is a familiar one.

Migration is highest in the Nordic countries, with Iceland at 29%, Sweden at 23%, Norway at 21%, Finland at 19% and Denmark at 18%, alongside Australia's 24% and the USA's 21%. It is lowest in Eastern Europe, with Poland, the Slovak Republic and Slovenia around 4%, the Czech Republic and Portugal around 5%, Greece and Spain on 8% and Italy at 9%. The very low rates in Eastern Europe are ascribed to a combination of labour market rigidities and the very specific features of housing markets there, including the small size of the private rented sector and the persistent undersupply of new dwellings. More generally, they find that residential mobility is lower among households living in social or subsidised housing and in countries where transaction costs are high, while they are higher where there is a more responsive housing supply and greater access to credit. These conclusions lead directly to policy recommendations aimed at increasing residential mobility in the cause of greater labour market efficiency.

Finally, in this review of previous cross-national analyses of internal migration intensities are two projects characterised by much greater global coverage. One is survey-based, using data from the Gallup World Poll, which in 2011–2012 asked adults in 139 countries whether they had moved from another city or area within their respective countries in the past five years. As reported by Esipova *et al.* (2013), the overall average came to 8%, with the highest rates for Australia, Scandinavian, North American and South American countries at 21–26%—as expected, but comforting given the rather small size of sample (around 1,700 adults per country) and the fact that the geography of 'another city or area' was not clearly specified. The main value added by this study, however, is its impressively full coverage of the less developed parts of the world, within which variation was found to be nearly as marked as has been seen for the more developed regions. Malawi and Syria were found to have rates of over 20%, but internal displacement through civil war and natural disaster respectively were thought to make these into special cases. Elsewhere, rates of 16% or more were found for Colombia in Latin America and Kenya in Africa alongside other countries in both those regions with rates of 5% or less. Even averaging across regions, the figure of 12% for both Sub-Saharan Africa and the Middle East was more than twice the 5% average for Developing Asia, with Latin America and North Africa in between at 10% and 8% respectively. So far, this Gallup survey has been just a one-off, affording no means of examining change over time in rates and with Esipova *et al.* (2013, p. 5) bemoaning the fact that 'Compared with the data on international migrants, the development of global estimates (of internal migration) is still in its nascent stages'.

The second such study, which goes a long way to addressing this perceived lacuna in demographic statistics, is the IMAGE (Internal Migration Around the GlobE) project, an international programme research which has established an inventory of migration data collections across the 193 UN member states. This project builds on the experience of a programme of work which compared Australia and Britain with respect to their internal migration patterns and trends over several decades and, for this purpose, devised a formidable suite of measures for doing this in a consistent and robust manner (see, for instance, Rees *et al.*, 2000; Bell *et al.*, 2002; Bell and Rees, 2006). The progress of the IMAGE project can be tracked through a series of publications, including Bell and Muhidin (2009), Bell and Charles-Edwards (2013), Bell *et al.* (2015), Bernard *et al.* (2014), Rees *et al.* (2016) and Stillwell *et al.* (2014, 2016). Most relevant in the present context is Bell *et al.* (2015), who provide a cross-sectional analysis of 96 countries for which there were suitable data relating to around the year 2000 on either one-year migration, five-year migration or both. The focus is on the totality of residential moves, which was recorded directly by 26 countries and estimated for the remainder using a modified version of the Courgeau method mentioned earlier. The IMAGE project has also collected data relating to around 1990 and 2010, permitting Bell and Charles-Edwards (2013) to report on trends in five-year intensities for 36 countries, alongside trends in lifetime migration for 62 countries.

It is not possible to include a full description of the IMAGE projects results in this review, nor is it necessary because the picture is updated in Chapter 4 of this book, following the treatment of the challenges of making robust international comparisons and the best ways around these that is provided in Chapter 3. Nevertheless, to whet the appetite for what follows, attention is drawn here to three key findings from Bell *et al.* (2015) and Bell and Charles-Edwards (2013). Firstly, the wide range of international variation in migration intensities found by previous, more limited studies is confirmed, with one-year address-changing rates ranging from 19% for Iceland to 1% for Macedonia and five-year rates ranging from 55% for New Zealand to 5% for India. Secondly, testing the idea of the link between migration intensity and modernisation discussed earlier in this chapter, there was found to be a strong correlation of both one-year and five-year rates with GDP per capita, the Human Development Index and the level of urbanisation. Thirdly, in terms of trends over time, the picture is a mixed one, even for the rather few developed world countries with the requisite data. In terms of lifetime migration intensity, the US rate is classified as stable between 1990 and 2000 and then fell in 2000–2010, while in 1990–2000 the rate for Spain rose and that for Portugal was stable (no data for 2000–2010). As regards five-year migration, Portugal's rate rose in 1990–2000, while the USA's fell, as did those for Australia and Canada across both decades. For the totality of countries covered, however, 'the dominant impression from the five-year data is a trend of declining migration intensities across the globe' (Bell and Charles-Edwards, 2013, p. 24)—which is at variance with the idea of continuously increasing migration though perhaps consistent with Zelinsky's (1971) anticipation that residential migration might in due course be substituted by increased circulation and telecommunication.

In sum, this section has reviewed the extensive literature that has compared countries with respect to their internal migration intensities at one

point in time or the more limited evidence on how these have been changing over time. The cross-sectional analyses paint a pretty clear picture, despite the majority of them dealing with data just on between-place moves
rather than the totality of address-changing and thus facing the problem—
recognised by Ravenstein onwards—caused by statistical areas varying
greatly in size between countries. As regards the developed world, these
studies highlight the contrast between the high frequencies of moving characteristic of North and South American countries as well as Australia and
the lower rates for Japan and Europe and, in relation to the latter, a general
gradation from highest in the Nordic countries to lowest in Southern and
Eastern Europe. Additionally, worldwide, there would appear to be a strong
cross-sectional relationship between level of migration and indicators of the
degree of modernisation. By contrast, the relatively few multi-national studies of trends over time suggest much greater complexity. In particular, while
the IMAGE project's global results suggest a prevailing tendency towards
declining intensities, its rather limited coverage of the developed world in
this respect reveals a variety of rising, stable and falling rates for single periods of change and also some countries switching between these three trends
from one change period to the next. To complicate matters further, there are
also some differences for individual countries depending on the type of migration measure used (e.g. whether lifetime or five-year or one-year) and on
the type of move being observed (e.g. whether all address-changes or local
ones or longer-distance inter-regional migration).

It is this variety of experience in trends that provided the initial rationale for this book. Certainly, as mentioned above, the US experience since
the 1980s has highlighted the fact that not all types of mobility are continuing to increase in all parts of the world, but then the finding by two of
the editors that inter-area migration rates were not falling in the UK long-
term (Champion and Shuttleworth, 2016) suggested that even countries in
the developed world might be following a different path from the USA despite many similarities in demographic, economic, social and technological
change. One possibility is that the countries are, in fact, following similar
trajectories, but that is not apparent because of differences in the definition
and measurement of migration. It is also the case that the main work on the
international comparative study of trends so far, notably Long's and the
IMAGE project, has been based almost entirely on comparing snapshots
taken once a decade, for which, as Long (1991) pointed out, the rates could
be temporarily affected by period factors like the business cycle. Admittedly, there is an increasing body of evidence that provides a continuous
record of annual migration rates going back over several decades, but such
studies invariably concentrate on just one country and adopt their own distinctive frames of reference, hence their not being included in the above review. In any case, a good number of these single-country studies relate to
the seven countries that have been selected as case studies for this book
and so will be discussed in the respective chapters in Part 2, where there is

also the opportunity to interpret their findings within the context of a more detailed consideration of these countries' migration records and particular national characteristics. The next section provides a fuller description of the book and justification for the approach we have adopted.

## The book: aims, approach and outline

### Aims and objectives

The material reviewed earlier in this chapter gives a clear indication that the US experience is not unique, even within the more developed part of the world. There are differences in migration rates between countries and also differences in trends through time. When considering migration rates cross-sectionally, countries like the USA and Australia are still ranked amongst the most residentially mobile, along with Nordic Europe, with most countries of Western Europe and Japan in an intermediate group and with Southern and Eastern Europe being amongst the least mobile. This reflects economic and labour market conditions but also less tangible factors such as culture and historical legacy and some broad structural differences such as whether a country was a settler society or not. Some variations in temporal trends were also noted: in recent years some countries have experienced decreases similar to the USA, whilst some have seen a fair degree of stability in internal migration rates and yet others have seen an increase.

In this context of the international variety of recent years and the limited knowledge about the past, the overriding aim of the book is to establish more clearly what the long-term trend in internal migration rates has been in the developed world and to achieve a better understanding of them. As hinted above, cross-national analyses of migration are by no means simple because of variations in data definitions and statistical systems, while adding temporal change complicates the task even further. The work of IMAGE cited in the previous section marks a considerable advance, but its time horizon has so far been largely confined to the latest couple of decennial census rounds. This is a limitation when trying to grapple with the structural forces that influence societies on multi-decade timescales and that form the wider social, historical, political and economic context for long-term changes internal migration rates.

The book therefore has three main objectives. The first is to find out what has been happening through detailed examination of national datasets on internal migration rates and patterns. The second is to identify explanations for the observed patterns and trends, requiring an in-depth understanding of national circumstances including demographic regimes, economic development, social inequalities, spatial inequalities and cultural and political history. The third and final objective is to relate internal migration, where possible, to other types of population mobility, such as international migration, temporary migration and everyday spatial mobility, and thereby engage critically, but constructively, with the NMP.

### The national case study approach

To achieve these objectives, a case-study approach was adopted, with the 'national' being selected as the level of analysis. While it is possible to make the case that the nation state has had its day and been superseded by supranational organisations and the forces of globalisation that have created spaces of networks and flows that no longer respect national boundaries, it can also be argued that internal migration within nation states remains a valid topic for study. These arguments crystallise around three major themes well represented in the writings of sociologists of globalisation like Michael Mann.

First, countries remain important as sites for the regulation of social and political life and the reproduction of labour. Education and legal systems remain predominantly national, for example. At the simplest and most basic level, laws and qualifications are often nation-specific. There are problems, for instance, in the validation and use of certification gained in one country being applied in another, making it harder for workers to gain employment there or at least to access jobs at the same occupational level.

Secondly, international migrants can be qualitatively different from people who move within countries. Quite apart from the lack of transferability of credentials and qualifications, state borders matter in structuring access to social welfare, housing and labour markets (Peck and Theodore, 2001; Shuttleworth, 2007). There is an ample and growing literature on how these differences separate international from internal migrants (May *et al.*, 2007; Standing, 2011) and these differences alone justify our analytical focus on internal migration within countries as a topic in its own right.

Thirdly, there is evidence of the growing importance of national borders in some places and in some contexts. During 2015, walls were constructed on Hungary's borders with Serbia and Croatia to keep out the thousands of refugees arriving daily from further east (see, for example, BBC, 2016). To this can be added the fears of international migration and terrorism that are prompting greater surveillance, as evidenced by the border checks proposed by the Trump Presidency. Perhaps most telling is that, in a supposed era of globalisation, we are witnessing a steady increase, rather than a decrease, in the number of nation states in the world.

Beyond these arguments, within-country migration is important in and of itself: residential location and relocation matter. Inter-regional migration has been, and continues to be, a major concern for economic policy, given its role as one of the chief mechanisms for labour market adjustments that are meant to balance supply with demand. The importance of more local residential migration is reflected in the extensive literature on neighbourhood effects and geographical context across various domains of social and economic life (van Ham and Manley, 2012; Thomas *et al.*, 2015) and by the historic and ongoing concerns of academics and policymakers with residential segregation (see the essays in Lloyd *et al.*, 2015). At the same time, there are no hard-and-fast boundaries between these two scales of migration—a point that is reflected by the fact that national statistical agencies almost invariably collect and publish internal migration data for the whole of their territories (and also normally separately from their data on international movement). There are thus very strong reasons for adopting the country level and for aiming to cover the totality of migration within their boundaries.

### Choice of case study countries

Three main criteria were applied in the choice of the countries to be put under the spotlight. In the first place, it was decided that the book's coverage should be restricted to the developed world, so as to be able to compare countries which are currently at roughly the same level of economic development and have followed similar trajectories in the past. Secondly, it was considered important to include a fairly wide range of national circumstances from within the developed world, including a range of migration behaviours, so as to help identify potential links between the two. Thirdly, the choice has necessarily been influenced by the availability of data and by the existence of willing researchers who are established scholars of their countries' internal migration, though not necessarily having previously had a prime focus on long-term trends in migration rates until asked to do so for this book.

In all, seven countries were selected for this purpose, as listed in Table 1.2 along with some of their key characteristics. They all justify the label of 'economically advanced', as recognised by bodies like the International Monetary Fund (IMF), and can be considered to have reached at least Stage IV of Zelinsky's (1971) mobility transition model after passing through Stages II and III by the mid-twentieth century. They are characterised by a significant history (and level) of industrialisation, high GDP per capita, and high scores on the Human Development Index. They also have ageing populations, although to varying degrees. In addition, all have experienced job losses from manufacturing, the growth of employment in financial and personal services, the casualisation of working conditions, wider financialisation, the general acceptance of neoliberal policies for economic and labour market management, and the expansion of internet technologies and air travel. The factors listed above are not exhaustive but give some idea of the suite of attributes in common which enable us to test the contention that the internal migration decline in the USA can be generalised to other similar economic, demographic, social, technical and political contexts.

At the same time, as shown by Table 1.2, they also differ in some important respects. First, they are drawn from different world regions: Europe, North America, East Asia and Australasia. They vary considerably in spatial size,

Table 1.2 Case-study countries: comparative indicators

| Indicator | Australia | Germany | Italy | Japan | Sweden | UK | USA |
|---|---|---|---|---|---|---|---|
| Surface area (000 km²)[1] | 7,682.3 | 348.4 | 294.1 | 364.6 | 407.3 | 241.9 | 9,147.4 |
| Population density (per km²)[1] | 3.1 | 232.1 | 208.5 | 348.7 | 23.8 | 269.2 | 34.9 |
| % urban population[2] | 89.0 | 75.3 | 69.0 | 93.0 | 86.0 | 82.0 | 81.0 |
| % population aged 65+[1] | 14.7 | 21.1 | 22.0 | 25.7 | 19.6 | 17.8 | 14.4 |
| GDP per capita, PPP (constant 2011 international $)[1] | 43,218 | 43,443 | 33,077 | 35,634 | 44,028 | 38,865 | 52,117 |
| GINI coefficient[3] | 30.3 | 27.0 | 31.9 | 37.9 | 24.9 | 32.4 | 45.0 |
| Strength of safety net for unemployed[4] | High | High | Low | Low | High | Medium | Low |
| Model of capitalism[5] | Liberal | Corporatist | Corporatist | Corporatist | Social democratic | Increasingly liberal | Liberal |
| Human Development Index 2013 rank[6] | 2/188 | 6/188 | 27/188 | 19/188 | 14/188 | 15/188 | 7/188 |
| Rank on one-year address-changing intensity, ca 2005[7] | 7/45 | 15/45 | 29/45 | 22/45 | 8/45 | 13/45 | 6/45 |

Sources: 1. World Bank, www.worldbank.org/en/country, accessed 29 October 2015; 2. UN (2014), Annex Table 1; 3. CIA, www.cia.gov/library/publications/the-world-factbook/rankorder/2172rank.html, accessed 6 March 2017; 4. OECD (2011), Table 1.1; 5. After Esping-Andersen (1990), 26–28; 6. UN (2013), Table 1; 7. Bell et al. (2015), Figure 4. Data for 2014 except where indicated otherwise.

settlement history, population density and level of urbanisation, and are characterised by specific inherited political, social and economic features, including the degree to which they have assimilated the Anglo-American free-market economic model that has been so dominant since 1980. They also provide representatives of four of Oláh *et al.*'s (2014) five European welfare regimes: Liberal (UK), Dual Earner (Sweden), General Family Support (Germany) and Familiaristic (Italy), omitting only the Transition post-socialist type. Finally, they cover a wide range of migration intensities ranging from USA, Australia and Sweden at the high end through to Italy at the low end of the developed country rankings made by Bell *et al.* (2015).

### *Plan of the book*

The book is organised into three parts, beginning with four scene-setting chapters of which this is the first. In Chapter 2, the book engages closely with the potential drivers of internal migration, looking at the factors that might be expected to alter the frequency of address-changing and taking the book into the area of political economy as it considers the hypothesis that common economic, social and demographic features across advanced societies are forcing internal migration down. Chapter 3 is statistical and methodological, but it is essential for establishing the evidence base on which the rest of the book is constructed. As was demonstrated earlier, international comparisons of migration are challenging, but this chapter discusses various solutions calling upon the expertise of book contributors. Chapter 4 draws on the experience of the IMAGE project to give a global overview of internal migration that sets in context the internal migration experiences of the nations of the developed world. It also allows the cross-sectional relationship between different stages of economic development and internal migration to be explored, permitting an initial assessment of conceptual frameworks that make a link between these two elements.

The second part, forming the core of the book, consists of seven chapters which are in-depth analyses of internal migration in the selected case-study countries: the USA, the UK, Australia, Japan, Sweden, Germany and Italy. The intention is to examine whether the migration slowdown seen in the USA is common across all advanced economies and whether the same structural forces suggested as causes for this in the USA are in play elsewhere. In each of these seven chapters, the authors provide more detail about their national context, describe and evaluate the data sources available there for studying internal migration patterns, go back as far as these sources allow in tracking their country's address-changing rates over time and attempt to identify the factors behind the trends that they observe. The latter is achieved both through a review of the existing literature and through their own manipulation of the migration data, whether by going into more detail about the different types of migration (such as classified by distance of move or patterns of flow) or by the statistical analysing of the migration patterns and trends.

The final part of the book comprises three chapters of reflection. Chapters 12 and 13 provide the opportunity for invited commentators to review the findings of the case study chapters from their own distinctive perspectives. Frey is able to draw on his long experience in American migration research and related policy matters, while Halfacree is a UK-based geographer sympathetic to the New Mobilities Paradigm and able to assess how far this book's observations on declining migration rate square with the main thrust of that paradigm. Lastly, in Chapter 14 the editors discuss the wider implications of the book's main conclusions and set future research directions in what we believe is an important and challenging field of study.

## References

Abel, G.J. and Sander, N. 2014. Quantifying international migration flows. *Science*, 343, 1520–1522.

Adey, P. 2010. *Mobility*. Abingdon, Oxon: Routledge.

Bauman, Z. 2000. *Liquid Modernity*. Cambridge: Polity Press.

BBC. 2016. Migrant Crisis: Hungary Police Recruit 'border hunters'. www.bbc. co.uk/news/world-europe-37259857, accessed 6 March 2017.

Bell, M., Blake, M., Boyle, P., Duke-Williams, O., Rees, P., Stillwell, J. and Hugo,G. 2002. Cross-national comparison of internal migration: Issues and measures. *Journal of the Royal Statistical Society A*, 165(3), 1435–1464.

Bell, M. and Charles-Edwards, E. 2013. Cross-national Comparisons of Internal Migration: An Update of Global Patterns and Trends. *Technical Paper* 2009/30. New York, NY: Population Division, United Nations Department of Economic and Social Affairs.

Bell, M., Charles-Edwards, E., Ueffing, P., Stillwell, J., Kupiszewski, M. and Kupiszewska, D. 2015. Internal migration and development: Comparing migration intensities around the world. *Population and Development Review*, 41(1), 33–58.

Bell, M and Muhidin, S. 2009. National Comparisons of Internal Migration: An Update of Global Patterns and Trends. *Technical Paper* 2013/1. New York, NY: Population Division, United Nations Department of Economic and Social Affairs.

Bell, M. and Rees, P. 2006. Comparing migration in Britain and Australia: Harmonisation through use of age-time plans. *Environment and Planning A*, 38(5), 959–988.

Bell, M. and Ward, G.J. 2000. Comparing permanent migration with temporary mobility. *Tourism Geographies*, 2(1), 97–107.

Bernard, A., Bell, M. and Charles-Edwards, E. 2014. Explaining cross-national differences in the age profile of internal migration: The role of life-course transitions. *Population and Development Review*, 40(2), 213–239.

Bonin, H., Eichhorst, W., Florman, C. *et al.*, 2008. Geographic Mobility in the European Union: Optimising its Economic and Social Benefits. *IZA Research Report* 19. Bonn: IZA.

Brooks, A.C. 2016. How to get Americans moving again. *New York Times*, 20 May.

Bunker, N. 2015. The consequences and causes of declining geographic mobility in the United States. Washington Centre for Equitable Growth. http://equitablegrowth.

org/equitablog/the-consequences-and-causes-of-declining-geographic-mobility-in the-united-states, accessed 17 January 2017.

Caldera Sanchez, A. and Andrews, D. 2011. To move or not to move: What drives residential mobility rates in the OECD? *OECD Economics Department Working Paper* 846. Paris: OECD.

Caletrio, J. 2016. Mobilities paradigm. Mobile Lives Forum. hhtp://en.forumvies mobiles.org/marks/mobilities-paradigm-3293, accessed 19 September 2016.

Castles, S., de Haas, H. and Miller, M.J. 2014. *The Age of Migration*. 5th edition. Basingstoke: Palgrave Macmillan.

Champion, A.G. and Shuttleworth, I. 2016. Is longer-distance migration slowing? An analysis of the annual record for England and Wales. *Population, Space and Place,* published online in Wiley Online Library. doi:10.1002/psp.2024

Charles-Edwards, E. 2004. *Have Work, Will Travel: Toward an Understanding of Work-related mobility in Australia.* Unpublished BA Honours thesis. Brisbane: School of Geography, Planning and Architecture, University of Queensland.

Cooke, T.J. 2011. It is not just the economy: Declining migration and the rise of secular rootedness. *Population, Space and Place*, 17(3), 193–203.

Cooke, T.J. 2013. Internal migration in decline. *The Professional Geographer*, 65(4), 664–675.

Courgeau, D., Muhidin, S. and Bell, M. 2012. Estimating changes of residence for cross-national comparison. *Population-E*, 67(4), 631–652.

Cresswell, T. 2006. *On the Move: Mobility in the Modern Western World.* London: Taylor and Francis.

Esipova, N., Puglieese, A. and Ray, J. 2013. The demographics of global internal migration. *Migration Policy Practice*, III(2), 3–5.

Esping-Andersen, G. 1990. *The Three Worlds of Welfare Capitalism.* Princeton, NJ: Princeton University Press.

Ferrie, J.P. 2006. Internal migration. Chapter Ac. In Carter, S.B. (ed) *Historical Statistics of the USA.* New York, NY: Cambridge University Press. http://hsus.cambridge.org/HSUSWeb/toc/showChapter.do?id=Ac, accessed 17 January 2017.

Gillies, A.D.S, Wu, H.W. and Jones, S.J 1997. The increasing acceptance of fly-in fly-out within the Australian mining industry. *Proceedings Aust. Inst. Min. Metall. Conference*, 87–95.

Green, A.E., Hogarth, T. and Shackleton, R.E. 1999. Longer distance commuting as a substitute for migration in Britain: A review of trends, issues and implications. *International Journal of Population Geography*, 5(1), 49–67.

Hannam, K., Sheller, M. and Urry, J. 2006. Editorial: Mobilities, immobilities and moorings. *Mobilities* 1, 1–22.

Houghton, D. 1993. Long distance commuting: A new approach to mining in Western Australia. *Geographical Journal*, 159(3), 281–290.

Jaffe, E. 2012. The mystery of declining mobility. *The Atlantic Cities*, December 17th.

Karahan, F. and Li, D. 2016. What caused the decline in interstate migration in the United States? New York, NY: Liberty Street Economics, October 17th. http://libertystreeteconomics.newyorkfed.org/2016/what-caused-the-decline-in-interstate-migration-in-the-united-states, accessed 17 January 2017.

Lee, E.S. 1966. A theory of migration. *Demography*, 3, 47–57.

Lennartz, C., Arundel, R. and Ronald, R. 2016. Young adults and homeownership in Europe through the Global Financial Crisis. *Population Space and Place*, 22, 823–35.

Lloyd, C.D., Shuttleworth, I.G. and Wong, D.W. 2015. *Social-spatial Segregation: Concepts, Processes and Outcomes*. Bristol: Policy Press.

Long, L. 1988. *Migration and Residential Mobility in the United States*. New York, NY: Russell Sage Foundation.

Long, L. 1991. Residential mobility differences among developed countries. *International Regional Science Review*, 14, 133–147.

Long, L. 1992. Changing residence: Comparative perspectives on its relationship to age, sex and marital status. *Population Studies*, 46, 141–158.

Lowery, A. 2013. Why are Americans staying put? *New York Times*, December 10th.

May, J., Wills, J., Datta, K., Evans, Y., Herbert, J. and McIlwaine, C. 2007. Keeping London working: Global cities, the British state and London's new migrant division of labour. *Transactions of the Institute of British Geographers*, 32, 151–167.

Miller, J. 2014. *The Decline in Geographic Mobility*. http://eyesonhousing.org/2014/01/the-decline-in-geographic-mobility, accessed 17 January 2017.

Molloy, R., Smith, C.L. and Wozniak, A. 2011. Internal migration in the United States. *Journal of Economic Perspectives*, 25(2), 1–42.

OECD. 1990. *OECD Employment Outlook 1990*. Paris: OECD.

OECD. 2005. *OECD Employment Outlook 1990*. Paris: OECD.

OECD. 2011. *OECD Employment Outlook 2011*. Paris: OECD.

Oláh, L.S., Richter, R. and Kotowska, I. 2014. The new roles of men and women and implications for families and societies. *EU FP7 Families and Societies Working Papers Series* 11. www.familiesandsocieties.eu/wp-content/uploads/2014/12/WP11OlahEtAl2014.pdf, accessed 17 January 2017.

Parish, W.L. 1973. Internal migration and modernization: The European case. *Economic Development and Cultural Change*, 21(4), 591–609.

Partridge, M. D., Rickman, D. S., Olfert, M. R. and Ali, K. 2012. Dwindling U.S. Internal migration: Evidence of spatial equilibrium or structural shifts in local labour markets? *Regional Science and Urban Economics*, 42(1), 375–388.

Peck, J. and Theodore, N. 2001. Contingent Chicago: Restructuring the spaces of temporary labour. *International Journal of Urban and Regional Research*, 25, 471–496.

Ravenstein, E.G. 1885. The laws of migration. *Journal of the Statistical Society of London*, 48(2), 167–235.

Ravenstein, E.G. 1889. The laws of migration. *Journal of the Statistical Society of London*, 52(2), 241–305.

Rees, P., Bell, M., Blake, M. and Duke-Williams, O. 2000. Harmonising databases for the cross national study of internal migration: Lessons from Australia and Britain. *Working Paper* 00–05. Leeds: School of Geography, University of Leeds.

Rees, P., Bell, M., Kupiszewski, M., Kupiszewska, D., Ueffing, P., Bernard, A., Charles-Edwards, E. and Stillwell, J. 2016. The impact of internal migration on population redistribution: An international comparison. *Population, Space and Place*, published online in Wiley Online Library. doi:10.1002/psp.2036

Rees, P. and Kupiszewski, M. 1999. Internal migration and regional population dynamics in Europe: A synthesis. *Population Studies* 32. Strasbourg: Council of Europe.

Robins, K. 2000. Encountering globalization. In Held, D. and McGrew, A.G. (eds) *The Global Transformation Reader*. Cambridge: Polity Press, pp. 239–245.

Samuelson, R.J. 2014. Our stay-put society. *The Washington Post*. 7th May.

Sennett, R. 1994. *Flesh and Stone: The Body and the City in Western Civilisation*. 1st edition. New York, NY: W.W. Norton.

Sheller, M. and Urry, J. 2006. The new mobilities paradigm. *Environment and Planning A*, 38, 80–93.

Shuttleworth, I. 2007. Reconceptualising local labour markets in the context of cross-border and transnational labour flows: The Irish example. *Political Geography*, 26, 968–981.

Standing, G. 2011. *The Precariat: The New Dangerous Class*. London: Bloomsbury Academic.

Stillwell, J., Daras, K., Bell, M. and Lomax, N. 2014. The IMAGE Studio: A tool for internal migration analysis and modelling. *Applied Spatial Analysis and Policy*, 7(1), 5–23.

Stillwell, J., Bell, M., Ueffing, P., Daras, K., Charles-Edwards, E., Kupiszewski, M. and Kupiszewska. D. 2016. Internal migration around the world: Comparing distance travelled and its frictional effect. *Environment and Planning A*, 48(8), 1657–1675.

Swarbrooke, J. and Horner, S. 2001. *Business Travel and Tourism*. Oxford: Butterworth-Heinemann.

Thomas, M., Stillwell, J. and Gould, M. 2015. Modelling multilevel variations in distance moved between origins and destinations in England and Wales. *Environment and Planning A*, 47, 996–1014.

UN. 2013. *Human Development Report 2013*. New York, NY: United Nations.

UN. 2014. World Urbanisation Prospects. New York, NY: United Nations Department of Economic and Social Affairs. https://esa.un.org/unpd/wup/publications/files/wup2014-highlights.Pdf, accessed 21 January 2017.

Urry J. 2007. *Mobilities*. Cambridge: Polity Press.

van Ham, M. and Manley, D. 2012. Neighbourhood effects research at a crossroads. Ten challenges for future research introduction. *Environment and Planning A*, 44, 2787–2793.

Vandenbrande, T. (ed) 2006. *Mobility in Europe: Analysis of the 2005 Eurobarometer Survey on Geographical and Labour Market Mobility*. Luxembourg: Office of the Official Publications of the European Communities.

Wolf, D.A. and Longino, C.F. 2005. Our 'increasingly mobile society'? The curious persistence of a false belief. *The Gerontologist*, 45, 5–11.

Zelinsky, W. 1971. The hypothesis of the mobility transition. *Geographical Review*, 61, 219–249.

# 2 Understanding the drivers of internal migration

*Anne Green*

Internal migration plays a key role in national well-being because of its effects on economic, social and demographic change. It can be a major factor in patterns of population and employment growth and decline within countries. It has been identified as fundamental to the efficient functioning of economies and housing markets, as well enabling individuals and families to achieve their goals and aspirations (Bell *et al.*, 2015; Bernard *et al.*, 2014). In so doing, individuals and families make individual, household, economic and non-economic trade-offs (Clark and Maas, 2015). Such decision-making is highly complex, with the role of different factors varying in importance according to the reason for moving and the distance moved. Classical theories of mobility tend to major on the broad distinction between moves over longer distances being motivated primarily by job-related considerations and those over shorter distances more by housing and neighbourhood ones (Boheim and Taylor, 2007; Coulter and Scott, 2015). In reality, however, the situation is much more varied than this.

This chapter draws on the existing literature to demonstrate the multiplicity of factors driving internal migration in its various forms and, in the context of this book's central aim of achieving a better understanding of long-term trends in migration intensities, provides an *a priori* assessment of whether changes in these factors might be expected to lead to an increase or a decrease in frequency of address-changing. It does this by examining five main groups of drivers relating to: (1) changing demography, (2) macro-economic and labour market factors, (3) technological developments, (4) societal and non-economic considerations and (5) other markets, regulatory and institutional structures. The first four of these may be considered generic, in that similar trends may be anticipated across the world, whereas in the case of the fifth group greater variability can be expected between countries due to their distinctive histories and cultures. Inevitably, some key trends and issues cut across the boundaries of these five groups and such intersections are highlighted below.

It should be noted that trends in internal migration are a function of two broad sets of factors: compositional changes, defined in terms of the changing distribution of the population between sub-groups associated with higher and lower migration propensities, and behavioural changes, defined as trends over time in the level of residential mobility of individual population sub-groups. Compositional change and behavioural change may reinforce each other in leading to an increase or decrease in internal migration, but this is not necessarily the case: compositional change may lead to an expectation of increased internal migration but behaviour change might indicate a decrease in internal migration—possibly leading to an aggregate outcome of no change. In this chapter, an attempt is made to separate out the implications for these two broad sets of changes for mobility trends.

It is also the case that internal migration is only one form of spatial mobility. Besides internal migration itself subsuming both long- and short-distance migration as just mentioned, other types of spatial mobility— notably commuting—that Zelinsky (1971) termed 'circulation' can act as a substitute for internal migration. Hence, where feasible and appropriate, a distinction is made in this chapter between whether the drivers and trends outlined might be expected to lead to an *increase* or a *decrease* in the intensity of movement for (a) long-distance internal migration, (b) short-distance internal migration, and (c) circulation. However, as we shall see, often the expected direction of change is not clear cut, in which case the outcome is labelled *unclear*. The next five sections of the chapter discusses each of the five different groups of drivers in turn, while the final section provides an overall assessment of how and whether the drivers identified might be leading to changes in internal migration propensities and volumes.

## Changing demography

This section considers four aspects of demographic change in turn: the age profile of the population, the changing ethnic make-up and diversity of the population, trends in international migration and household structures and living arrangements.

### Age

Age is a key determinant in migration and is sure to remain so. Age may be considered a proxy for the life course, given that key migration events (e.g. moves to higher education, first jobs, family formation, birth of children, retirement and ill-health and housing equity release) tend to occur progressively through the life course. At the same time, variations in cultural context, social norms, economic conditions and institutional structures mean that there are detailed differences in the timing and sequencing of life-course events by age (Bernard *et al.*, 2014).

Propensities to migrate, and reasons for migration, vary over the life course (Coulter and Scott, 2015). The key feature of the age profile of migration highlighted in the migration literature is that rates of migration are highest for young adults, a substantial proportion of whom are students or graduates, associated with participation in post-compulsory education (Lundholm, 2007) and first job (Lomax and Rees, 2015). The propensity to move then declines with age as individuals accumulate 'commitments'— such as children or an employed partner—which tend to make migration decisions more complex and costly (Coulter and Scott, 2015). However, an aggregate migration schedule by age disguises considerable heterogeneity of experience within age groups—and volumes and patterns of internal moves are related also to household structures (discussed in a later sub-section) and economic circumstances (considered in the next section).

The key feature of change across all case study countries in this volume is the ageing of the population, characterised by a greater share of the population in older age groups and a reduced proportion in younger age groups. In compositional terms, this would be expected to be associated with an overall decrease in internal migration because individuals in older age groups tend to be less migratory than those in younger age groups.

In terms of behaviour change amongst young adults, a trend towards increased participation in higher education might be expected to lead to an increase in long-distance migration. Evidence from Sweden indicates a long-term increase in inter-regional migration amongst students (Lundholm, 2007). However, a greater tendency towards attending more local higher education institutions (Christie, 2007) in a country such as the UK, with a tradition of going away to university, may counter this trend.

A broader tendency is for young people to stay in the parental home for longer (Lennartz *et al.*, 2016) and/or move back and forth from the parental home before establishing a more stable independent household, which means that young people may be more migratory (over both long- and short-distances) over an extended period. Evidence from Southampton, England, highlights that the migration experience of students moving into employment is often complex, involving multiple temporary moves between place of study, parental home and a residence close to a new place of employment (Sage *et al.*, 2013). However, in the USA there is evidence that the younger age groups migrated less often in 2009 than in 1999 (Cooke, 2011). There is some evidence that car usage amongst young adults is in decline in Europe, as they wait longer to get a driving licence and use greener modes of transport. This might lead to a decrease in circulation.

Amongst adults in the middle of the age range, a desire for 'rootedness' (in part to provide stability for children's education) may be expected to lead to a reduced propensity for internal migration, and within this a predominance of short-distance housing related moves amongst internal moves— with labour market adjustments made through increased circulation.

With the lengthening of the lifespan, amongst older adults it seems appropriate to distinguish the younger ones who, if they are owner-occupiers, might move to release equity in property and/or to adjust to new lifestyle in a longer physically active old age. Indeed, evidence from Sweden indicates that migration propensity for these was higher in 2001 than in 1970

(Lundholm, 2007). In addition, they may have caring responsibilities for grandchildren and/or for elderly parents—involving travelling to (mainly local) locations. Amongst this sub-group these trends might be expected to lead to increases in both long- and short-distance internal migration and also to increased circulation. For the very elderly, moves may take place over short distances primarily to adjust to the vulnerabilities of the ageing body and associated care requirements (Lundholm and Malmberg, 2009; Findlay *et al.*, 2015). While this might be expected to lead to more short-distance moves amongst the very elderly, trends toward a greater emphasis on independent living may mean fewer movers, so leading to an overall decrease in internal migration for this group.

### Ethnicity and diversity

Changes in the ethnic composition and diversity of populations may impact on the volume and types of internal migration. Evidence from the UK suggests that, although there are differences in migration propensities by ethnic group (e.g. see Hussain and Stillwell, 2008) and in internal migration distances (e.g. with Chinese people tending to move over longer distances more than people from South Asian groups), in general ethnic minorities have higher rates of internal migration than the white population. This suggests that any increase in the population from ethnic minority groups would lead to an increase in internal migration—although there may be differences between countries here.

With globalisation, the ethnic origins of the population are generally becoming more diverse. There are pronounced ethnic variations in household structures, socio-economic structures and age profiles (for instance, non-White ethnic groups tend to have a younger age profile than the host population in Europe, North America and Australasia), which in turn will have implications for internal migration. Hence, it would be expected that many of the differences in volumes and patterns of internal migration by ethnic group would be captured by such other variables, rather than by a separate ethnicity dimension *per se*. Indeed, analyses of internal migration data for the UK from the 2001 Census by Finney and Simpson (2007) suggest that differences in levels of migration result from differing socio-economic and age compositions of ethnic groups. Moreover, different factors influencing migration may vary in importance between ethnic groups; for instance, Raymer and Giulietti (2009) suggest that education level is an important factor influencing migration patterns for the white population, whereas employment status is a more important factor for the ethnic minority population.

### International migration

Related to ethnic composition and diversity, and also to political systems/ institutional structures and policies, is international migration. There is increasing recognition that international migration can impact on internal migration, and also that conventional binary divides between international and internal migration (King, 2002) and between international migration and circulation are becoming less clear cut. In a cross-national analysis, Bell *et al.* (2015) point to a positive relationship between internal migration intensity and the international migration rate.

National and supra-national policy plays an important role in countries' openness to international migration. For instance, there are important distinctions between free movers (i.e. those with a right to move and reside freely within a territory, as between member states of the European Union) and managed migration (i.e. where governments seek to control the types and volumes of international migrants such as for high-skilled workers who may be targeted to address labour shortages and skills deficiencies). There are also other types of international migration flows; for example, asylum seekers and refugees. In general, individuals who have been mobile across international borders might also be expected to be more mobile (over long and short distances within destination countries), such that the compositional change may be expected to lead to increased internal migration.

Highly skilled immigrants on visas associated with managed migration policies might be expected to be more migratory than average (given the associations between skill level and internal migration discussed in the next section). In the UK, for instance, such policy means that there is an increase in intra-corporate transferees—coming to the destination country for short periods—with an increase in circulation at the expense of migration within destination countries. National policy differences are likely to impinge on the behaviour as well as the composition of managed migrants, such that it is unclear what overall trends to expect.

Likewise, free movers are a diverse group, but often tend to work in occupations below their skills levels and so seek to move between locations and jobs once in the destination country. Hence they tend to be characterised by higher than average internal migration rates (over both short and long distances) and also have higher than average levels of circulation as they seek better jobs/different experiences. However, as they become more established in the destination country, it might be expected that such rates of migration and mobility will decline, though still relatively high.

International migration may also impinge on the internal migration propensities of the host population. Since international migration tends to reduce labour and skill shortages, it may serve to obviate the need for longer-distance migration of the host population to address them. However, internal migration over short distances may increase as some members of the host population move out from migrant-dense areas.

### *Household structures and living arrangements*

Although individual characteristics are significant determinants of migration, household structures and living arrangements are important too because internal moves often involve individuals moving together as members of a household. There are some important differences in household structures cross-nationally; for instance, between Anglo-American welfare states and Southern European ones, with the latter tending to have a history of young people living in the parental home for longer than in the former.

In compositional terms, two key features and trends are a rising share of single-person households and an increase in more complex and fluid household structures at the expense of nuclear households comprising of parents with children. An increase in household formation and dissolution would be expected to lead to an increase in internal migration associated with such transitions. A growth in single-person households amongst younger and middle-age adults would be expected to be associated with increases in both long- and short-distance internal migration as they satisfy their own needs and desires, without needing to consider other household members. On the other hand, more single-person households amongst older adults may be expected to be associated with a decrease in long-distance migration.

Other developments in household structures, such as the increase in dual-earner (with two adults in employment) and dual-career households (with two adults pursuing careers in high-skill occupations), have important overlaps with labour market change. Such households need to be in locations where there are employment opportunities for more than one household member and so these types of household are considered in the next section. A growth in the share of more 'fractured' households resulting from divorce—sometimes associated with complex living arrangements, including children living in more than one household—might also be expected to be associated with a decrease in long-distance internal migration as parents living in different households seek to remain physically close to each other to enable dual-residence arrangements for children. Such arrangements might necessitate an increase in short-distance internal migration and also circulation in order to achieve the desired configuration of residential arrangements. In aggregate, the expected compositional impact of these changes in household structure might be a decrease in long-distance internal migration taking place alongside increases in short-distance migration and circulation.

In terms of behaviour change, internal migration amongst single-person households comprised of younger adults is likely to be motivated by attraction to large cities for employment-related and cultural reasons. Migration for lifestyle reasons, including 'discovery migration' linked to wanting new experiences, may also involve long-distance moves. The associated expected behaviour change is an increase in long-distance moves, while housing and affordability constraints (in some countries and some local contexts) may prompt an increase in short-distance moves. Amongst single-person households in the oldest age groups, advances in healthcare technology and lower costs involved in home care for older age groups are both likely to lead to older people remaining in single-person households for longer, rather than moving into residential care. This leads to fewer long-distance moves, but perhaps to more short-distance ones into more appropriate accommodation. There may be an opposing trend for some older single persons to move into extended households, involving either long-distance or short-distance migration.

Turning to fractured households, it seems likely that divorce and other household transitions will continue to culminate in moves to new accommodation. These are generally short-distance so as to enable partners to maintain family ties and stay in contact with children (Mulder and Cooke, 2009). For complex households, there are likely to be advantages in being located in close physical proximity, meaning a continuing trend for reduced long-distance migration. Likewise, Mulder (2007) has suggested that the long-term decline in the importance of family context as an influence on residential choice might now be reversing, given the fluidity of family structures: single parents may be increasingly reliant on family members outside the household because they do not have a partner for support. Similarly, extended family relationships involving three (or more) generations are also likely to impact on migration decision-making (Clark and Maas, 2015). This may help tie households to a particular place, so leading to an expectation of reduced long-distance migration. In relation to short-distance moves in an intra-urban context, the presence of extended family has been shown to be an important factor in determining neighbourhood choice, particularly for immigrants and individuals with low socio-economic status (Hedman, 2013). Proximity to wider family members can also be important in providing care for the elderly.

## Macro-economic and labour market factors

This section considers two features of the macro economy, namely labour demand and the changing organisation of work, and changes in labour supply. Neo-classical models of migration place economic determinants of migration centre-stage, foregrounding the importance of human capital factors in migration and assuming that individuals move for economic gain, typically measured in terms of income (Greenwood, 1975; Mincer, 1978; van Ham, 2002). While recent migration literature has placed enhanced emphasis on non-economic determinants of migration, it remains the case that securing continued employment is of paramount importance in migration decisions for the majority of working-age migrants (Morrison and Clark, 2011).

### *Labour demand and the changing organisation of work*

Five topics are considered here in turn: the state of the macro economy, the changing sectoral profile of employment, the changing occupational structure of employment, changing spatial patterns of economic opportunity and changes in work organisation.

Neo-classical economics suggests that, during times of economic growth, rising demand for labour will result in greater opportunities for job-related mobility and hence long-distance migration, whereas recessionary conditions tend to be associated with below-average migration propensities. This is likely to reflect both a reduction in job openings and the fact that economic downturns tend to be associated with risk aversion. Macro-economic impacts on migration propensities can be substantial, as exemplified by a decomposition analysis of inter-county moves in the USA between 1999 and 2009 in which 63% of the decline in migration rates was attributed to the direct effects of the Great Recession from 2007 (Cooke, 2011).

Throughout recent periods of economic growth and decline there have been clear medium-term trends in the sectoral profile of employment. At a broad level, there has been a decline in the share of employment in primary and manufacturing sectors and an increase in service sector employment (Wilson *et al.*, 2014). In general, since employment in the primary and manufacturing sectors has historically been tied to specific geographical locations to a greater extent than services (although there is an important distinction between producer services and customer services, with the latter being more tied to local populations), sectoral trends may be expected to be associated with diminished sectoral diversity in employment across space and perhaps a decrease in internal migration. Technological developments may serve to reinforce this trend towards a reduction in sectoral diversity as boundaries between sectors become less clear and enable new decentralised production processes.

The key feature of medium-term change in the occupational profile of employment is 'professionalisation' of occupational structures, characterised by increasing demand for labour in high-skilled non-manual occupations. However, there is some evidence for 'polarisation', along with growth in low-skilled service occupations and shrinkage in intermediate occupations (including administrative, secretarial and skilled manual workers) in the middle of the skill and income range. In aggregate, however, the increase in high-skilled non-manual occupations would be expected to lead to an increase in internal migration (albeit there are counter trends in relation to dual-career households, as discussed below).

In terms of spatial patterns of economic opportunity, according to a conceptualisation of migration as a labour market adjustment process, individuals in employment would be expected to move to more dynamic areas. Where employment opportunities are relatively abundant across space, other non-employment goals may play a greater role in migration decision-making (Morrison and Clark, 2011). As noted above, over the long-term, sectoral and occupational structures tend to have become more similar, such that a decrease in long-distance migration might be expected due to there being fewer potential economic gains to be made by moving. However, analyses of medium-term occupational advancement show that there are advantages (as measured by so-called 'elevator' and 'escalator' effects) to individuals of migration to and residence in dynamic city regions, such as London and the Greater South East in the UK context. This underscores the economic advantages of urban agglomerations, with their 'thick' labour markets offering a greater quantity and quality of opportunities. These effects are particularly pronounced in knowledge-intensive sectors and for young people (Gordon *et al.*, 2015). This could lead to an expectation of an increase in long-distance migration to, and circulation involving, the largest urban agglomerations in order to benefit from the more advantageous spatial opportunity structures there.

In terms of hours of work, types of employment contracts and employment status, there are varying traditions between countries in the extent of full-time and part-time working, and in self-employment. However, a general trend in work organisation is a shift towards more flexible working. In general, this has different consequences for those in high-status/high-skill and low-status/low-skill occupations, with the former tending to be able to exercise greater discretion in when and where they work than the latter. Indeed, low-hours flexible contracts for some low-skilled workers place a premium on living close to the workplace, and so for this group of workers the expectation might be one of a decrease in long-distance migration and circulation—but not necessarily an increase in short-distance moves closer to the workplace because of the costs of moving and the precarious nature of some low-skilled employment.

It remains the case that the majority of workers in employment are employees and only a minority are self-employed. The demise of large bureaucracies outlined above opens up increased opportunities for self-employment. Although self-employment tends to be associated with an emphasis on entrepreneurial skills, there has been some growth in low-hours self-employment following recession. This highlights the heterogeneity of the self-employed and the difficulty of identifying what changes in the proportion of self-employed mean for internal migration trends. Traditionally, the literature has tended to suggest that the self-employed tend to be rooted in place and to be less migratory than employees because of the importance of family ties and contacts in the local business community. However, some recent analysis using data from Germany suggests that the self-employed are not more immobile than employees, and indeed those that flow into self-employment are positively associated with inter-regional moves (Reuschke, 2014). This suggests that behaviour change of the self-employed may be associated with an increase in long-distance migration. While there might be nationally specific factors at work here, this latter finding is in line with Florida's (2002) creative class theory that self-employment may be associated with lifestyle preferences about living in (and so migration to) certain geographical locations.

At an organisational level, the key features of change are the demise of large bureaucracies and the rise of a new organisational paradigm in which companies are defined increasingly as 'network orchestrators', in which collaboration in value creation networks is enabled by the virtualisation of business processes, fuelled by the rise of the digital economy (Störmer *et al.*, 2014)—so highlighting the importance of technological change (as discussed in the next section). As a result, work has become less location-specific, more network-oriented, more project-based and increasingly technology-intensive. Workplaces and modes of working are under pressure to increase flexibility and to adapt to business volatility through outsourcing, leading to jobs and organisations becoming increasingly fluid. Businesses are increasingly able to create and disband corporate divisions rapidly, as they shift tasks between slimmed-down pools of long-term core employees, international colleagues and outsourced external service providers (Störmer *et al.*, 2014). From a migration perspective, this means that career paths are less clearly defined than formerly, such that the positive association between social mobility and spatial mobility is less clear cut than formerly, both because of the demise of large organisations and because the changing organisation of work places greater emphasis on individuals (as opposed to large organisations) in career advancement.

### *Changing labour supply*

The main long-term developments in labour supply revolve around ageing and increasing ethnic diversity, alongside changes in the gender and qualifications profiles and—related to these—the changing configuration of labour supply at household level. What impact can these changes be expected to have on internal migration intensities?

Key features of changing labour supply by age are a general shift to later entry into the labour market and also an increase in older workers in the labour market, as older people participate in the workforce for longer (in part because of a rise in state pension ages). Given the association between younger age groups and higher migration propensities, in compositional terms this might be expected to lead to a reduction in internal mobility. In regards to increasing immigration and ethnic diversity, it is unclear what this might mean for internal migration, given ethnic variations in migration propensities and differences in the timing of arrival and characteristics between the various groups of immigrants.

One of the most important changes in labour supply over the medium-term has been the long-term increase in women (including mothers with young children) in employment. In the USA, for instance, the proportion of women participating in the labour force rose from around 40% in the 1960s to around 60% by the end of the first decade of the twenty-first century (US Department of Labor, 2011). In terms of labour market participation, hours of work and wages, women and men have become more equal over time, and women account for a substantial proportion of projected growth in high-skilled non-manual occupations (Störmer *et al.*, 2014). Nevertheless, it is well established that there are important differences between women and men in sectoral and occupational profiles of employment and these have implications for career prospects and financial gain. Analyses of data from the British Household Panel Survey and the UK Labour Force Survey show that female-dominated occupations have lower potential for earnings progression and greater geographic ubiquity (for example, in teaching and health-related occupations). These characteristics make individuals working in these occupations less likely to progress their careers through geographic mobility, leading to a lower propensity to become a 'lead mover' and a higher propensity to become a 'tied mover' in household migration decisions (Perales and Vidal, 2013). Moreover, the gendering of migration serves to reinforce occupational segregation by gender (Halfacree, 1995). Hence, although the increased labour market participation of women and their increased penetration into high-skilled occupations might lead to an expectation in compositional terms of increased long-distance internal migration, patterns of occupational segregation and the presence of many women in dual-earner/dual-career households may militate against such an increase.

A further clear trend in labour supply is the increase in qualification levels, resulting primarily from increased participation in higher education. It is clear that university students now comprise one of the most important migration streams within countries: in Sweden, for example, Lundholm (2007) reveals that students accounted for 40% of inter-regional migrants aged 18–65 years in 2001, compared with just 10% in 1970. Individuals with higher educational attainment working in occupations associated with higher level skills search over geographically more extensive areas than those with lower level qualifications (Van Ham *et al.*, 2001; Fielding, 2012). There is a clear positive association between level of educational attainment and geographical mobility (Greenwood, 1997; Molloy *et al.*, 2011; Brandén, 2013). Moreover, given that individuals with higher qualifications typically have access to greater financial resources to offset the costs of internal and migration and commuting than less qualified workers (Thomas *et al.*, 2015), an increase in the highly qualified proportion of the workforce would be expected to manifest itself in more long-distance migration and commuting. Implications for short-distance migration are less clear. However, given the greater heterogeneity of those with higher-level qualifications arising from the massification of higher education, a reduction in differentials by qualification in migration propensity may be expected and the positive association between education and migration propensity may weaken. The fact that highly educated men tend to be more mobile than highly educated women serves to reinforce this direction of change (Brandén, 2013).

A key feature of the changing configuration of labour supply at household level is the increase in dual-earner and dual-career households and decline in single-earner households. Dual-earner and dual-career households are disproportionately drawn from younger people, the highly educated

and those in higher level occupations (Mulder, 2007)—all factors sugges-
tive of a compositional increase in mobility. Yet research has indicated that
dual-earner couples are less likely to migrate than single-earner couples
(Smits *et al.*, 2003; Lundholm, 2007) and, given the greater complexity of
migration decisions that they face, are more likely to substitute commuting
for migration in order to mitigate disruption to one partner's career that
might result from internal migration to benefit the other partner (Green,
1997). Hence, the compositional effect of an increase in the proportion
of dual-earner and dual-career households is likely to be a decrease in
longer-distance migration. There is no reason to expect a particular change
in behaviour regarding short-distance moves (prompted mainly by non-
economic factors). It is possible that technology could to some extent substitute for commuting (at least for some of the time) amongst dual-career
households, changing the nature of circulation to take on an electronic
rather than a physical form.

## Technological change

Two aspects of technology are pertinent to changes in mobility: physical
travel and transport and the role of the internet and information and communications technologies (ICT). Developments in the former have impacted
on the position of the threshold where trade-offs between internal migration
and circulation take place, and so have a role in redefining the desirability
of internal migration vis-à-vis commuting. ICT developments have enabled
virtual mobilities to substitute for physical movement, both internal migration and commuting (Findlay *et al.*, 2015).

### *Physical travel and transport*

There has been a general trend over the long term towards easier and cheaper
travel, especially over longer distances by land and air, so facilitating circulation. This might be expected to lead to a reduction in internal migration,
as increasingly commuting can substitute for it—over both short and longer
distances. Improvements in the ease of travel can enable an increasing separation of places of residence and workplaces (Fielding, 2012) through longer
(in terms of both time and distance) commutes. The fact that there is a trend
for local labour market areas to become larger over time is indicative of
longer average commuting trips. However, it is possible that more frequent
travel in terms of commuting, business and leisure trips, facilitated by easier
physical travel, transport and virtual contacts over larger geographies enabled by mobile phones and ICT, might lead to greater internal migration, as
mobility begets further mobility (Cohen, 2011; Cohen and Gossling, 2015).
In a similar vein, Bell *et al.* (2015) identify a positive association between the
proportion of the population with a mobile phone subscription and internal
migration propensity.

There is some evidence that, at least in some countries and especially in large cities, fewer young people have driving licences—partly because of the rising costs of running a car. This restricts their travel and commutes to work to locations that can be reached by public transport or by walking or cycling. This behaviour change might lead to an expectation of an increase in internal migration (over both long and short distances) to accessible locations, and a concomitant reduction in circulation. The scheduling and fare structures on public transport (and so the cost of travel) is likely to be influenced by national and sub-national level transport and accessibility policies, which are likely to have a particular impact on groups with less access to cars such as young adults, older adults and people on low incomes.

### The internet and ICT

The internet and ICT have increased access to information about potential travel and migration destinations. Individuals born since the 1980s have grown up in the so-called 'digital age' and this could mean that there are particular cohort effects pertaining to this generation that mark them out from previous generations. In a qualitative study of young Swedish adults, Vilhelmson and Thulin (2013) presented evidence that use of the internet and social media could increase individuals' understanding, awareness and curiosity about other places, so broadening their spatial horizons. Theoretically this might lead to an expectation of increased internal (and international) migration and circulation. Conversely, the fact that use of the internet and ICT enables people to maintain contact with each other more easily and frequently across space might lead to an expectation of reduced migration and circulation. However, the significance of the internet for internal migration might rest more as an enabler, than a key driver. The greater availability of information via the internet negates the need for speculative long-distance moves to find out about opportunities available at the destination; consequently, a greater proportion of moves that are made are likely to be pre-planned and researched. Indeed, modelling of international bi-lateral migration flows by Winkler (2016) suggests that an increase in internet adoption among migrant-sending countries reduces migration from these locations.

As noted above in the section on labour demand, developments in ICT have facilitated the rise of network organisations and have been a key factor in changes in ways of working for individuals. The result is that some types of work are more footloose geographically. Also previously strong links between the 'times' and 'spaces' of work have become fractured, so allowing greater flexibility in where and when work is undertaken. An ability to work virtually can obviate the need for long-distance migration and circulation. However, working from home might prompt an increase in short-distance moving in order to achieve a 'housing space' more commensurate with full- or part-time home working. Evidence suggests that part-time virtual working (i.e. working remotely for part but not all of the time) has increased more than full-time virtual working (Felstead, 2012), and such working arrangements may be associated with a change in the nature of circulation, with fewer longer journeys substituting for more frequent shorter journeys as the internet and ICT facilitate more spatially and temporally flexible working arrangements. Hence, although at face value the diffusion of the internet and ICT might be expected to enable a reduction in mobility, on further investigation the picture is less clear cut.

## Societal and non-economic considerations

Reference has been made above to increasing emphasis in the migration literature on the role of non-economic considerations in migration. Fielding (1992) called for a more culturally informed understanding of migration, and subsequently Halfacree (2004) made a case for a greater appreciation of non-economic issues (including life-course, cultural and spatial factors) to balance work undertaken in the economic tradition. This section highlights the part played by socio-cultural factors and a desire for rootedness in changing mobilities and identifies selected other non-economic considerations impinging on internal migration and circulation.

### *Socio-cultural factors and rootedness*

Social networks and exchanges of social support can influence and be configured by mobility (Mulder, 2007). In research focusing on individuals in challenging economic circumstances in six deprived neighbourhoods in Britain, a need to live close to family and friends emerged as the most important factor relating to mobility and immobility (Hickman, 2010). A desire to live close to family and friends was a key factor in short-distance moves to their current place of residence as well as being a reason for not moving over long distances. The research suggested that, for many workless residents of deprived localities, their ability to 'get by' in difficult and challenging circumstances depended on their immobility, as their neighbourhood provided them with support from friends and family in both material terms and in other ways that helped to maintain working arrangements (e.g., helping with transport and care responsibilities). This location-specific capital is particularly important at times of austerity, when it acts as a deterrent to long-distance migration (Mulder, 2007). Where physical care-giving relationships are involved, geographical proximity is very important. There is a strong distance-decay effect in such circumstances (Knijn and Liefbroer, 2006, quoted in Mulder, 2007), so such relationships may influence migration over short distances. Religious and ethno-cultural factors may also play a similar role in residential mobility decisions.

   The fact that developments in communication infrastructures have facilitated greater mobility and flexibility may in turn fuel a desire for 'spatial anchoring' and 'rootedness'. Cohen and Gossling (2015, p. 1673) suggest that there is "a darker dimension of hypermobility" and one key feature of this is a scaling back of individuals' local and community social networks and a reduced ability to participate in family life. The implications for internal migration are unclear. Over a certain threshold long-distance daily and weekly commuting can disrupt local and family connections, but frequent long-distance moves can have the same effect. It is likely that the location of the threshold for substitution of migration with circulation (and vice versa) varies between individuals and households. There is some evidence, however, that availability of flexible working is becoming an increasingly important factor in employer choice over time, especially for parents and workers with caring responsibilities, as well as the highly qualified who are more likely to commute and migrate over long distances (Störmer *et al.*, 2014).

On balance, a desire for spatial rootedness may lead to an expectation of a reduction in long-distance migration, with a mix of increased virtual and physical circulation substituting for it, but there is no clear reason why an enhanced importance for place-based social capital and rootedness should impact on short-distance migration.

### *Selected non-economic factors*

Compositional factors may play some part in increasing the relevance of non-economic factors in understanding internal migration. Noting the higher-than-average internal migration propensities amongst young people with lower employment rates, Lundholm (2007) contends that it is reasonable to assume that fewer internal migration decisions are motivated by labour market factors than was the case formerly. Nevertheless, economic considerations may be expected to play a key part in mobility decisions for younger adults. Moreover, analyses of longitudinal data by Coulter and Scott (2015) show that people are more likely to move residence for 'targeted' reasons (such as employment opportunities) than 'diffuse' reasons (such as area characteristics). The same analysis of reasons reported by individuals for wishing to move residence indicated that area-related quality of life increases in importance for people from their mid-thirties to mid-sixties (Coulter and Scott, 2015).

It is important, however, not to underplay the role of economic factors in internal migration. As noted by Halfacree and Boyle (1993), a long-distance move for a new job among people of working age may be such a taken-for-granted component in migration decision-making that survey respondents do not think to mention this. Indeed, in many circumstances, it is so important that it must be addressed before a residential move is made (Morrison and Clark, 2011). Clark and Maas (2015) suggest that availability of jobs may be the context within which migration occurs, and that lifestyle choices, access to amenities, housing considerations and family change are important non-economic motivations for movement therein. Hence, it is appropriate (within a broader economic context) to consider migration as a social and consumption decision (Morrison and Clark, 2011). Glaeser *et al.* (2001) emphasise the attractiveness of cities as centres of consumption and associated cultural amenities. In contrast, Partridge (2010) highlights that long-term growth patterns in the USA are consistent with natural amenity-led migration to locations endowed with natural amenities such as pleasant climates and attractive landscapes, mountains and oceans. More generally, green issues have risen up the agenda, leading to increasing concern with environmental considerations in ways of living, working and moving.

In the context of decisions about where to live, considerations about social, cultural and consumption amenities and environmental features of place attractiveness appear to have risen in importance. But the literature on these subjects suggests that such factors may impinge more on the spatial patterns of movement than on its volume.

## Other markets, regulatory and institutional structures

Whereas previous sections dealt with drivers that are largely generic in nature, this section considers factors that are shaped to a greater extent by national regulatory and institutional structures, with nation-specific impact on migration intensity. It begins by discussing housing market factors and then the factors associated with labour market policy (including labour market regulation and welfare provision) and education and training policy. Other national-level policies (e.g. regarding transport) are also important, as highlighted in the section on the role of technology, but space constraints preclude detailed discussion here.

### *Housing market factors*

Housing satisfaction and housing costs are important factors in shaping residential mobility (Morrison and Clark, 2011), with residential moves being prompted by changes in family and household structures in accordance with life course events and also by the cost of housing. Particular emphasis has been placed on variations in migration propensities by housing tenure. In the UK, for instance, social housing tenants have faced particular restrictions in moving between local authority areas and so this tenure is associated with low migration propensities over long distances (Hughes and McCormick, 2000). By contrast, private-sector tenants display greater migration propensities, reflecting greater ease of movement within this tenure. Contrastingly, owner-occupation is an important source of local ties, and so owner-occupiers tend to be less migratory than renters (Mulder, 2007), and are likely to be more inclined than private renters to substitute commuting for internal migration. Macro-economic factors also play a role in the propensity for residential mobility amongst owner-occupiers. While some owner-occupiers in some circumstances might move residence to make financial gains, the state of the macro-economy can also act to repress migration amongst owner-occupiers, as exemplified by a decomposition analysis by Cooke (2011) which showed a significant reduction in the propensity of owner-occupiers to move between 1999 and 2009, attributable either directly or indirectly to the impact of the Great Recession.

Analyses of cross-sectional statistics across the European Union confirm a general trend towards diminishing access to owner-occupation amongst younger adults (aged 18–34 years), resulting in a larger rented sector in many countries (Lennartz et al., 2016). In compositional terms, this might lead to an expectation of increased internal migration (over long and short distances), given higher migration propensities amongst renters than owner-occupiers. A shift in the profile of the rental sector from social to private renting might also lead to an expectation of increased long-distance migration. However, Lennartz et al. (2016) suggest that the focus on the rise of 'Generation Rent' in the popular media may be over-exaggerated, given the trend for adult children to live with their parents for longer (as indicated in the section above on changing demography). Adverse labour market conditions facing young people in the Great Recession and its immediate aftermath, as well as volatility in more financialised housing markets, highlight the role of macro-economic and national financial policy factors in impinging on access to different tenures. Socio-demographic factors also play a role in access to owner-occupation, given the role of financial transfers to younger adults by parents and grandparents.

Housing costs in the owner-occupied and private-rented sectors are also a function of the volume of housing supply relative to demand. Higher rates of internal migration might be expected in circumstances where housing supply is relatively generous relative to demand *ceteris paribus*. Conversely, constrained housing supply relative to demand is likely to mean that for some people a desire to migrate remains unfulfilled. Indeed, analyses across 23 European countries by Caldera Sanchez and Andrews (2011) show that the probability of moving is facilitated by the responsiveness of housing supply and access to credit and constrained by transaction costs and rental regulations.

### *Labour market regulation, welfare and education and training policy*

Differences in labour market regulation, institutional structures, welfare and education and training policies between countries would be expected to lead to national differences in the frequency and nature of mobilities. The openness of national labour markets has implications for international migration and also for the ease of sectoral and occupational mobility associated with geographical mobility within countries. In general, a negative association would be expected between labour market regulation and long-distance migration, alongside increased short-distance migration and circulation amongst labour market 'insiders'. Limitations on pension portability might be expected to lead to an increase in long-distance migration at the behest of an employer, but a reduction in long-distance migration otherwise.

Where active labour market policies are relatively punitive and benefit levels are low (as tends to be the case in Anglo-American liberal economies), the general expectation would be for higher rates of internal migration and circulation to access job opportunities. However, the tendency for greater reliance on family and friends to 'get by' in a context of austerity runs counter to this. By contrast, in countries with more generous benefit entitlements, such as in the Nordic welfare state model, the unemployed are likely to have low mobility rates.

Because post-compulsory education and training policy is typically directed at young adults, the funding arrangements associated with different types of provision and the spatial organisation of education and training establishments can have important implications for the mobility of young people and—given the relational nature of migration—for subsequent migration too. There are marked national differences in the organisation of, and participation in, vocational education and training at initial and advanced levels, but typically participation tends not to be associated with internal migration. By contrast, as outlined in the earlier section on labour supply, there has been a general trend across countries for increased participation in higher education, but there are variations between countries in the extent to which this has been associated with moves away from the home area. These national differences are likely to have implications for variations in future migration propensities and spatial patterns of migration.

## Concluding comments

It is clear from this chapter that understanding the impact on internal migration of recent, current and likely future trends in its drivers is a challenging task. At least five overarching trends can be seen to emerge from reviewing the literature on the various forms of internal migration and their links with the socio-demographic, economic, political-institutional and technological contexts in which address-changing takes place.

In the first place, there is a trend towards increasing *heterogeneity* within demographic and economic sub-groups. This is exemplified by the increasing diversity of older people in terms of income (in part depending on their work histories), health and family situations, and by the greater variability within the population qualified to at least degree level associated with the expansion of higher education. Furthermore, a polarisation in employment structures may fuel ever greater heterogeneity.

Secondly, increased *fragmentation* is evident in demographic and economic spheres. For instance, rising divorce rates and family break-up have led to fissuring in household structures. In terms of work organisation, trends such as short working-hours contracts in some service activities and the technology-enabled separation of economic activities across local areas and national boundaries are indicative of fragmentation. This impacts on how individuals organise their lives in time and space, and so has implications for internal migration.

Thirdly, the form and nature of *networks* is changing. Historically, the migration literature has highlighted the role of social networks in understanding migration decisions and patterns. Social networks remain important and the growth in access to technology has extended the reach and penetration of networks. A key issue in relation to the role of social networks in understanding internal migration relates to when and how electronic networks can substitute for physical networks. Also on the theme of networks, in economic terms businesses may be increasingly seen as 'network orchestrators', using technology to link to non-local resources to meet their objectives (Störmer *et al.*, 2014).

Fourthly, the literature highlights the increasing *complexity* of migration decision-making. Alongside the economic considerations that have dominated neo-classical models of migration, there is increasing emphasis on both the range of motivations for migration and the role of non-economic factors. While these are not necessarily new considerations, arguably they have risen in prominence and so complexity is increasingly important in understanding migration.

Lastly, and with links to both the themes of networks and complexity, a strong emerging theme in the migration literature in recent years is the *relational nature of movement,* such that rather than being a 'one off' event, individuals' life courses and behaviours are linked to those around them (in their own households and beyond) in the broader time-space context of economic, social and cultural change (Findlay *et al.*, 2015). Taken together, these trends suggest that understanding how drivers of internal migration play out in practice is far from simple. The aggregate picture of internal migration is the function of the interplay of short-term cyclical economic processes, medium-term restructuring and long-term shifts in socio-cultural and economic values.

Turning to the more detailed account of the nature of and recent developments in the key drivers of internal migration intensities provided in the preceding sections, Table 2.1 attempts to summarise their expected implications for long-distance migration, short-distance migration and circulation. No attempt has been made here to quantify the relative importance of different drivers and trends. It is evident that different drivers operate in different directions, with some expected to lead to a decrease in long-distance internal migration and others to an increase, and likewise for short-distance migration and circulation. It is also apparent that in several instances it is unclear whether the expectation is for an increase or decrease in mobility.

The attempt to separate out compositional factors from behavioural change in Table 2.1 indicates that, in some instances, expected changes operate in

*Table 2.1* Summary of the changing nature of the drivers of internal migration and their implications for three types of spatial mobility

| | Key features and trends | Long-distance migration | Short-distance migration | Circulation |
|---|---|---|---|---|
| **Demography** | | | | |
| Composition | Ageing of population | **Decrease** | INCREASE | Unclear |
| | Greater diversity in ethnic composition of the population | INCREASE | INCREASE | Unclear |
| | Increase in international migration | INCREASE | INCREASE | INCREASE |
| | More single person households and more complex and fluid household structures | **Decrease** | INCREASE | INCREASE |
| Behaviour change | Young adults | INCREASE | Unclear | **Decrease** |
| | Middle age adults | **Decrease** | INCREASE | INCREASE |
| | Older adults—younger third agers | INCREASE | INCREASE | INCREASE |
| | Older adults—older third agers | **Decrease** | **Decrease** | Not applicable/ Unclear |
| | Ethnicity | Unclear | Unclear | Unclear |
| | High skilled managed migrants | Unclear | Unclear | INCREASE |
| | Free movers | **Decrease** | **Decrease** | **Decrease** |
| | Host population | **Decrease** | INCREASE | Unclear |
| | Single person households amongst younger adults | INCREASE | INCREASE | **Decrease** |
| | Single person households amongst oldest age groups | **Decrease** | INCREASE | **Decrease** |
| | Fractured and complex households | **Decrease** | INCREASE | INCREASE |
| **Macro-economic and labour market factors** | | | | |
| Composition | Sectoral shift from primary and manufacturing to services | **Decrease** | Unclear | Unclear |
| | Occupational change | INCREASE | INCREASE | INCREASE |
| | Changing spatial opportunity structures—in aggregate: | Unclear | Unclear | INCREASE |
| | –to largest agglomerations | INCREASE | | |
| | –to other destinations | **Decrease** | | |
| | Increase in employment precarity for low-skilled | **Decrease** | **Decrease** | Unclear |

| | Col 1 | Col 2 | Col 3 | Col 4 |
|---|---|---|---|---|
| Increase in network organisations and lean management | **Decrease** | Unclear | **Decrease** | Unclear |
| Ageing of workforce | **Decrease** | Unclear | **Decrease** | Unclear |
| Increasing proportion of women in the workforce | **Decrease** | Unclear | **Decrease** | INCREASE |
| Increase in proportion of highly educated in workforce | INCREASE | Unclear | INCREASE | INCREASE |
| Increase in dual-career and dual-earner households | **Decrease** | Unclear | **Decrease** | INCREASE |
| Behaviour change — Occupational change | **Decrease** | **Decrease** | **Decrease** | Unclear |
| Highly educated workers | **Decrease** | Unclear | **Decrease** | Unclear |
| *Technological change* | | | | |
| Composition — Easier and cheaper travel | **Decrease** | **Decrease** | **Decrease** | **Decrease** |
| Increased use of internet and ICT | Unclear | Unclear | Unclear | **Decrease** |
| Behaviour change — Fewer young people with driving licenses (in some countries) | INCREASE | INCREASE | INCREASE | **Decrease** |
| *Societal and non-economic considerations* | | | | |
| Composition — Increased concern about 'green' issues | Unclear | Unclear | **Decrease** | **Decrease** |
| Behaviour change — Increased desire for spatial 'rootedness' | **Decrease** | Unclear | **Decrease** | INCREASE |
| *Other markets, regulatory and institutional structures* | | | | |
| Composition — Medium-term rise in owner-occupation | **Decrease** | **Decrease** | **Decrease** | INCREASE |
| Recent rise in proportion of private renters—and also decline in social renters | INCREASE | INCREASE | INCREASE | **Decrease** |
| Decrease in labour market regulation | INCREASE | Unclear | INCREASE | Unclear |
| Increase in labour market regulation | **Decrease** | INCREASE | **Decrease** | INCREASE |
| Spread of labour market activation policies to more sub-groups | INCREASE | Unclear | INCREASE | INCREASE |
| Massification of higher education | INCREASE | INCREASE | INCREASE | INCREASE |
| Behaviour change — Higher education students | **Decrease** | Unclear | **Decrease** | INCREASE |

Source: Compiled by the author.

different directions. This is particularly evident in the case of occupational change where recent and projected trends would suggest increases in internal migration and circulation as a result of compositional effects. By contrast, behaviour change would indicate a decrease in internal migration. In part, this is linked to an increase in dual-earner/-career households, and an associated tendency towards substitution of commuting for migration. This latter point about occupational and household change highlights the intersecting nature of the various drivers of mobility, which in turn means that the aggregate impacts on mobilities cannot readily be determined.

The challenges in understanding the drivers of internal migration are further compounded by the destandardisation of the life course, which has affected the timing, frequency and geography of moves and the relational nature of mobilities (Findlay *et al.*, 2015). In turn, these trends emphasise the complexity of migration decision-making (Clark and Maas, 2015) and the importance of understanding not only what prompts mobility for movers but also the decisions (which are neither made unconsciously nor just once without renegotiation) surrounding immobility for stayers (Hjalm, 2014).

The assessment presented in this chapter supports the significance of factors such as population ageing, rising immigration, the increase in dual-earner/career households, greater geographical uniformity in the structure of employment and the growth in a desire for socio-spatial rootedness identified in Chapter 1 as explaining the decline in internal migration rates in the USA. Yet other factors—such as technological change—can enable both immobility and mobility, such that the aggregate outcome is less clear. Moreover, as outlined in the preceding section of this chapter, national level institutional and regulatory structures and policies may lead to national differences in the aggregate impact of the some of the drivers of internal migration. Hence, the value of the national empirical studies presented in the second part of this book.

## References

Bell, M., Charles-Edwards, E., Ueffing, P., Stillwell, J., Kupiszewski, M. and Kupiszewska, D. 2015. Internal migration and development: Comparing migration intensities around the world. *Population and Development Review*, 41, 33–58.

Bernard, A., Bell, M. and Charles-Edwards, E. 2014. Life-course transitions and the age profile of internal migration. *Population and Development Review*, 40, 213–239.

Boheim, R. and Taylor, M. 2007. From the dark end of the street to the bright side of the road? The wage returns to migration in Britain. *Labor Economics*, 14, 99–117.

Brandén, M. 2013. Couples' education and regional mobility—the importance of occupation, income and gender. *Population, Place and Space*, 19, 522–536.

Caldera Sanchez, A. and Andrews, D. 2011. To move or not to move: What drives residential mobility rates in the OECD? *OECD Economics Working Paper* 846. Paris: OECD Publishing.

Christie, H. 2007. Higher education and spatial (im)mobility: Non-traditional students and living at home. *Environment and Planning, A* 39, 2445–2463.

Clark, W.A.V. and Maas, R. 2015. Interpreting migration through the prism of reasons for moves. *Population, Place and Space*, 21, 54–67.

Cohen, S.A. 2011. Lifestyle travellers: Backpacking as a way of life. *Annals of Tourism Research*, 38, 1535–1555.

Cohen, S. and Gossling, S. 2015. A darker side of hypermobility. *Environment and Planning A*, 47, 1661–1679.

Cooke, T.J. 2011. It is not just the economy: Declining migration and the rise of secular rootedness. *Population, Place and Space*, 17, 193–203.

Coulter, R. and Scott, J. 2015. What motivates residential mobility? Re-examining self-reported reasons for desiring and making residential moves. *Population, Place and Space*, 21, 354–371.

Felstead, A. 2012. Rapid change or slow evolution? Changing places of work and their consequences for the UK. *Journal of Transport Geography*, 21, 31–38.

Fielding, T. 1992. Migration and culture. In Champion, T. and Fielding, T. *(eds) Migration Processes and Patterns*, Volume 1. *Research Progress and Prospects. London: Belhaven Press,* 201–212.

Fielding, T. 2012. *Migration in Britain: Paradoxes of the Present, Prospects for the Future. Cheltenham: Edward Elgar.*

Findlay, A., McCollum, D., Coulter, R. and Gayle, V. 2015. New mobilities across the life course: a framework for analysing demographically linked drivers of migration. *Population, Place and Space*, 21, 390–402.

Finney, N. and Simpson, L. 2007. Internal migration and ethnic groups: Evidence for the UK from the 2001 Census. *CCSR Working Paper* 2007–04. Manchester: Cathy Marsh Centre for Census and Survey Research, University of Manchester.

Florida, R. 2002. *The Rise of the Creative Class. New York, NY: Basic Books.*

Glaeser, E., Koklo, J. and Saiz, A. 2001. Consumer city. *Journal of Economic Geography*, 1, 27–50.

Gordon, I., Champion, T. and Coombes, M. 2015. Urban escalators and interregional elevators: The difference that location, mobility and sectoral specialisation make to occupational progression. *Environment and Planning A*, 47, 1661–1679.

Green, A.E. 1997. A question of compromise? Case study evidence on the location and mobility strategies of dual-career households. *Regional Studies*, 31, 641–657.

Greenwood, M. 1975. Research on internal migration in the United States: A survey. *Journal of Economic Literature*, 13, 397–433.

Greenwood, M. 1997. Internal migration in developed countries. *In* Rosensweig, M.R. and Stark, O. *(eds) Handbook of Population and Family Economics. Amsterdam: Elsevier,* 647–720.

Halfacree, K. 1995. Household migration and the structuration of patriarchy: Evidence from the USA. *Progress in Human Geography*, 19, 159–182.

Halfacree, K. 2004. A utopian imagination in migration's terra incognita? Acknowledging the non-economic worlds of migration decision-making. *Population, Place and Space*, 10, 239–253.

Halfacree, K. and Boyle, P. 1993. The challenge facing migration research: The case for a biographical approach. *Progress in Human Geography*, 17, 333–348.

Hedman, L. 2013. Moving near family? The influence of extended family on neighbourhood choice in an intra-urban context. *Population, Place and Space*, 19, 32–45.

Hickman, P. 2010. Understanding residential mobility and immobility in challenging neighbourhoods. Sheffield: CRESR, Sheffield Hallam University.

Hjalm, A. 2014. The 'stayers': Dynamics of lifelong sedentary behaviour in an urban context. *Population, Place and Space*, 20, 569–580.

Hughes, G. and McCormick, B. 2000. *Housing Policy and Labour Market Performance*. London: Department of the Environment, Transport and the Regions.

Hussain, S. and Stillwell, J. 2008. *Internal Migration of Ethnic Groups in England and Wales by Age and District type*. School of Geography Working Paper 08/3 Leeds: School of Geography, University of Leeds.

King, R. 2002. Towards a new map of European migration. *International Journal of Population Geography*, 8, 89–106.

Knijn, T.C.M. and Liefbroer, A.C. 2006. More kin than kind: Instrumental support in families. In Dykstra, P.A., Kalmijn, M., Knijn, T., Komter, A., Liefbroer, A. and Mulder. C.H. *(eds) Family Solidarity in the Netherlands. Amsterdam: Dutch University Press*, 89–106.

Lennartz, C., Arundel, R. and Ronald, R. 2016. Younger adults and homeownership in Europe through the global financial crisis. *Population, Place and Space*, 22, 823–835.

Lomax, N. and Rees, P. 2015. *UK Internal Migration by Ethnicity*. Presentation at the GISRUK2015 Conference, University of Leeds, Leeds, 16 April. www. ethpop.org/Presentations/NewETHPOP/NLGISRUK%20UK%20Internal%20 Migration%20by%20Ethnicity%20NL.pdf

Lundholm, E. 2007. Are movers still the same? Characteristics of interregional migrants in Sweden. *Tijdschrift voor Economische en Sociale Geografie*, 98, 336–348.

Lundholm, E. and Malmberg, G. 2009. Between elderly parents and grandchildren—geographic proximity and trends in four-generation families. *Journal of Population Ageing*, 2, 121–137.

Mincer, J. 1978. Family migration decisions. *Journal of Political Economy*, 86, 749–773.

Molloy, R., Smith, C.L. and Wozniak, A. 2011. Internal migration in the United States. *Journal of Economic Perspectives*, 25, 173–196.

Morrison, P.S. and Clark, W.A.V. 2011. Internal migration and employment: Macro flows and micro motives. *Environment and Planning A*. 43, 1948–1964.

Mulder, C.H. 2007. The family context and residential choice: A challenge for new research. *Population, Place and Space*, 13, 265–278.

Mulder, C.H. and Cooke, T.J. 2009. Family ties and residential locations. *Population, Place and Space*, 15, 299–304.

Partridge, M. 2010. The dueling models: NEG vs. amenity migration in exploring U.S. engines of growth. *Papers in Regional Science*, 89, 513–536.

Perales, F. and Vidal, S. 2013. Occupational characteristics, occupational sex segregation, and family migration decisions. *Population, Place and Space*, 19, 487–504.

Raymer, J. and Giulietti, C. 2009. Ethnic migration between area groups in England and Wales. *Area*, 41, 435–451.

Reuschke, D. 2014. Self-employment, internal migration and place embeddedness. *Population, Place and Space*, 20, 235–249.

Sage, J., Evandrou, M. and Falkingham, J. 2013. Onwards or homewards? Complex graduate migration pathways, well-being and the 'parental safety net'. *Population, Place and Space*, 19, 738–755.

Smits, J., Mulder C.H. and Hooimeijer, P. 2003. Changing gender roles, shifting power balance and long-distance migration of couples. *Urban Studies*, 40, 603–613.

Störmer, E., Patscha, C., Prendergast, J., Daheim, C. and Rhisiart, M. 2014. The future of work jobs and skills in 2030. *Evidence Report* 84. Wath-upon-Dearne and London: UK Commission for Employment and Skills.

Thomas, M., Stillwell, J. and Gould, M. 2015. Modelling multilevel variations in distance moved between origins and destinations in England and Wales. *Environment and Planning A*, 47, 996–1014.

US Department of Labor. 2011. *Women in the Workforce: A Data Book*, U.S. Bureau of Labor Statistics. Washington DC: US Department of Labor.

van Ham, M. 2002. *Job Access, Workplace Mobility, and Occupational Achievement.* Delft: *Eburon.*

van Ham, M., Mulder, C.H. and Hooimeijer, P. 2001. Spatial flexibility in job mobility: macrolevel opportunities and microlevel restrictions. *Environment and Planning A*, 33, 921–940.

Vilhelmson, B. and Thulin, E. 2013. Does the internet encourage people to move? Investigating Swedish young adults' internal migration experiences and plans. *Geoforum*, 47, 209–216.

Wilson, R., Beaven, R., May-Gillings, M., Hay, G. and Stevens, G. 2014. Working Futures 2012–2022. *Evidence Report* 83. Wath-upon-Dearne and London: UK Commission for Employment and Skills.

Winkler, H. 2016. How does the internet affect migration decisions? *Applied Economics Letters*. doi:10.1080/13504851.2016.1265069

Zelinsky, W. 1971. The hypothesis of the mobility transition. *Geographical Review*, 61, 219–249.

# 3 Studying internal migration in a cross-national context

*John Stillwell, Martin Bell*
*and Ian Shuttleworth*

This book draws together empirical material on temporal trends in internal migration in selected countries across the developed world in order to explore whether the decline in migration intensity observed in the USA is evident in other developed societies. If the trends in the USA are observed across a suite of comparator nations, then it becomes plausible to contend that the structural economic and social changes that have taken place across the advanced nations of Europe, North America, Asia and Australasia have acted to reduce the propensity for internal migration. If, on the other hand, each country has experienced different, and possibly unique, temporal trends in migration rates in recent decades, the opportunity for grand theory formulation becomes less attractive. Alternatively, it may be that declining migration intensity is not confined to countries that are economically advanced, but that it is more widely spread across nations at earlier stages of development, as indeed seems to be the case with the evidence presented in Chapter 1, which suggests that there is a general period effect that is acting on all countries to a greater or lesser extent.

The question which is the focus of the book clearly requires a cross-national comparative perspective. Given the weight of research that has focussed on internal migration in all its guises across several disciplines, it is tempting to assume that cross-national analysis is straightforward and analyses of migration behaviour in different countries abound. This, however, is not the case; our basic question is easy to pose but much harder to answer as countries—even those with well-established population data systems—differ in the way that data are collected, in their definitions of migration and in the spatio-temporal coverage of the data that are available. Consequently, previous cross-national studies of internal migration have tended to focus on comparisons between a relatively small number of selected countries rather than confront the challenges of data collection and harmonisation associated with a more comprehensive set of countries. This book falls primarily into the former category by asking individual experts to produce case studies for seven countries that answer the same general questions without being rigidly prescriptive about the means of doing this. In this sense, it differs from studies designed to compare migration using a standard set of migration indicators in two countries (e.g., Bell *et al.*, 2002) or to compare one dimension of migration amongst a relatively small set of countries (e.g., Long *et al.*, 1988). However, it does also embrace the work of a research project that involved comparative analysis using consistent indicators of migration intensity, impact and distance for a much larger number of countries across the world—the IMAGE project (Bell *et al.*, 2014, 2015c; see also Chapter 4 of this book and www.imageproject.com.au).

Whilst the statistical issues associated with data harmonisation are particularly important when it comes to making consistent comparisons between countries, as in the IMAGE project, it is also essential to discuss these issues to better inform the limitations and possibilities of the empirical approaches in the country case study chapters that form the core of this book. There are also other conceptual and methodological issues that underlie the comparative focus of the book. Implicit in the structure of the volume are assumptions about the desirability and feasibility of cross-national comparisons, the focus on nation states as the units of analysis and the importance of internal migration versus other types of population mobility. In assessing the material that is presented in the country case studies, it is therefore also important to explore these issues further so as to understand the limits (and benefits) of our comparative approach.

The purpose of this chapter is to examine these issues in more detail and thereby help readers interpret the statistical material and the arguments that are presented in subsequent chapters. It begins by discussing some of the benefits and the problems of the comparative case study approach in social science, before going on to justify the focus on selected countries (rather than other spatial units) and the migration of people within them. The section that follows thereafter presents an account of the statistical problems of conceptualising and measuring migration in a way that is comparable between countries, drawing primarily on the work of the IMAGE project but also on the experience of preparing the country case studies for Chapters 5–11. It includes a discussion of the challenges faced not only in measuring time series internal migration but also in undertaking quantitative analyses with the data that are available.

## The comparative approach

As mentioned in Chapter 1, the comparative approach that has been adopted in this book is often used in social science and has been commonly (although not exclusively) applied by political scientists. It is a useful approach for generating and testing hypotheses (Collier, 1993) and can assist in the analysis of similarities and differences between societies. It is usually based on the systematic analysis and discussion of a small number of cases and it is therefore suitable when it is not feasible to work with a large number of observations or when it is impossible to design and conduct an experiment, something which is often precluded by the open-ended and complex nature of many social science questions. A comparative approach can throw light on how differences and similarities occur, whether they are between groups of individuals or geographical areas, but important for its success are the units selected for analysis. Lijphart (1971) argues for analytical units that have as much as possible in common—otherwise they would not be comparable—but which differ on some key dimensions that are the focus of exploration. The strengths of comparative approaches are that they are suitable for answering many social science questions and they are relatively easy to implement, but their weaknesses lie in the many variables and characteristics which it is impossible to devise controls for. There is thus sometimes a danger of overplaying the exceptional and the unique at the expense of the general, but these tensions mean that the method is suitable to tease out how general structural forces are mediated and modified by national circumstances. This type of approach is thus well suited to cope with the issues that are the concern of the book.

A comparison between a selection of more advanced countries permits assessment of the extent to which declines in internal migration are common across the developed world, thereby allowing an assessment of the argument that common forces are operating there. However, although this approach can be used to challenge and invalidate the central hypothesis of the book, it cannot be used to prove that it is true. In fact, we are not in the business of accepting and discarding hypotheses in these statistical scientific terms but instead operate on the border between quantitative and qualitative approaches, making informed judgements on the basis of the evidence presented in each of the country case study chapters and attempting to draw conclusions from them.

Plausible arguments might be made that challenge this focus on countries (or states) and also on migration flows within them at the expense of other types of mobility and other units used for analysis. For example, it is entirely feasible to conceptualise migration along a geographical continuum from the very local to the international/global, differentiated only by the prime motives for moving and with a distinction often drawn for intra-national movement between shorter-distance residential mobility and longer-distance internal migration (see, for instance, Niedomysl, 2011; Niedomysl *et al.*, 2017). Likewise, the same type of argument can apply to the temporal dimension of migration with long-distance commuting shading into long-term business trips and other types of transience which involve moves of varying degrees of permanence over differing spatial scales (see, for instance, Bell and Ward, 2000). One classic example of a population subgroup whose status as internal migrants is questionable is higher education (HE) students in the UK, living some of the year at an address in the vicinity of their chosen university but returning on regular intervals to the parental domicile (Duke-Williams, 2009). The picture is made more complex by the fact that different forms of mobility are often closely connected. International moves are often preceded or followed by migration within a country, and international migration can influence internal migration by placing pressure on labour and housing markets. In a similar way, temporary forms of movement such as long-distance commuting or seasonal migration may substitute for or morph into more permanent changes of residence (Bell and Ward, 2000).

Despite these complications, internal migration remains important in its own right. It is one of the prime mechanisms by which labour demand and supply are matched at local and regional spatial scales, and where someone lives is an important determinant of their life chances, educational success and life expectancy. However, the distinction between internal and international migration can sometimes be blurred by the regulations operated by national governments or supra-national organisations to manage migrant flows as well as by motives. Job-related reasons, for instance, are important for both international and longer-distance internal migrants. Whereas international migration frequently requires some form of documentation or permission (commonly a passport and often a visa), there are situations where this is not a formal requirement, such as migration between the countries of the UK, or where freedom of movement across borders is accepted, such as movement between those countries in the European Union (EU) Schengen zone. The role of the nation state in protecting its borders has therefore been diminished by supra-national organisations such as the EU and the United Nations (UN) in the interests of allowing individuals freedom of movement for economic reasons or to seek sanctuary as refugees. However, in recent years, some parts of the world have witnessed a backlash against this sort of arrangement, as evidenced by the reaction of some EU member states to the mass refugee migration from the Middle East and Africa, which many commentators suggest underpinned the 2016 Brexit referendum vote in the UK. Increasingly, nation states are responding to the erosion of control of their political territories (e.g. by globalisation and supranational political organisations) by establishing or strengthening border controls and by reclaiming sovereignty in areas such as labour law, housing market regulation, welfare provision and education. These are all arenas which influence internal migration. Moreover, despite the extensive attention given to international migration in the media and public debate, migrations between regions *within* countries outnumber movements *between* countries by a factor of four to one (Bell and Charles-Edwards, 2013), and the difference would be even greater if within-region moves and residential mobility were included. The scale of this movement alone, coupled with the changing international context, therefore readily justifies the national focus adopted in this volume—'internal migration' still matters.

This section has provided a justification for the subject matter of the book. However, it is one thing to set out this framework for the comparative study of internal migration and quite another to measure it in ways that permit cross-national comparisons. The latter is by no means easy to achieve. Indeed, while some statistical or methodological difficulties can be overcome or lessened, others are more intractable. The remainder of this chapter is therefore, as mentioned above, devoted to examining these statistical and technical issues in more detail, so as to make readers more critically aware when approaching the material presented in subsequent chapters.

## Studying internal migration in an international comparative framework

The chief obstacle to achieving a fully comparable analysis is the difference between national statistical systems for the provision of population data. In the UK, for example, data on internal migration can be obtained from the population census administered by three separate national statistical agencies and from administrative sources such as the National Health Service Central Register (NHSCR) and the Patient Register Data Service (PRDS). In Sweden it can be procured from a population register administered by the Swedish Tax Agency, and in the United States from a decennial population census and from surveys such as the American Community Survey (ACS) and the Current Population Survey (CPS) conducted by the US Census Bureau. There is considerable diversity, therefore, between just three of our case study countries, never mind the entire sample.

The types of issues that cross-national comparison raises for migration researchers are categorised into four general groups of problems by Bell *et al.* (2002): (i) temporal comparability—the interval over which migration is measured, commonly 1 or 5 years, but also the availability of comparable time-series data; (ii) the way in which migration is measured—for instance as a transition or an event; (iii) data coverage and quality—for instance, certain population groups can be undercounted or omitted entirely from some collections; and (iv) differences in spatial units—this is the well-known modifiable areal unit problem (MAUP) which makes it difficult to compare meaningfully countries with different administrative geographies. An example of the latter is a comparison of inter-state migration in the USA with inter-regional migration in the UK. These are very different geographies at different spatial scales, and hence the type of migration and the distance of move might also be very different. These problems make comparative measurement and analysis a difficult exercise and are discussed in turn below. Some of the problems are important for comparisons between places and through time, whereas some are more important for one or the other. Certain limitations can be addressed and their effects ameliorated, whereas others are more intractable and, in making comparisons through time and between places, our only solution is to be aware of the potential difficulties and to make an informed judgement about the evidence that is presented.

*Measuring migration*

One important difference between population data systems is the way in which they record and conceptualise migration. Population registers and administrative data systems usually record migration as an *event*. These systems are designed to capture every move that is made by an individual, though the data are generally assembled into discrete length periods prior to release, or designed to capture changes of residence by extracting information at specified annual intervals. An example of this is Statistics Sweden (SCB, *Statistiska Central Byrån*) which compiles data collected from various public agencies. The migration registers are created by combining the Total Population Register (*Rikets Totalbefolkning*—the country's total population) with the Property-Tax Register (*Fastighets Taxerings Registret*) with matched information downloaded annually on the last day of December. This means it is possible to follow individuals and their moves on a yearly basis.

In the UK, NHS Digital (formerly the Health and Social Care Information Centre) maintains a demographic database of all registered patients called the Patient Register Data Service (PRDS). A version frozen at the end of each July is used in conjunction with a similar version from one year earlier to generate records of patients who have changed address (recorded as postcodes). These postcode-to-postcode records, classified by age and gender, are used to produce counts of migrants between local authorities; that is, transition data. These transition counts are converted to movement counts by applying ratios of moves to transitions available from a legacy database that counts patient re-registrations for current health administration areas, the National Health Service Central Register (NHSCR). The local authority to local authority counts are supplied to ONS for use in creating mid-year population estimates (ONS, 2012). The NHSCR was not created as a migration registration system, so it provides an indirect method of deriving estimates of relatively long-distance moves taking place between health areas by capturing patient re-registrations with their doctors (Champion and Shuttleworth, 2016a). Although data on all postcode to postcode changes are captured from the PRDS, shorter distance relocations (transitions) are not released for confidentiality reasons. The only moves not captured are those of addresses within a postcode, which are likely to be very rare, as well as those of patients who fail to inform the NHS of their new address when they move (Barr and Shuttleworth, 2012). Medicare registers which provide migration data from the Australian national health system are subject to similar problems and, like British NHSCR data, fail to capture the movements of military personnel. Indeed, selective coverage is endemic to most types of administrative by-product statistics, including such sources as electoral rolls, which are commonly confined to citizens aged 18 and over. Such issues highlight the complexities involved in using administrative data to measure migration and the need for users to be aware of the procedures adopted to generate the data sets that are released. The creation of a migration data time series is complicated further in the UK because different registers are maintained and different methods of estimation are used by the respective national statistical agencies in Scotland and Northern Ireland (see Lomax *et al.*, 2013, for more detail).

Population censuses, in contrast, generally measure migration as a *transition*. A good example of this is the UK Census. Since 1961, UK population censuses have asked a question on address one year before the census, which yields counts of the number of migrants over the 12-month period. Of course, this approach provides only a single snapshot of migration in a particular year and omits moves earlier in each intercensal decade. The Office for National Statistics Longitudinal Study (ONS-LS) provides a longer-term perspective by linking the locations of a sample of individuals from one census to the next. This, in effect, generates a 10-year transition measure which is dependent, of course, on accurately measuring residential locations in successive censuses (Champion and Shuttleworth, 2016b). The weakness of this type of data is that there will be at least some multiple or return moves over a decade about which we know nothing at the start and end of a decade. In the most extreme case, a person might be located in the same place in 2001 and 2011, but have made multiple moves in the intervening period before returning to their original residence. Nevertheless, these longitudinal data prove very useful in providing an indication of the distance over which migrants travel, suggesting a significant decline in the propensity to move relatively short distances (<10 km) in England and Wales over the last four decades, as summarised in Chapter 6 and in Champion and Shuttleworth (2016b).

Bell *et al.* (2002) note that analytically these distinctions between data types are important. Transition data capture *migrants* whereas event data capture *migrations*. Differences between the two types of measures are relatively small over short periods, such as a single year, but increase exponentially as the observation interval lengthens. Even over single years, careful harmonisation is needed since transition data measure ages at the end of the migration interval whereas event data capture ages at the time of migration (see Bell and Rees, 2006). A comprehensive global inventory of the types of internal migration data collected across the 183 of the 193 member states of the United Nations is provided by Bell *et al.* (2014), who also assess their comparative strengths and weaknesses in detail.

### *Temporal comparability*

The theme of temporal comparability is complex. One very obvious problem concerns the interval over which migration is measured. It is not possible to compare a question in one country on previous address five years ago with data from elsewhere on address one year ago. In particular, it is not possible to create comparable rates simply by multiplying the one year rate by five because five year rates are affected by return and repeat migration, and there is no simple empirical multiplier to place the estimates on a consistent base. As a result, cross-national comparisons must be made separately for countries that collect one-year or five-year data (Bell *et al.*, 2002). Moreover, many countries collect information by referring simply to the 'last move', irrespective of timing, while others measure only 'lifetime migration', comparing place of residence at the census with place of birth (Bell *et al.*, 2015c). In this situation, reliable comparisons between countries, or over time, are largely out of reach.

There are also differences in the temporal depth of various data sources. For example, the Swedish Population Register used in Chapter 9 runs from 1990 to 2014 and other Swedish data from 1900, whilst the annual NHSCR time series for England and Wales used in Chapter 6 starts in 1975. Chapter 6 also uses a set of annual estimates of inter-district migration intensity for the UK running from 2001 to 2013 as well as the ONS-LS data in England and Wales from 1971 to 2011, whereas the United States Population Survey of Income Dynamics (PSID), referred to in Chapter 5, started in 1968 and ran annually until 1997 and every other year thereafter. Clearly, considerable caution must therefore be exercised when collating information on temporal trends given these considerable differences between just three countries, all with well-established population data and statistical systems.

Differences in the timing of population censuses also hinder comparability, because they are not necessarily synchronised with fluctuations in the labour and housing markets which often shape migration (Bell *et al.*, 2002). In these situations, it is important to rely on country-specific knowledge and a combination of sources to get a 'best picture' of a reality that might only be seen with difficulty. These problems are compounded by the sensitivity of internal migration to economic cycles which mean that the start and end point of an analysis cannot be ignored (Champion and Shuttleworth, 2016a).

## *Data coverage and quality*

One problem with migration data, especially from censuses, is that migrants are hard to enumerate. These are 'hard-to-count' populations because they are mobile—and hence difficult to tie down in statistics—and fall into just those demographic groups (for example, the young, students, and those in private rented and communal accommodation) which are problematic for other reasons. It is likely, therefore, that there is an undercount of migrants to a greater or lesser extent. Administrative systems that record migration events are particularly susceptible to these problems. The NHSCR is based on health identification numbers and internal migration is measured by de- and re-registrations with doctors. However, we know that some groups (such as the younger, healthier and more mobile) tend to lag in re-registering (or even not register at all), which means some migrants are left in the wrong place, as far as the registration system is concerned, and some moves are unobserved (Stillwell *et al.*, 1992; Barr and Shuttleworth, 2012). This is also a problem faced by the Swedish population register which may undercount the internal migratory moves of young people who leave the parental home if they refuse to register for certain services. At the same time, groups such as military personnel, overseas visitors and recent immigrants may be omitted entirely from population registers and administrative systems in certain countries since the criteria for inclusion and for registration vary widely between countries and data collections (see Bell *et al.*, 2015c).

There are other analytical issues with data quality and coverage. These are readily exemplified by the problems experienced when trying to analyse internal migration in the UK that arise from changing questions and definitions used in the census. Questions asked at each census vary through time and some topics, for example education, undergo major changes that reflect fast-moving changes as new qualifications start and others end. New questions are also asked: ethnicity, for instance, was introduced to the England & Wales Census in 1991. Furthermore, the population base also changed, with students before 2001 being recorded at their vacation (normally parental home) address but in 2001 and 2011 recorded at their term-time address. All these changes make it difficult to make reliable comparisons of internal migration through time. These problems are not restricted to censuses but can also apply to surveys where changed methodologies can lead to discontinuities, as is the case with the United States PSID, and which mean that care must be taken in interpreting changes from the start of the data series in 1968.

### Differences in spatial units

Internal migration is an inherently spatial phenomenon. It is therefore bedevilled by the problems that are common to all spatial statistics in which measurement is conditioned by the geographies used to capture, output and represent data. The problem of the MAUP is well known and applies across many fields of study (see, for instance, Openshaw, 1984; Fotheringham and Wong, 1991; Openshaw and Rao, 1995; Holt *et al.*, 1996; Manley, 2014). The gold standard is to have *x,y* coordinates for individuals or households as is the case in the Swedish population register. This is geo-referenced to a 100-m resolution, allowing local and longer-distance moves to be accurately measured. Something similar is possible with the ONS-LS, where detailed address information collected by the census permits internal migration/housing moves to be defined at a very fine spatial resolution, although it should be noted that the accuracy of this geo-referencing varies between censuses and care must be taken in considering whether all moves or just some over a certain threshold should be considered (Champion and Shuttleworth, 2016b). It is very important to have this finely grained data since most address-changes occur over short distances and the majority of internal migration is thus relatively local. In the UK, data from a consumer survey by a commercial company known as the Acxiom Research Opinion Poll has provided this level of geographical granularity for migrants in Britain over three years in the mid-2000s (Stillwell and Thomas, 2016).

However, except in a few national cases, the migration analyst must work with migration statistics based on pre-defined geographies such as states, regions, parishes, *Länder* or other statistical/administrative areas. These are inconsistent in size and shape between countries and this means that estimates of migration distances based on moves between population or geometrical centroids are not comparable. Moreover, administrative and statistical geographies can and do change within countries, making comparisons through time even within the same state problematic (Champion and Shuttleworth, 2016a). One solution is to compile data on the lowest common denominator, but more analytically sophisticated approaches have also been devised. Building on the work of Courgeau (1973; see also Courgeau *et al.*, 2012), Bell *et al.* (2015a) have used migration intensities measured at a range of different spatial scales to make estimates of *all residential moves* which are comparable across countries, circumventing the problems caused by differences in spatial frameworks. Coupled with data from the few countries that collect this information directly in the census, this method provided the basis for robust comparisons of migration intensity across 96 countries representing 80% of the global population (see Chapter 4). In a similar manner, Stillwell *et al.* (2016) have shown how spatial interaction models can be fitted to inter-zonal migration flows to generate distance decay (beta) parameters that capture the effects of distance on migration in a single index that is largely independent of the spatial scale of which migration was measured. Rees *et al.* (2016) describe another new index, the Index of Net Migration Impact, which allows reliable cross-national and temporal comparisons of the extent to which internal migration operates to shape the redistribution of population within countries.

### Data availability

Notwithstanding this progress in the development of analytical techniques, cross-national comparisons are fundamentally constrained by data availability. The IMAGE inventory has catalogued what internal migration data are available and from which sources for all countries across the world, indicating that, of the 193 UN member states, 82% collected data from censuses, 26% from administrative sources, 57% from surveys and 56% from multiple sources (Bell *et al.*, 2015b, Table 1). The inventory also indicates that countries collect transition data based on different observation periods (one-year, five-year, other fixed interval, lifetime, last move) and, of particular importance, the variety of forms in which the data collected are released.

If we consider only aggregate migration, there are some countries whose national statistical agencies release origin-destination matrices of migration flows at a number of spatial scales. This is particularly beneficial when directional migration patterns or migration distances are the focus of comparison between countries. Nevertheless, caution is needed in interpreting the diagonal cells of the matrix in some countries since these may contain either counts of intra-zonal flows, counts of intra-zonal movers and stayers or even flows between zones at a lower level in the geographical hierarchy than that for which the data are released. In some countries, this component of the matrix is missing altogether, preventing a figure for total migration in the country from being derived from the matrix. In other countries, only the marginal totals of zonal in-migration and out-migration are available, especially in the case of flows disaggregated by gender or age group (Bell *et al.*, 2015b).

Another common form in which migration data are made available is as simple counts of total internal movement at various spatial scales, such as between states, between counties or between municipalities. These are often referred to as *migration status* data, and are the most regular form in which information on population mobility is reported on national statistical websites. Sample surveys also commonly provide data of this type, often accompanied by details of the characteristics of movers or the reasons for migration. This form of count data provides no information on the spatial pattern of migration flows, but it does provide a crucial measure of the overall intensity of migration at different spatial scales. A small number of countries also collect information on all changes of address irrespective of spatial scale (see Bell *et al.*, 2015b, 2015c), but even where this is unavailable, migration counts at multiple spatial scales provide the essential building blocks to estimate aggregate migration intensities in a form that is comparable between countries, as mentioned above (see Courgeau *et al.*, 2012; Bell *et al.*, 2015a). By the same logic, such counts provide a basis to estimate the trend in migration intensities through time, as explained below, even where regional and local area boundaries have undergone considerable change.

Not all the data collected by countries using various instruments are published in readily available tables or accessible from online information systems, so the task of gathering data for comparative analysis is often less than straightforward. The IMAGE inventory is one attempt to accomplish this: it contains internal migration data of various types and forms extracted from repositories (such as the Integrated Public Use Microdata Series-International (IPUMS), the Centro Latinoamericano y Caribeno de Demografia (CELADE), the EUROSTAT database, or the USAID's Demographic and Health Surveys (DHS)) or supplied by national statistical offices in countries around the world, together with the relevant aggregate populations at risk and the boundaries of the geographical zones at each geographical scale for which spatial migration data are available. Many of these data sets, together with GIS boundary files, are now freely available on GitHub (https:// github.com).

Finally, we have to recognise that migration data collected by national agencies using census or survey instruments may go through extensive processing before being released as 'official statistics'. The UK Census is a particular case in point, with a range of pre-tabulation and post-tabulation adjustments made to create a set of estimates from the raw statistics that meet the confidentiality requirements imposed by the current legislation. In the case of England and Wales, adjustment methods have changed from one census to the next, creating further uncertainty over the legitimacy of comparison between censuses. One specific example of this is the use of small cell adjustment methods (SCAM) in the 2001 Census to ensure that all flows of one or two individuals in the migration tables were changed probabilistically to values of zero or three, consequently rendering the matrices of flows between small areas such as output areas much less useful and limiting the opportunity for consistent comparison with flows at this spatial scale in 2011 when the SCAM was not applied. In a similar way, new coding procedures were introduced in the 1996 Australian Census which resulted in a major disruption to the five yearly census-based time series of migration that dates back to 1976, in this case resulting in a marked upwards shift in the apparent level of local residential mobility (see Chapter 7).

## Pathways to comparability

As indicated in the previous section, there are real problems which limit the reliability of comparisons of internal migration between countries. Temporal analysis within countries confronts similar obstacles. The individual country contributions which comprise the heart of this volume are all faced with the impediments described above and have adopted a number of approaches to solve them, or at least minimise their effects.

### *Creating consistent definitions*

Some of the solutions are relatively straightforward, such as coding variables to the lowest common denominator and compiling data on a single consistent geography. This was the approach adopted by Champion and Shuttleworth (2016a) in constructing a time series of aggregate migration flows between health areas and regions in England and Wales from the NHSCR from 1975 to 2011. In a similar manner, the chapter on Australia draws on the Australian Internal Migration Database which was carefully constructed using GIS overlays of the basic building blocks (Statistical Local Areas) to create a consistent geography of 69 functional regions (Temporal Statistical Divisions— TSDs) that are spatially harmonised across seven censuses to produce a time series spanning 35 years (Blake *et al.*, 2000). Stillwell *et al.* (2000, 2001) used TSDs to create a hierarchical structure built around six types of city regions that allowed robust comparisons to be made with migration flows through a similar spatial system constructed from districts in the UK.

Even where concerted attempts are made to harmonise the data, however, there is little basis on which to compare particular forms of migration from one country to another. The supposed distinction between residential mo-bility and migration is a particular case in point. As Niedomysl (2011) makes patently clear, there is no obvious or easily defined empirical cut off between local and long-distance migration, and in most countries the distinction is based simply on readily available data for convenient administrative bound-aries, which inevitably differ between countries in size, extent and relevance. Thus, what is defined as local in the USA or Australia is likely to bear little correspondence to data which are similarly described in Japan or the UK. Stillwell *et al.* (2016) have proposed creative solutions to the problem of com-paring countries with respect to migration distance but, for the purposes of this volume, particular care is needed in making comparisons in regard to 'local' or 'long-distance' mobility, because these will likely measure quite different things across our sample of countries.

All our sample countries suffer errors and inconsistencies due to under-counts and undercoverage to a greater or lesser degree depending on the type of data they use, as spelt out in the relevant chapters. In some cases, consid-erable data manipulation has been needed to create a useable time series. In the case of the UK, for example, a nation state that contains four component home nations (England, Wales, Scotland, Northern Ireland) with three na-tional statistical agencies (NSAs), construction of a consistent time series was particularly problematic. The Office for National Statistics (ONS) col-lects data from the other two NSAs—National Records of Scotland (NRS) and the Northern Ireland Statistics and Research Agency (NISRA)—which are compiled to produce aggregate mid-year population estimates for local authority districts across the UK together with estimates of the components of change using a common methodological approach (ONS, 2011). There are, however, various availability and consistency problems associated with the internal migration data used in the population estimation process by each NSA.

*Modelling data*

Chapter 6 reports the temporal changes in migration propensities within the UK that are evident from a time series of estimates that connect the two census periods, 2000–2001 and 2010–2011, and which has been assembled from data collected from administrative sources used by the three NSAs. An important distinction is drawn between migrant flows between Local Authority Districts (LADs) that occur within the same nations and flows between LADs that cross the boundaries between England & Wales and Scotland and Northern Ireland and are described by Lomax *et al.* (2013) as internal 'international' flows. Whilst administrative sources provide infor-mation about all the flows in the former group with the exception of flows be-tween districts in Northern Ireland, the latter migration flows are unknown and have to be estimated from data on known marginal totals of migration flows between the countries. A number of methods are available to solve the problem of estimating missing data in origin-destination matrices, including log-linear models, gravity models, spatial interaction models, entropy and information maximisation models and Iterative Proportional Fitting (IPF). Different techniques have been compared by Willekens (1980, 1983) and, after modelling a multidimensional dataset using different methods, van Imhoff *et al.* (1997) conclude that IPF is the most efficient in terms of the time taken to generate a solution. IPF was probably first applied to fit a contingency table using marginal constraints by Deming and Stephan (1940) and a comprehensive history of the methodology is provided by Záložnik (2011). Details of the methodology underpinning the estimates used in Chapter 6 are available in Lomax *et al.* (2013) which also reports strong correlations between IPF-derived estimates and observed annual data for the districts of England and Wales derived from PRDS data.

### Selecting robust migration indicators

Another issue in making time series comparisons is selection and application of the most appropriate statistical indicators. Bell *et al.* (2002) suggested that four discrete domains could be recognised for comparison of migration within countries (intensity, impact, distance and connectivity) and specified a set of indicators in each domain which could be used for cross-national comparison. While each of these domains provides a particular perspective on the nature of migration, it is migration intensity and, to a lesser extent, migration impact and migration distance that are of primary relevance to the focus of the current volume. Each of these can be captured using a number of different indicators, but by far the simplest and most basic measure for comparison between countries or over time is the Crude Migration Intensity (CMI) computed simply as:

$$CMI = M / P \tag{3.1}$$

where $M$ represents the number of migrants or migrations at a particular level of spatial scale (e.g., between states or districts) and $P$ is the population at risk (PAR). Following van Imhof *et al.* (1997), the term intensity is used to encompass both rates and probabilities (see below). While measurement of migration is a primary concern, care is also needed in selection of the appropriate denominator for computation of the CMI.

Whilst a count of those usually resident in a country is generally available from national censuses, this statistic may refer to different points in time. In comparing migration intensities between Britain and Australia, Rees *et al.* (2000) clarify that different forms of the PAR are needed for event data and for transition data. Event data are distributed throughout the observation interval and require a midpoint population to generate *occurrence-exposure rates*. Transition data, on the other hand, capture only those who were alive and in the country at the start and end of the interval. In Australia,

the PAR for both one- and five-year transition probabilities is readily derived from census migration matrices, since these include non-movers as
well as those who moved within and between zones, and following Rees
*et al.* (2000) this is used to compute migration intensities. In contrast, the
UK Census reports only an end-of-period population for the one-year migration data it collects and no start-of-period population is readily available. Theoretically, since a migration event is, on average, likely to take
place halfway through the period, the PAR most appropriate for use in the
intensity calculation is the mid-period population, which, for census data,
requires estimation. This can be obtained by interpolating between mid-
year estimates. One benefit of using NHSCR-based event data in the UK
is that the time period is from mid-year to mid-year, and the start and end
points coincide with date (30 June) for which the ONS produces mid-year
sub-national population estimates. Countries vary in regard to the types of
data available to represent the PAR, but, provided a consistent approach is
used, temporal consistency should be maintained.

In making comparisons between countries, Bell *et al.* (2015a) argued that
the only reliable comparison was in terms of all changes of address, irrespective of spatial scale, since countries differed widely in the number of
zones or regions for which the CMI was measured. In some countries, data
on all changes of permanent addresses are available from the periodic census. However, this is often not the case, and is rarely true for population
registers or data from administrative sources such as the NHS, when only
flows between LADs and between areas used for administering the NHS
are available.

To address this problem, the IMAGE project developed an alternative
mechanism to estimate the intensity of migration at all spatial scales (the
Aggregate Crude Migration Intensity or ACMI), using available information on flows at a range of different spatial scales. The estimates leverage
the linear relationship which is found between crude migration intensities and the average number of households per zone at each spatial scale
(see Courgeau, 1973; Courgeau *et al.*, 2012). Where data are available at
three spatial scales (such as regions, provinces and municipalities), for instance, the crude migration intensity, CMI($n$) can be plotted against the
natural log of the average number of households per zone and a regression
line can be defined as:

$$\text{CMI }(n) = a + b \, (\ln (H \,/\, n)) \tag{3.2}$$

where $H$ is the aggregate number of households in the country as a whole,
$n$ is the number of zones at which the value of the crude migration intensity
(CMI) is recorded, $b$ is the slope of the linear association and $a$ is the inter-
cept. When the number of households is the same as the number of zones,
then $H/n = 1$, so that any move represents a migration from one household
(or dwelling) to another. The log of $H/n$ then equals zero, and the equation

defaults to the value of the *Y*-intercept, *a*, which provides an estimate of all changes of address—the ACMI.

Chapter 4 of this book makes use of this technique to provide time-series estimates of the ACMI for 20 countries around the world, including developing countries as well as the developed nations that are the main focus of this volume. However, the data needed to calculate the ACMI are not widely available. Additionally in Chapter 4, therefore, comparisons of trends in migration intensity are made on a between-area-flow basis for a much larger set of countries, 56 in all, using the CMI calculated for the geographies for which comparable time-series data are available in each country.

Like all crude measures, the CMI is influenced by age composition effects, so Bell *et al.* (2002) set out a number of more sophisticated measures of intensity including age-standardised rates and migration expectancies. Age standardisation is an important consideration when comparing migration trends over time, since population ageing will lead to a reduction in the CMI even if the underlying propensity to move remains unchanged. This effect is considered explicitly in several of the chapters that follow, since it is one of the fundamental potential causes of the observed decline in migration intensity in those countries undergoing population ageing.

Standardisation might also be applied to allow for other compositional effects, such as shifts in educational attainment, or occupational mix, since increasing proportions of highly mobile educated and professional classes are likely to exert upward pressure on mobility. In most analyses such factors are generally introduced as explanatory rather than as structural variables (see Bell *et al.*, 2015a). As Bernard *et al.* (2014a, 2016) show, however, compositional effects also affect the age at which migration occurs, because the age profile of migration is shaped by the timing of key transitions in the life course, both in the family sphere (partnership and family formation) and in the economic domain (educational participation and labour force entry). Economic development therefore acts not only to change the overall level of migration intensity in a national population, but also the ages at which migration occurs. This impacts not only on young adults, but also on those reaching retirement age (see Sander and Bell, 2016).

Bernard *et al.* (2014b) demonstrate that age and intensity at the peak are the optimum measures to capture cross-national differences in the age profile of migration, since these measures encompass both the breadth and symmetry of the peak among young adults. They also demonstrate (Bernard and Bell, 2015) that considerable care is needed in the choice of technique for data smoothing, since conventional model migration schedules may obscure subtle but important shifts over time in the age at which peak migration occurs. It follows that careful attention to the choice and computation of migration indicators is needed if shifts over time in the overall level of migration within a country, and their underlying causes, are to be properly understood.

Robust indicators are also needed to measure changes over time in the extent to which migration is generating population redistribution within a country. Migration is ultimately a form of spatial behaviour and movements over longer distances commonly take place in response to regional differentials in economic opportunities. As countries develop and become progressively more urbanised, the extent of redistribution between regions rises, slowly at first, then more rapidly and finally falls away at more advanced stages of economic development (Rees *et al.*, 2016). We know that the extent of population redistribution arising from internal migration can be captured by the Aggregate Net Migration Rate (ANMR) and this in turn is a product of two distinct components: migration intensity and migration effectiveness (Bell *et al.*, 2002), the latter (captured in the Migration Effectiveness Index or MEI) simply measuring the extent to which movements in one direction are more or less balanced by movements in the other. Thus:

$$ANMR = CMI \times MEI \tag{3.3}$$

where CMI is defined in Equation (3.1) and:

$$MEI = N / M \tag{3.4}$$

where $N$ represents the sum of the net migration balances for all gaining regions and $M$ is the total migration between regions.

If longer-distance migration is indeed declining, as appears to be the case in at least some of our sample countries, then it begs the question as to how this decline is interacting with migration effectiveness, and how this then plays out in terms of changes in the level of population redistribution. The MEI, CMI and ANMR provide the tools to trace this interaction over time in those countries where the requisite data are available, as is the case for Australia (see Chapter 7 in this volume).

## Conclusions

Individually and collectively, the sample countries in this volume confront all of the data deficiencies and methodological problems outlined in this chapter, though the specifics vary widely from one country to the next. As argued elsewhere (Bell *et al.*, 2002, 2014), there is a strong case to be made for greater uniformity and harmonisation in the way migration data are collected, analysed and made available. In the case of international migration, significant progress has already been made in this regard (see Bilsborrow *et al.*, 1997; Nowok *et al.*, 2006). For internal migration, however, the task for scholars and statisticians remains largely ahead.

The individual chapter contributions outline the particular challenges of tracing mobility trends in the country of interest and draw on a range of different data types and analytical methods, as well as covering a variety of temporal frameworks. What they all have in common is the endeavour to identify trends in internal migration over an extended period, to distinguish the forms of migration that have changed and to attempt to account for the underlying causes of the observed patterns. To establish the broader context for the particular case studies that follow, the next chapter reviews trends in a broader, global sample of countries that include examples drawn from all continents and reflect a range of cultural settings and stages of economic development. This enables the case study countries to be compared with each other as well as being set in their wider context, identifying issues and arguments which will be returned to in the final part of the book.

## References

Barr, P. and Shuttleworth, I. 2012. Reporting address changes by migrants: The accuracy and timeliness of reports via health card registers. *Health and Place*, 18, 595–604.

Bell, M., Blake, M., Boyle, P., Duke-Williams, O., Rees, P., Stillwell, J. and Hugo, G. 2002. Cross-national comparison of internal migration: Issues and measures. *Journal of the Royal Statistical Society A*, 165(3), 1435–1464.

Bell, M, and Charles-Edwards, E. 2013. Cross-national Comparisons of Internal Migration: An Update of Global Patterns and Trends. *Technical Paper* 2013/1. New York, NY: Population Division, United Nations Department of Economic and Social Affairs. www.un.org/en/development/desa/population/publications/pdf/technical/TP2013-1.pdf

Bell, M., Bernard, A., Ueffing, P. and Charles-Edwards, E. 2014. *The IMAGE Repository: A User Guide*. QCPR Working Paper No 2014/01. Brisbane: University of Queensland. https://imageproject.com.au/framework/repositoryuserguide/

Bell, M., Charles-Edwards, E., Ueffing, P., Stillwell, J., Kupiszewski, M. and Kupiszewska, D. 2015a. Internal migration and development: Comparing migration intensities around the world. *Population and Development Review*, 41(1), 33–58.

Bell, M., Bernard, A., Charles-Edwards, E., Kupiszewska, D., Kupiszewski, M., Stillwell, J., Zhu, Y., Ueffing, P. and Booth, A. 2015b. *The IMAGE Inventory: A User Guide*. QCPR Working Paper 2015/01. Brisbane: Queensland Centre for Population Research, University of Queensland. http://imageproject.com.au/framework/inventoryuserguide/

Bell, M., Charles-Edwards, E., Kupiszewska, D., Kupiszewski, M., Stillwell, J. and Zhu, Y. 2015c. Internal migration data around the world: Assessing contemporary practice. *Population, Space and Place*, 21(1), 1–17.

Bell, M. and Rees, P. 2006. Comparing migration in Britain and Australia: Harmonisation through use of age-time plans. *Environment and Planning A*, 38(5), 959–988.

Bell, M. and Ward, G.J. 2000. Comparing permanent migration with temporary mobility. *Tourism Geographies*, 2(1), 97–107.

Bernard, A. and Bell, M. 2015. Smoothing internal migration age profiles for comparative research. *Demographic Research*, 32 (33), 915–948.

Bernard, A., Bell, M. and Charles-Edwards, E. 2014a. Explaining cross-national differences in the age profile of internal migration: the role of life-course transitions. *Population and Development Review*, 40(2), 213–239.

Bernard, A., Bell, M. and Charles-Edwards, E. 2014b. Improved measures for the cross-national comparison of age profiles of internal migration. *Population Studies*, 68(2), 179–195.

Bernard, A., Bell, M. and Charles-Edwards, E. 2016. Internal migration age patterns and the transition to adulthood: Australia and Great Britain compared. *Journal of Population Research.* doi:10.1007/s12546-016-9157-0

Bilsborrow, R., Hugo, G., Oberai, A.S. and Zlotnik, H. 1997. *International Migration Statistics: Guidelines for Improving Data Collection Systems.* Geneva: International Labour Organization.

Blake, M., Bell, M. and Rees, P. 2000. Creating a temporally consistent spatial framework for the analysis of interregional migration in Australia. *International Journal of Population Geography*, 6, 155–174.

Champion, A.G. and Shuttleworth, I. 2016a. Is longer-distance migration slowing? An analysis of the annual record for England and Wales. *Population, Space and Place,* published online in Wiley Online Library. doi:10.1002/psp.2024

Champion, A.G. and Shuttleworth, I. 2016b. Are people moving address less? An analysis of migration within England and Wales, 1971–2011, by distance of move. *Population, Space and Place*, published online in Wiley Online Library. doi:10.1002/psp.2026

Collier, D. 1993. The comparative method. In Finifter, A.W. (ed) *Political Science: The State of the Discipline.* Washington, D.C.: American Political Science Association. https://papers.ssrn.com/sol3/papers.cfm?abstract_id=1540884

Courgeau, D. 1973. Migrations et découpage du territoire. *Population*, 28, 511–536.

Courgeau, D., Muhidin, S. and Bell, M. 2012. Estimating changes of residence for cross-national comparison. *Population (English edition)*, 67(4), 631–652.

Deming, W.E., and Stephan, F.F. 1940. On a least squares adjustment of a sampled frequency table when the expected marginal totals are known. *The Annals of Mathematical Statistics*, 11(4), 427–444.

Duke-Williams, O. 2009. The geographies of student migration in the UK. *Environment and Planning A*, 41(8), 1826–1848.

Fotheringham, A.S. and Wong, D.W.S. 1991. The modifiable areal unit problem in multivariate statistical analysis. *Environment and Planning A*, 23, 1025–1044.

Holt, D., Steel, D.G. and Tranmer, M. 1996. Area homogeneity and the modifiable areal unit problem. *Geographical Systems*, 3, 181–200.

Lijphart, A., 1971. Comparative politics and the comparative method. *American Political Science Review*, 65(03), 682–693.

Lomax, N., Rees, P., Norman, P. and Stillwell, J. 2013. Subnational migration in the United Kingdom: Producing a consistent time series using a combination of available data and estimates. *Journal of Population Research*, 30, 265–288.

Long, L., Tucker, C.J. and Urton, W.L. 1988. Migration distances: An international comparison. *Demography*, 25(4), 633–640.

Manley, D. 2014. Scale, aggregation, and the Modifiable Areal Unit Problem. In Fischer, M.M. and Nijkamp, P. (eds) *Handbook of Regional Science.* Berlin: Springer-Verlag, 1157–1171.

Niedomysl, T. 2011. How migration motivations change over migration distance: Evidence on variations across socioeconomic and demographic groups. *Regional Studies*, 45(6), 843–855.

Niedomysl, T., Ernston, U. and Fransson, U. 2017. The accuracy of migration distance measures. *Population, Space and Place*, 23(1), e1971. doi:10.1002/psp.1971

Nowok, B., Kupiszewska, D. and Poulain, M. 2006. Statistics on international migration flows. In Poulain, M., Perrin, N. and Singleton, A. (eds) *Towards Harmonised European Statistics on International Migration* (THESIM). Louvain-la-Neuve: Presses Universitaires de Louvain, 203–231.

ONS. 2011. Mid-year Estimates Short Methods Guide. Titchfield, UK: Office for National Statistics. www.ons.gov.uk/ons/guide-method/method-quality/specific/population-and-migration/pop-ests/mid-year-population-estimates-short-methods-guide.pdf, accessed 14 August 2012.

ONS. 2012. A Conceptual Framework for UK Population and Migration Statistics. Titchfield, UK: Office for National Statistics. http://webarchive.nationalarchives. gov.uk/20160105160709/http://www.ons.gov.uk/ons/guide-method/method-quality/imps/latest-news/conceptual-framework/index.html

Openshaw, S. 1984. *The Modifiable Areal Unit Problem.* CATMOG 38. Norwich: GeoBooks.

Openshaw, S. and Rao, L.1995. Algorithms for re-engineering 1991 census geography, *Environment and Planning A,* 27, 425–446.

Rees, P., Bell, M., Blake, M. and Duke-Williams, O. 2000. *Harmonising Databases for the Cross National Study of Internal Migration: Lessons from Australia and Britain.* Working Paper 00–05. Leeds: School of Geography, University of Leeds.

Rees, P., Bell, M., Kupiszewski, M., Kupiszewska, D., Ueffing, P., Bernard, A., Charles-Edwards, E. and Stillwell, J. 2016. The impact of internal migration on population redistribution: An international comparison. *Population, Space and Place,* published online in Wiley Online Library. doi:10.1002/psp.2036

Sander, N. and Bell, M. 2016. Age, period and cohort effects of migration on the baby-boomers in Australia. *Population, Space and Place,* 22(6), 617–630.

Stillwell, J., Bell, M., Blake, M., Duke-Williams, O. and Rees, P. 2000. A comparison of net migration flows and migration effectiveness in Australia and Britain, 1976–96. Part 1: Total migration patterns. *Journal of Population Research,* 17(1), 17–41.

Stillwell, J., Bell, M., Blake, M., Duke-Williams, O. and Rees, P. 2001. A comparison of net migration flows and migration effectiveness in Australia and Britain, 1976–96: Part 2, Age-related migration patterns. *Journal of Population Research,* 18(1), 19–39.

Stillwell J., Bell M., Ueffing P., Daras K., Charles-Edwards E., Kupiszewski M. and Kupiszewska, D. 2016. Internal migration around the world: Comparing distance travelled and its frictional effect. *Environment and Planning A,* 48(8), 1657–1675.

Stillwell, J., Rees, P.H. and Boden, P. (eds) 1992. *Migration Processes and Patterns Volume 2: Population Redistribution in the United Kingdom.* London, Belhaven Press.

Stillwell, J. and Thomas, M. 2016. How far to internal migrants really move? Demonstrating a new method for the estimation of intra-zonal distance. *Regional Studies, Regional Science,* 3, 28–47.

van Imhoff, E., van der Gaag, N., van Wissen, L. and Rees, P. 1997. The selection of internal migration models for European regions. *International Journal of Population Geography,* 3(2), 137–159.

Willekens, F.J. 1980. Entropy, multiproportional adjustment and the analysis of contingency tables. *Systemi Urbani,* 2(3), 171–201.

Willekens, F.J. 1983. Log-linear modelling of spatial interaction. *Papers of the Regional Science Association,* 52, 187–205.

Založnik, M. 2011. *Iterative Proportional Fitting: Theoretical Synthesis and Practical Limitations.* Unpublished PhD thesis. Liverpool: University of Liverpool.

# 4 Global trends in internal migration

*Martin Bell, Elin Charles-Edwards,
Aude Bernard and Philipp Ueffing*

Population mobility is widely regarded as integral to the process of national development (United Nations, 2009). It enables individuals, families and households to meet their goals and aspirations, and it is essential to the efficient functioning of labour and housing markets (World Bank, 2009). In an 'age of migration' (Castles and Miller, 2009), mobility itself has become synonymous with development. Following Zelinsky (1971), high levels of mobility are therefore seen as indicative of advanced stages of development, while nations where mobility is lower are taken to be at an earlier point on the development ladder. In practice, however, population movement takes a wide variety of forms, and low levels of permanent migration may be compensated by high levels of seasonal or circular mobility (Skeldon, 1990, 1997; Bell and Ward, 2000). Types of movement within a country also vary and alter over time, as exemplified by rural-urban migration which appears to follow a logistic function (Rees *et al.*, 2016). Assessment of trends in migration therefore calls for careful attention to the types and forms of mobility as well as its overall intensity.

The case studies presented in this volume are drawn primarily from developed nations and provide a solid testbed for the proposition of a decline in mobility among countries at the advanced end of the development spectrum. If the results are to be meaningful, however, what is further needed is a clear understanding of the broader context within which the selected countries are situated. Where on the international spectrum of mobility do they sit? And what is the trajectory among other countries? Is mobility decline confined only to the highly mobile, advanced economies in the western hemisphere, or is it found more broadly across other regions and among countries at different levels of development?

This chapter endeavours to establish the context of mobility decline by drawing on data assembled as part of the IMAGE project, an international collaborative programme of research that aims to make comparisons of Internal Migration Around the GlobE (see www.imageproject.com.au). Building on an inventory of migration data collections for 193 UN member states (Bell *et al.*, 2014), the project assembled internal migration data covering 135 countries (Bell *et al.*, 2015a) and built a bespoke software platform, the

IMAGE Studio, to compute multiple migration indicators using flexible geographies (Stillwell *et al.*, 2014; Daras, 2014). Various papers have explored methodological issues (Bell *et al.*, 2013) and made cross-national comparisons of overall internal migration intensities (Bell *et al.*, 2015b), migration age profiles (Bernard *et al.*, 2014a, 2014b; Bernard and Bell, 2015), migration impacts (Rees *et al.*, 2016) and migration distance (Stillwell *et al.*, 2016) globally as well as for selected regions and group of countries (Charles-Edwards *et al.*, 2016).

In this chapter, we draw on the IMAGE database to examine the trajectory of migration intensities in a global sample of countries representing all continents, extending earlier work by Bell and Charles-Edwards (2013, 2014). We begin by reviewing the nature of the available data and underlining the constraints involved in comparing migration between countries and over time. We then seek to position the sample countries examined in this volume in the global league table of aggregate migration intensities established by the IMAGE project (Bell *et al.*, 2015b). The following section examines trends in migration intensity at a range of spatial scales in those countries for which data enable us to compute migration over a sequence of time intervals. We seek to identify commonalities and differences among countries and link these to regional patterns and stages of development. We conclude by summarising the global evidence on contemporary trajectories in internal migration.

## Global data on internal migration

Fully 179 of the 193 member states of the United Nations collect some form of data on internal migration, but the mechanisms used vary widely, as do the types of data collected. Despite predictions of their imminent demise (Coleman, 2013), censuses remain the most common source, with 158 countries collecting some form of internal migration data in the 2000 Census round (Bell *et al.*, 2015a). A further 50 countries draw on population registers or administrative data (such as national health registers, or electoral rolls), while 110 countries employ national surveys in a range of forms.

Despite their widespread coverage, censuses differ markedly in the type of data collected, particularly in regard to the interval over which migration is measured. As shown in Table 4.1, lifetime migration, measured by comparing place of residence with place of birth, is the most common form of internal migration data, but is of limited use for analysis of contemporary trends because it reflects the cumulative redistribution of population over many decades. Abel and Sander (2014) have shown that it is possible to estimate temporal trends in international migration using similar data on migrant stocks, but comparisons of internal migration between countries, and over time, using lifetime data are further prejudiced by differences in age composition (which affect lifetime exposure) and by the relatively coarse spatial frameworks used to collect place of birth within countries (Bell *et al.*, 2015a).

*Table 4.1* Internal migration data collection by data type, interval and region

| Continental region | Census transition interval | | | | | Migration events | Total countries |
|---|---|---|---|---|---|---|---|
| | *1 year* | *5 years* | *Other fixed* | *Lifetime* | *Latest move* | | |
| Africa | 9 | 8 | 8 | 29 | 13 | 0 | 54 |
| Asia | 1 | 13 | 8 | 26 | 18 | 15 | 47 |
| Europe | 14 | 4 | 12 | 26 | 10 | 32 | 43 |
| Latin America | 2 | 16 | 2 | 28 | 12 | 0 | 33 |
| North America | 1 | 3 | 0 | 3 | 0 | 2 | 2 |
| Oceania | 2 | 8 | 2 | 10 | 2 | 1 | 14 |
| Total | 29 | 52 | 32 | 122 | 55 | 50 | 193 |

Source: IMAGE Inventory of Internal Migration.

Notes: Census data refer to 2000 Census round; migration events refer to data drawn from population registers and administrative collections.

Measuring migration over a fixed interval provides a more robust basis for cross-national comparisons. In the 2000 Census round, 52 countries collected data on migration over a five-year interval but, as shown elsewhere, it is possible to supplement this count using data collected in the form of the 'latest move' in combination with data on duration of residence (Bell *et al.*, 2015b). In a similar way, the count of countries which measure migration as a transition over a single year can be supplemented using migration events generated from population registers and administrative systems. While there are important differences between these types of data (see, for example, Bell and Rees, 2006), these combinations significantly increase the global coverage on which comparisons between countries can be made. As Table 4.1 reveals, there is nevertheless a distinct spatial bias, with one-year transition and event data confined primarily to Europe and parts of Africa, while five-year transition data are more common in other parts of the world. This complicates interpretation because there is no simple translation between data measured over one- and five-year intervals. Nevertheless, since some countries are represented in both groups, it provides a viable framework for interpreting cross-national differences in overall levels of mobility.

Making comparisons of mobility over time presents a more significant challenge, because suitable data are available for few countries, and rarely in a format that delivers robust time series. Censuses are conducted infrequently and often involve changes in the format of migration questions. Population registers and administrative collections generally provide more frequent data, usually annual, but rarely offer a lengthy history. Both sources are compromised by changes over time in the geographical boundaries used to capture migration—that is, in the number of zones and their spatial configuration—giving rise to the well-known modifiable areal unit problem (Wrigley *et al.*, 1996; Bell *et al.*, 2002). Lack of geographic detail and changing definitions commonly undermine the utility of survey data as a source of information on migration trends.

For comparative purposes, migration should ideally be measured at similar levels of spatial aggregation across any sample of countries, since trends at the local level may differ from those at the regional level. In practice, administrative and statistical geographies vary widely between countries and there is no simple method by which to harmonise migration observed at different spatial scales. Recent work based on flexible spatial aggregation procedures has demonstrated that certain measures of migration, notably distance-decay and migration effectiveness, remain remarkably stable irrespective of spatial scale, at least when countries are divided into 30 zones or more (Rees *et al.*, 2016; Stillwell *et al.*, 2016). Where such data are not available, however, some measure of harmonisation between countries can be achieved using the commonly recognised division between Major Regions (e.g., States, Provinces) and Minor Regions (e.g., municipalities). Comparisons based on zones of similar average area or similar average population have also proved instructive.

For the purposes of this chapter, we draw on a range of data from censuses and administrative sources that provide internal migration intensities at one or more spatial scales for at least two time intervals. For the USA, we also draw on the American Community Survey and the Current Population Survey. In total, we have data for 66 countries, with the most extensive coverage in Europe and Latin America, and somewhat lower representation from Asia and Africa (Table 4.2). The counts for North America and Oceania are small but include all the most populous nations. There are census data for 48 countries and, for 29 of these, observations are available for three intervals or more. Most of the census time series are for countries which measure migration over five-year intervals, but for five countries (Mozambique, Australia, Canada, Portugal and Greece) we have data for both one- and five-year migration intervals. Register and administrative data are available for 20 countries, but these are confined mainly to Europe, together with South Korea (for which census data are also available) and Japan.

*Table 4.2* Time series on internal migration from censuses and registers in the IMAGE repository

| Continental region | Census observations | | | Register observations | | | Total countries with data |
|---|---|---|---|---|---|---|---|
| | *Two* | *Three* | *Four* | *<10* | *10–20* | *>20* | |
| Africa | 4 | 3 | 1 | 0 | 0 | 0 | 8 |
| Asia | 5 | 3 | 1 | 0 | 1 | 1 | 10 |
| Europe | 1 | 1 | 3 | 3 | 8 | 6 | 22 |
| Latin America | 9 | 6 | 6 | 0 | 0 | 0 | 21 |
| North America | 0 | 1 | 1 | 0 | 1 | 0 | 2 |
| Oceania | 0 | 0 | 3 | 0 | 0 | 0 | 3 |
| Total | 19 | 14 | 15 | 3 | 10 | 7 | 66 |

Source: IMAGE Repository of Internal Migration.

For a handful of countries, census-based migration data can be assembled back to the 1960s. However, in order to enhance comparability, we restrict the focus here to the period since the 1980 round of censuses (1975–1984) and concentrate particularly on shifts in migration trends over the latest decade, measuring migration at dates corresponding to the 2000 and 2010 census rounds. Register and administrative data provide a finer temporal grain and capture more recent trends but generally offer less historical context. With both datasets we endeavour to differentiate movements between Major Regions and between Minor Regions, where possible, in order to distinguish the distinctive suite of forces operating to shape migration at each spatial level. For a handful of countries, we are also able to make comparisons over time with respect to overall levels of migration intensity, as explained below.

## Cross-national differences

To contextualise the analysis of temporal trends, it is important to first establish the relativity between countries in terms of overall intensities of internal migration. As argued elsewhere (Bell *et al.,* 2002), the only reliable statistic for comparing the level of migration between countries is the Aggregate Crude Migration Intensity, computed as ACMI = $M/P$ where $M$ represents *all* changes of residence within the country, irrespective of distance moved, and $P$ is the population at risk. This measure effectively combines local residential mobility with longer distance migration, since arguably there is no clear or consistent distinction between the two. Unfortunately, few countries collect data on all changes of address directly, but Courgeau *et al.* (2012) show that reliable estimates of the ACMI can be derived if data are available on the crude intensity measured at two or more levels of scale. Coupling this technique with the flexible geographies implemented through the IMAGE Studio (Stillwell *et al.*, 2014), Bell *et al.* (2015b) compared the ACMI across 96 countries representing more than 80% of the global population.

Figures 4.1 and 4.2 extend and update the analysis reported by Bell *et al.* (2015b). What is most striking is the extent of variation between countries. Migration intensities are consistently high in Oceania and in North America, but other parts of the world are characterised by massive diversity. In Europe ACMIs vary from a low of just 1% per annum in Macedonia to a high of more than 19% in Iceland. In Asia, where five-year data provide the greatest coverage, South Korea stands out as the most mobile country while the lowest intensity is found in India. Similarly, in Africa, intensities are highest in Senegal and Cameroon and lowest in Egypt and Mali, while in Latin America the peaks are found in Panama and Chile, and the troughs in Haiti and Venezuela. Underpinning these variations, Bell *et al.* (2015b) identified evidence of clear spatial patterning. In Europe, migration intensities are highest in Scandinavia and decline steadily moving south and east. In Latin America, mobility rates are highest among the Andean countries and fall away to the east of the continent and in Central America and the Caribbean. In Asia, it is in the south and southeast that the lowest migration intensities are found, with somewhat higher rates to the east and west of the continent. The evidence for Africa is less complete and presents a more varied mosaic.

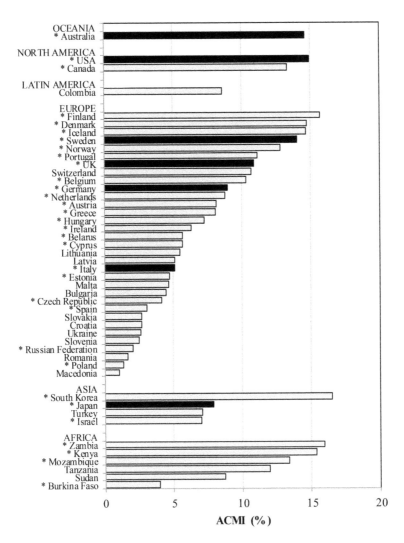

*Figure 4.1* Aggregate Crude Migration Intensities (ACMI) for countries collecting one-year migration data.

Source: Modified after Bell *et al.* (2015b) and extended and updated using data from the IMAGE Repository.

*Notes*: Asterisks identify countries for which we analyse time series data at various scales; the countries indicated with black bars are those for which detailed analyses are contained within this volume.

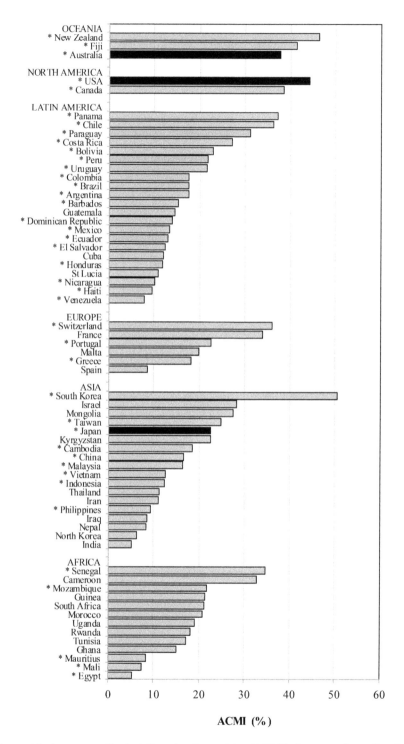

*Figure 4.2* Aggregate Crude Migration Intensities (ACMI) for countries collecting
five-year migration data.

Source and notes: As for Figure 4.1.

Bell *et al.* (2015b) reported strong associations between aggregate migration intensity and a range of national socio-economic indicators including urbanisation, GDP per capita, net international migration, and the Human Development Index. These associations offer *a priori* evidence of close links between development and migration. What is yet to be established is whether or not there is a general tendency for migration intensity to decline at advanced stages of development. As revealed in Figures 4.1 and 4.2, the countries selected for analysis in other chapters of this volume are all drawn from economically advanced countries, but they vary widely in terms of their level of migration intensity. In the following sections, we examine trends in the intensity of migration in selected countries for which data are available, focusing first on aggregate migration, then turning to trends at varying geographic levels.

## Trends in migration intensity

As noted earlier, only a small number of countries collect data on all residential moves, and time series data are available for just 20 of these. Table 4.3 sets out the recorded intensities centred on the end of each decade from 1980 to 2010 and assesses the trend as either rising, falling or stable over the 10 year interval. The change is classified as stable if the recorded intensity varies by less than 5% from one interval to the next, since this is likely to signify a fluctuation rather than a trend. The dominant impression across the dataset is of declining migration intensity. Of the 20 countries listed, only in the small island nation of Cyprus is there evidence of a clear rise in migration. Finland and Iceland also recorded a rise in the 1990s, but this halted (Finland) or reversed (Iceland) in the following decade. Similarly, Israel registered a rise in the 1980s but a fall in the 1990s. Across the remaining European countries, the trend in the 2000s is either stability (Austria, Ireland, Norway, Sweden and UK) or a decline (Denmark, Hungary and the Netherlands). In Asia we have time series of aggregate migration intensities for just three countries, Japan, South Korea and Taiwan, and in Africa for just one, Mauritius: all show evidence of decline. In Australia, New Zealand and the USA, the picture is unequivocal, and while Canada is categorised as stable, the tendency also appears to be towards declining mobility.

When the analysis is extended to distinguish migration at different levels of scale, the picture takes on added complexity. Table 4.4 summarises trends in migration intensity over the first decade of the twenty-first century, differentiating movements between Major and between Minor Regions. As noted earlier, national statistical agencies vary in the administrative geographies used to capture migration and there is no simple basis for comparison. For some countries, data are coded to three or four different spatial levels, while for others only a single level of geography is made available. Table 4.4 adopts an average population of around 250,000–500,000 to distinguish between Major and Minor Regions and, to avoid double counting, it admits a maximum of two levels per country, even where migration data for multiple geographies are available. Similarly, where migration is measured

*Table 4.3* Trends in aggregate internal migration intensity, selected countries

| Country | Interval (years) | Aggregate crude migration intensity (%) | | | | Trend | | |
|---|---|---|---|---|---|---|---|---|
| | | 1980 | 1990 | 2000 | 2010 | 1980– 1990 | 1990– 2000 | 2000– 2010 |
| *Africa* | | | | | | | | |
| Mauritius | 5 | | 13.36 | 11.98 | 8.33 | | Fall | Fall |
| *Asia* | | | | | | | | |
| Japan | 5 | | | 27.75 | 22.45 | | | Fall |
| South Korea | 1 | 21.70 | 22.10 | 19.00 | 16.50 | Stable | Fall | Fall |
| South Korea | 5 | | | 52.79 | 50.59 | | | Stable |
| Taiwan | 5 | | 29.75 | 27.86 | 24.83 | | Fall | Fall |
| *Europe* | | | | | | | | |
| Austria | 1 | | | 8.03 | 8.15 | | | Stable |
| Cyprus | 1 | | | 3.83 | 5.69 | | | Rise |
| Denmark | 1 | | | 15.96 | 14.72 | | | Fall |
| Finland | 1 | | 11.36 | 15.22 | 15.66 | | Rise | Stable |
| Hungary | 1 | | | 8.09 | 7.25 | | | Fall |
| Iceland | 1 | | 16.31 | 17.29 | 14.63 | | Rise | Fall |
| Ireland | 1 | | | 6.07 | 6.30 | | | Stable |
| Israel | 1 | 6.57 | 8.42 | 7.55 | | Rise | Fall | |
| Netherlands | 1 | | 10.38 | 9.96 | 8.80 | | Stable | Fall |
| Norway | 1 | | | 12.78 | 12.83 | | | Stable |
| Sweden | 1 | | | 14.06 | 14.04 | | | Stable |
| UK | 1 | | | 10.95 | 10.93 | | | Stable |
| *North America* | | | | | | | | |
| Canada | 1 | | | 13.43 | 13.29 | | | Stable |
| | 5 | | | 39.82 | 38.50 | | | Stable |
| USA (ACS) | 1 | | | 14.48 | 14.90 | | | Stable |
| USA (CPS) | 1 | | | 15.51 | 12.23 | | | Fall |
| USA (Census) | 5 | 45.40 | 46.72 | 44.27 | | Stable | Fall | |
| *Oceania* | | | | | | | | |
| Australia | 1 | 16.44 | 18.34 | 17.55 | 14.58 | Rise | Stable | Fall |
| | 5 | 40.77 | 40.38 | 42.38 | 37.73 | Stable | Stable | Fall |
| New Zealand | 5 | | | 48.96 | 46.30 | | | Fall |

Source: Analysis of data from the IMAGE Repository.

Notes: Intensities are interpreted as stable if the change over the interval is less than 5%.

over both one- and five-year intervals, only the latter are included. The results provide a synthesis of trends in migration at levels that broadly correspond to longer distance moves, such as between provinces, state or regions (Major Regions), and those that more closely match migration over short or intermediate distances (Minor Regions).

*Table 4.4* Trends in internal migration intensity, 2000–2010, Major and Minor Regions

| Continental region | Trend in migration intensity | | | Total countries |
|---|---|---|---|---|
| | *Fall* | *Stable* | *Rise* | |
| *Major Regions* | | | | |
| Africa | 4 | 0 | 1 | 5 |
| Asia | 4 | 0 | 3 | 7 |
| Europe | 3 | 5 | 6 | 14 |
| Latin America | 8 | 2 | 2 | 12 |
| North America | 2 | 0 | 0 | 2 |
| Oceania | 1 | 1 | 0 | 2 |
| Total | 22 | 8 | 12 | 42 |
| *Minor Regions* | | | | |
| Africa | 2 | 1 | 2 | 5 |
| Asia | 4 | 1 | 2 | 7 |
| Europe | 6 | 5 | 9 | 20 |
| Latin America | 5 | 1 | 1 | 7 |
| North America | 2 | 0 | 0 | 2 |
| Oceania | 1 | 2 | 0 | 3 |
| Total | 20 | 10 | 14 | 44 |

Source: Analysis of data from the IMAGE Repository.

Notes: Major Regions generally have populations of 250,000 or more. Countries are catego-rised as having stable migration intensity if the level altered by less than 5% over the decade. Membership of the two panels differs since data are not always available for both Major and Minor Regions. See text.

Trends over the period 2000–2010 can be computed for 56 of the 66 countries for which we have data. Migration between Major Regions can be tracked for 42 countries, and between Minor Regions for 44 countries. Thirty countries are represented in both panels of Table 4.4. Both datasets reveal a similar story, although the trends for individual countries are not always consistent. Overall, two-thirds of countries recorded either stability or a decline in migration intensity over the decade, while just one in three registered a rise. Falls outnumbered rises in a ratio of three to two. Of the six world regions, however, only in North America was there a consistent pattern of mobility decline. Stability or decline also predominates across Oceania and Latin America, and this pattern is sustained, although less pro-nounced, across Africa and Asia. It is Europe where the balance is reversed, with a majority of countries recording rising migration intensities over the decade, and this is true at both levels of spatial scale. In the following sec-tions we explore these patterns by country within each continental region.

### Trends in migration intensities: Africa

The eight African countries for which we have data are scattered widely across the continent. For five of these—Burkina Faso, Egypt, Kenya, Mauritius and Mozambique—there is a distinctive downward trend, and this seems

*Table 4.5* Trends in internal migration intensity, 1980–2010, countries in Africa

| Country | Observation interval (years) | Number of regions | Average population per region | Migration intensity (%) 1980 | 1990 | 2000 | 2010 |
|---|---|---|---|---|---|---|---|
| Burkina Faso | 1 | *45 | 300,563 | | 2.25 | 1.35 | 1.34 |
| Egypt | 5 | 27 | 2,403,554 | | | 1.79 | 1.28 |
| Kenya | 1 | 8 | 3,406,953 | 3.66 | 3.86 | 3.29 | 2.16 |
| | 1 | 69 | 395,009 | 5.12 | 5.12 | 5.07 | |
| Mali | 5 | 9 | 808,864 | | | 1.84 | 4.01 |
| | 5 | 47 | 154,889 | | | 2.46 | 5.06 |
| Mauritius | 5 | 10 | 113,129 | | 4.68 | 4.48 | 3.58 |
| Mozambique | 5 | 11 | 1,459,585 | | | 4.60 | 2.55 |
| | 5 | 141 | 113,868 | | | 12.18 | 7.09 |
| | 1 | 11 | 1,749,754 | | | 1.46 | 1.12 |
| | 1 | 141 | 136,506 | | | 4.50 | 4.41 |
| Senegal | 5 | 11 | 764,357 | | 6.29 | 3.70 | |
| | 5 | *34 | 247,292 | | 5.88 | 8.10 | |
| Zambia | 1 | 10 | 956,954 | | 2.46 | 2.94 | 2.55 |
| | 1 | *74 | 129,318 | | 4.28 | 5.40 | 5.73 |

Source: Analysis of data from the IMAGE Repository.

Notes: Asterisks indicate observations with significant changes in number of regions over time. Dates for intensities refer to the mid-point of the decennial census intervals, so 1980 refers to 1975–1984, or exact years for register-based data.

most pronounced for long distance migration, that is, for migration between major regions (Table 4.5). In Kenya, for example, movement between eight regions fell by 45% between the 1990 and 2010 censuses. Egypt, Mozambique and Zambia registered similar precipitous falls in mobility over the latest decade, while in Senegal the downward trend began in the 1990s. Similarly, Burkina Faso, Kenya, Mauritius and Mozambique also registered declining mobility over short or intermediate distances. In Senegal and Zambia, however, shorter distance migration was rising, while in Mali migration intensities more than doubled over the 2000–2010 decade at both spatial scales. African migration data sometimes lack reliability and mobility itself may fluctuate widely in response to natural disasters or regional conflicts, such as the long running conflict in Senegal and the civil war in Mozambique. Nevertheless, it is salutary to note that declining mobility is by no means confined to the advanced economies of the western world.

### Trends in migration intensities: Asia

In Asia, it is two of the giants, China and Indonesia, together with Vietnam that stand out with rising mobility since the turn of the century (Table 4.6). For Indonesia and Vietnam, this represented a reversal of earlier trends,

*Table 4.6* Trends in internal migration intensity, 1980–2010, countries in Asia

| Country | Observation interval (years) | Number of regions | Average population per region | Migration intensity (%) | | | |
|---|---|---|---|---|---|---|---|
| | | | | 1980 | 1990 | 2000 | 2010 |
| Armenia | 5 | 11 | 269,637 | | | 11.22 | 11.15 |
| Cambodia | 5 | 24 | 558,384 | | | 2.99 | 2.79 |
| China | 5 | 31 | 42,993,899 | | 1.11 | 2.74 | |
| | 5 | 347 | 3,840,954 | | 3.42 | 6.70 | 8.99 |
| Indonesia | 5 | *33 | 6,453,453 | 2.85 | 2.87 | 2.19 | 2.41 |
| | 5 | *456 | 467,026 | | | 3.98 | 4.43 |
| Japan | 1 | 47 | 2,684,681 | 2.88 | 2.58 | 2.24 | 1.85 |
| Malaysia | 5 | 15 | 1,195,397 | | 7.17 | 4.75 | |
| | 5 | 133 | 134,819 | | 13.33 | 8.00 | |
| Philippines | 5 | 81 | 1,010,691 | | | 3.28 | 1.76 |
| | 5 | 1490 | 54,944 | | | 4.55 | 3.35 |
| South Korea | 5 | 17 | 2,831,681 | | | 12.84 | 11.32 |
| | 5 | 263 | 183,036 | | | 22.83 | 18.27 |
| | 1 | 17 | 2,966,115 | | | 3.80 | 3.81 |
| | 1 | 263 | 191,726 | | | 6.99 | 6.81 |
| Taiwan | 5 | 25 | 865,181 | | 10.21 | 8.06 | 6.29 |
| | 5 | 369 | 58,617 | | 17.75 | 15.38 | 14.08 |
| Vietnam | 5 | 8 | 9,801,889 | | 2.49 | 1.94 | |
| | 5 | 63 | 1,244,684 | | 3.35 | 2.90 | 4.33 |
| | 5 | 660 | 118,811 | | 4.47 | 4.55 | 6.51 |

Source and notes: As for Table 4.5.

and may prove short lived. In China, on the other hand, rising mobility is well-entrenched and reflects the steady loosening of restrictions on movement within the country, as well as the rapid pace of urbanisation associated with economic development (Bell and Charles-Edwards, 2013; Fan, 2005). Elsewhere in East Asia, the data for Japan, Taiwan and South Korea document a sustained decline in both long and short distance migration, confirming the trend in the estimates of aggregate migration intensity (Table 4.4). Similarly, across Southeast Asia, mobility levels in the Philippines, Malaysia and Cambodia appear to be declining. Armenia is the sole representative from Western Asia where mobility levels appear to be relatively stable.

### Trends in migration intensities: Europe

The IMAGE Repository contains time series data for fully half of the 43 countries in Europe and, for all but one of these, the trend in migration intensity can be computed for the early years of this century (Table 4.7). In 11 of these countries, the trend is unequivocally upwards. Austria,

Belarus, the Czech Republic, Greece, Ireland, Norway and Spain all registered increasing mobility at both spatial scales. Trend data are only available at one level of geography in Belgium, Estonia, Finland and the UK, but these countries also show a clear rise in migration intensity (though the more comprehensive analysis in Chapter 6 reveals a more complex picture for the UK). In three other countries, Poland, Portugal and Sweden, the trend is mixed, with stable or declining intensities over longer distances but rising mobility at the local or intermediate level. Just seven countries display a contrary trend. In Germany, Hungary, Iceland, the Netherlands and the Russian Federation, migration intensities declined at all spatial scales. The same was true of Italy, except for movements at the local level (between the 8,100 communes), while Denmark too registered a decline in migration between its 99 municipalities. Data for Switzerland are not available beyond 2000 but the 1990s suggest a mixed trend, with falling long distance mobility offset by rising intensities over shorter distances, as found in a number of other jurisdictions.

*Table 4.7* Trends in internal migration intensity, 1980–2010, countries in Europe

| Country | Observation interval (years) | Number of regions | Average population per region | Migration intensity (%) | | | |
|---|---|---|---|---|---|---|---|
| | | | | 1980 | 1990 | 2000 | 2010 |
| Austria | 1 | 9 | 929,008 | | | 0.93 | 1.16 |
| | 1 | 99 | 84,455 | | | 2.00 | 3.63 |
| | 1 | 2,354 | 3,552 | | | 3.23 | 4.91 |
| Belarus | 1 | 7 | 1,353,286 | | | 0.76 | 1.00 |
| | 1 | 145 | 65,331 | | | 1.69 | 1.95 |
| Belgium | 1 | 589 | 17,735 | | | 4.32 | 5.24 |
| Czech Republic | 1 | 14 | 751,019 | 0.73 | 0.69 | 0.56 | 0.88 |
| | 1 | 77 | 136,549 | 1.64 | 1.58 | 0.88 | 1.33 |
| | 1 | 6,234 | 1,687 | 2.75 | 2.59 | 1.95 | 2.30 |
| Denmark | 1 | 99 | 56,654 | | | 5.39[5] | 4.94 |
| Estonia | 1 | 15 | 88,765 | | | 0.48 | 1.31 |
| Finland | 1 | 336 | 16,037 | | 3.56 | 4.70 | 5.05 |
| Germany | 1 | 16 | 5,118,887 | | | 1.40[3] | 1.30 |
| | 1 | 412 | 198,792 | | | 3.23 | 3.14[6] |
| | 1 | 12,227 | 6,698 | | | 4.73 | 4.37 |
| Greece | 5 | 54 | 175,387 | 7.67 | 5.42 | 6.04 | |
| | 1 | 54 | 175,387 | | 2.14 | 2.80 | |
| | 5 | *947 | 10,518 | 11.97 | 8.66 | 10.29 | 10.80 |
| Hungary | 1 | 7 | 1,424,532 | | 1.38[2] | 1.36 | 1.35 |
| | 1 | 3,152 | 3,164 | | 4.58 | 4.04 | 3.80 |
| Iceland | 1 | 8 | 40,090 | | 3.01 | 3.18 | 2.70 |
| | 1 | 79 | 4,060 | | 6.05 | 6.55 | 5.69 |
| Ireland | 1 | 8 | 498,674 | 1.32 | 1.07 | 1.79 | 2.21 |

| Country | Observation interval (years) | Number of regions | Average population per region | Migration intensity (%) 1980 | 1990 | 2000 | 2010 |
|---|---|---|---|---|---|---|---|
| Italy | 1 | 20 | 3,039,133 | | $0.51^2$ | 0.63 | 0.55 |
| | 1 | 107 | 568,062 | | $0.78^2$ | 0.92 | $0.84^6$ |
| | 1 | 8,100 | 7,504 | | $1.95^2$ | 2.23 | 2.27 |
| Netherlands | 1 | 40 | 415,385 | | | 2.24 | 2.12 |
| | 1 | 431 | 38,551 | | 3.79 | 3.86 | 3.55 |
| Norway | 1 | 7 | 725,738 | | | $1.65^4$ | 1.76 |
| | 1 | 19 | 267,377 | | | $2.43^4$ | 2.57 |
| Norway | 1 | 428 | 11,870 | | | $4.24^4$ | 4.42 |
| Poland | 1 | 16 | 2,391,166 | | $0.28^2$ | 0.26 | 0.26 |
| | 1 | 3,095 | 12,361 | | $1.09^2$ | 1.03 | 1.11 |
| Portugal | 5 | 7 | 1,439,769 | | 1.78 | 1.92 | |
| | 5 | 30 | 345,865 | | 2.34 | 2.78 | 2.70 |
| | 1 | 30 | 345,865 | 1.10 | 0.83 | 0.99 | 0.99 |
| | 5 | 308 | 33,688 | | 5.11 | 6.55 | 6.83 |
| | 1 | 308 | 33,688 | 2.28 | 1.69 | 2.19 | 2.64 |
| Russia | 1 | 80 | 1,785,618 | | | 0.69 | 0.61 |
| | 1 | 2,585 | 55,261 | | $2.01^1$ | 1.57 | 1.34 |
| Spain | 1 | 52 | 789,140 | | | 1.01 | 1.44 |
| | 1 | 8,108 | 5,061 | | | 2.56 | 3.45 |
| Sweden | 1 | 21 | 453,300 | | | 2.17 | 2.12 |
| | 1 | 290 | 32,825 | | | 4.43 | 4.63 |
| | 1 | 2,512 | 3,790 | | | 8.22 | 8.40 |
| Switzerland | 5 | 26 | 245,943 | 7.22 | 6.62 | 6.13 | |
| | 5 | 2,896 | 2,208 | 19.80 | 19.13 | 20.23 | |
| United Kingdom | 1 | 404 | 156,386 | | | 4.24 | 4.44 |

Source and notes: As for Table 4.5; also, 1. data for 1993; 2. data for 1995; 3. data for 2003; 4. data for 2005; 5. data for 2006; 6. data for 2009.

## Trends in migration intensities: Latin America

For the analysis presented here, we group together 10 countries in South America and 11 in Central America and the Caribbean. Data for the last decade are not available at all spatial scales, but migration statistics date from the 1980s in many countries, and these show a long-standing trend of falling mobility (Table 4.8). In South America, migration intensities declined in Argentina, Brazil, Colombia, Ecuador, Paraguay, Peru, Uruguay and Venezuela. Only Bolivia and Chile show a contrary trend, though data for the latter have been withdrawn due to concerns about data accuracy. The picture is similar in Central America and the Caribbean. Antigua and Barbados registered rising intensities but mobility declined in Mexico, Panama, Nicaragua, Honduras, Haiti, El Salvador and Belize. Costa Rica and the Dominican Republic display more stable levels of mobility.

*Table 4.8* Trends in internal migration intensity, 1980–2010, countries in Latin America

| Country | Observation interval (years) | Number of regions | Average population per region | Migration intensity (%) | | | |
|---|---|---|---|---|---|---|---|
| | | | | 1980 | 1990 | 2000 | 2010 |
| Antigua | 5 | 8 | 6,792 | | 11.13 | 13.01 | |
| Argentina | 5 | 24 | 1,553,057 | 8.55 | 6.33 | 3.55 | 3.40 |
| Barbados | 5 | 11 | 17,556 | | 6.94 | 6.38 | 7.72 |
| Belize | 5 | 6 | 32,301 | | 6.20 | 5.71 | |
| Bolivia | 5 | 9 | 963,230 | 6.09 | 5.64 | 5.98 | 8.04 |
| | 5 | *112 | 77,402 | 10.82 | 9.30 | 10.02 | |
| Brazil | 5 | 27 | 6,463,144 | | 3.86 | 3.40 | 2.64 |
| | 5 | 588 | 296,777 | | | 7.11 | 5.23 |
| Chile | 5 | 15 | 985,423 | 5.91 | 6.05 | 5.82 | 5.32 |
| Chile | 5 | 44 | 306,007 | | | 9.62 | 9.59 |
| | 5 | 304 | 44,291 | 18.11 | 19.46 | 18.24 | |
| Colombia | 5 | 33 | 1,308,607 | | 8.15 | | 4.21 |
| | 5 | 532 | 81,173 | | 13.45 | | 6.42 |
| Costa Rica | 5 | 7 | 584,834 | 6.60 | | 5.53 | 5.56 |
| | 5 | 81 | 50,541 | 13.16 | | 10.67 | 10.43 |
| Dominican Rep | 5 | 9 | 1,112,978 | | | 4.17 | 4.89 |
| | 5 | *155 | 64,625 | | | 6.43 | 6.01 |
| Ecuador | 5 | 25 | 521,202 | 8.47 | 6.16 | 5.55 | 4.73 |
| | 5 | *128 | 101,797 | | 8.64 | 8.25 | |
| El Salvador | 5 | 14 | 363,603 | | 4.79 | | 3.13 |
| | 5 | 262 | 19,429 | | 7.58 | | 5.70 |
| Haiti | 5 | 10 | 899,623 | 5.52 | | 2.39 | |
| | 5 | 41 | 219,420 | 6.90 | | 6.00 | |
| Honduras | 5 | 18 | 353,613 | 3.07 | 4.92 | 4.24 | |
| Mexico | 5 | 32 | 3,683,950 | | 4.86 | 4.44 | 2.48 |
| Nicaragua | 5 | 17 | 320,895 | | | 3.54 | 2.45 |
| | 5 | 153 | 35,655 | | | 5.24 | 4.08 |
| Panama | 5 | 12 | 254,568 | 4.36 | 4.38 | 6.35 | 5.32 |
| | 5 | *76 | 40,195 | | 9.33 | 12.56 | 10.30 |
| Paraguay | 5 | 18 | 309,521 | 10.78 | 9.11 | 7.61 | |
| | 5 | *236 | 23,608 | 16.77 | 12.56 | 11.51 | |
| Peru | 5 | 25 | 1,133,136 | | 8.59 | | 5.40 |
| Uruguay | 5 | 19 | 178,078 | 6.65 | 7.52 | 6.54 | 5.08 |
| Venezuela | 5 | 25 | 979,731 | | | 5.07 | 1.82 |

Source and notes: As for Table 4.5.

### Trends in migration intensities: North America and Oceania

North America and Oceania are considered together here, since each of their most populous countries, the USA and Australia, are examined closely in later chapters. Moreover, as can be seen from Tables 4.9 and 4.10, the trends in each continent are crystal clear. In North America, the data point to long-established declines in migration intensities in both Canada and the United States, and these are apparent at all spatial scales. In Oceania, the same is true for Australia, with the decline starting in the 1990s. Mobility rates in New Zealand and Fiji have shown greater stability, but they too display a general downward trend from the 1980s.

*Table 4.9* Trends in internal migration intensity, 1980–2010, countries in North America

| Country | Observation interval (years) | Number of regions | Average population per region | Migration intensity (%) | | | |
|---|---|---|---|---|---|---|---|
| | | | | 1980 | 1990 | 2000 | 2010 |
| Canada | 5 | 11 | 2,579,420 | 5.27 | 4.04 | 3.37 | 2.89 |
| | 1 | 11 | 2,579,420 | | 1.24 | 0.99 | 0.95 |
| | 5 | 288 | 98,519 | 15.70 | 14.85 | 12.46 | 11.34 |
| | 1 | 288 | 106,225 | | | 3.81 | 3.59 |
| | 5 | *4,916 | 5,772 | | 20.69 | 16.57 | 14.96 |
| USA | 5 | 9 | 28,317,625 | | 6.97 | 6.57 | |
| | 5 | 51 | 4,997,228 | 9.86 | 9.60 | 8.94 | |
| | 1 | 51 | 5,956,280 | | | 3.14 | 1.45 |
| | 1 (ACS) | 51 | 5,325,385 | | | 2.42 | 2.25 |
| | 1 (CPS) | 3,144 | 96,619 | | | 6.42 | 3.54 |

Source and notes: As for Table 4.5; ACS data from American Community Survey; CPS data from Current Population Survey.

*Table 4.10* Trends in internal migration intensity, 1980–2010, countries in Oceania

| Country | Observation interval (years) | Number of regions | Average population per region | Migration intensity (%) | | | |
|---|---|---|---|---|---|---|---|
| | | | | 1980 | 1990 | 2000 | 2010 |
| Australia | 5 | 8 | 2,688,465 | 5.19 | 5.37 | 4.82 | 4.40 |
| | 1 | 8 | 2,177,554 | 2.09 | 2.02 | 1.63 | 1.50 |
| | 5 | 38 | 565,993 | 16.90 | 16.52 | 14.90 | 12.61 |
| | 5 | 69 | 311,706 | 18.85 | 18.19 | 16.47 | 13.99 |
| | 5 | 1,379 | 15,597 | 28.64 | 28.29 | 28.06 | 25.06 |
| | 1 | 1,379 | 14,298 | 10.89 | 11.01 | 10.45 | 9.11 |
| New Zealand | 5 | 16 | 204,895 | 11.58 | 10.36 | 10.02 | 9.71 |
| | 5 | 73 | 44,908 | | | 20.51 | 20.51 |
| Fiji | 5 | 10 | 74,227 | 15.95 | 15.14 | 14.24 | 14.75 |

Source and notes: As for Table 4.5.

## Understanding migration trends

As foreshadowed in Chapters 1 and 2, trends over time in migration intensities are the product of a complex interplay of forces. While some of these operate at a global or regional level, such as the global and Asian financial crises, trends in individual countries are also shaped by local contingencies. Teasing out their relative effects is a task for the conclusion to this volume, rather than for a chapter such as this that aims to sketch the global context. Nevertheless, it is useful to draw attention to four factors that inevitably exert some influence on temporal trends.

The first of these is inconsistencies in zonal boundaries over time. As noted earlier, the IMAGE project attempted to establish temporally consistent boundaries where possible, but this is only feasible where migration data are available in the form of inter-regional flow matrices, with associated digital boundaries. Blake *et al.* (2000) review the issues and solutions. If, however, migration intensities are derived from tabulations classified by migration status, as is often the case with census reports that do not provide any data on origins or destinations, such adjustments are not achievable. Even where detailed data are available, complex revisions may seriously impede harmonisation, and hence prejudice reliable comparisons between census observations a decade apart. Since the general tendency is for the number of zones to increase over time, the effect is to artificially inflate apparent migration intensities, since more moves are likely to cross zonal boundaries. This effect almost certainly underpins the rise in migration intensity in Zambia (where the number of regions increased from 55 in 1990 to 74 in 2010). Similar effects may have occurred in Senegal and Bolivia, as well as in Indonesia where the number of both Major and Minor regions increased sharply between the 2000 and 2010 Census rounds. In a similar vein, the downward trend in migration intensity in Ecuador, Panama and Paraguay may have been sharper, but for the increased geographic fragmentation over successive decades. Countries in which comparison of intensities is potentially seriously affected by boundary changes are asterisked in Tables 4.5–4.10, and full details of the data are available on the IMAGE Repository website (Bell *et al.*, 2014).

Allied to this is the issue of temporal harmonisation. The 2000–2010 interval encompassed the global financial crisis of 2008–2009, and migration intensities measured over a five-year interval at the 2010 or 2011 Census are more likely to have been affected than intensities measured over a single year. Five-year data might therefore be more susceptible to showing a decline over the decade. In practice, the countries which measure migration over both one- and five-year intervals show little discernible difference in trend (see Tables 4.5–4.10). Going beyond the evidence shown in these tables and looking at countries which collect annual data from registration systems, for some of these (notably Norway, Netherlands and Hungary) there is a discernible dip in migration intensities in 2008–2010 compared with 2007,but this is by no means a universal phenomenon. More notable, perhaps, among these predominantly European countries, is a distinctive upward trend in intensities post 2011, maintaining the rising trajectory described earlier.

A third and more widespread impetus to declining intensities is population ageing, which moves larger cohorts into older age groups, where the propensity to move is lower than among young adults (Rogers and Castro, 1981; Bernard *et al.*, 2014a). *Ceteris paribus*, migration intensities can be expected to decline as populations age. Correcting for changes in age composition calls for standardisation, which requires data on the age profile of migration. Such data are not widely available for historical time series, but some indication of the significance of age differences can be gained from the work of Bell *et al.* (2015b) who age-standardised migration intensities for 12 countries widely spaced along the development continuum. They found that age standardisation reduced the range between high and low intensities by around 20%, suggesting that age composition accounted for about one fifth of the variation between countries. In a similar way, Charles-Edwards *et al.* (see Chapter 7 in this volume) show that population ageing accounted for about one quarter of the fall in migration intensities between 69 regions of Australia over the 35 year period from 1976 to 2011. The pace of ageing varies across the world and its effects on crude migration intensities are complicated by shifts over time in the age profile of migration itself (Bernard *et al.*, 2014b). In general, however, it seems clear that the force of ageing will have intensified the downward trend in countries which recorded a fall in migration intensities, while reducing the apparent increase in those where intensities are rising. It also suggests the need for caution in interpreting shifts at the margin, since any change in crude intensities may result from age structure effects rather than from fluctuations in the underlying propensity to migrate.

Another general factor thought to influence migration is the level of national development. Bell *et al.* (2015b) reported moderate to strong and statistically significant associations between overall migration intensity (as measured by the ACMI) and the Human Development Index (HDI), whether computed across countries that collect migration data over a single year (mainly the more developed countries ($r = 0.62$, $n = 40$)), or across those that utilise a five-year measurement interval ($r = 0.48$, $n = 59$). While migration intensity itself is therefore closely linked to development, Table 4.11 provides little support for the notion that declining migration is confined to a particular segment of the development spectrum. Here, countries have been classified into four categories of development using the fixed boundaries re-introduced in the 2014 HDI, with cut-offs at 0.55, 0.7 and 0.8 (United Nations, 2014). It is patently clear from Table 4.11 that falls, stability and rises in intensity over the decade 2000 to 2010 were distributed quite evenly across countries in all four categories of HDI. Correlation coefficients confirm the absence of any association between HDI and change in migration intensity, with coefficients (Pearson $r$) of $-0.011$ for movement between for the 42 countries reporting migration between Major regions, and $-0.201$ for the 44 countries with data for Minor regions.

*Table 4.11* Trends in internal migration intensity, 2000–2010, major and minor regions, by HDI level

| HDI | Trend in migration intensity | | | Total countries |
|---|---|---|---|---|
| | *Fall* | *Stable* | *Rise* | |
| *Major regions* | | | | |
| Low | 3 | 0 | 1 | 4 |
| Medium | 4 | 0 | 4 | 8 |
| High | 6 | 2 | 2 | 10 |
| Very high | 9 | 6 | 5 | 20 |
| Total | 22 | 8 | 12 | 42 |
| *Minor regions* | | | | |
| Low | 1 | 1 | 2 | 4 |
| Medium | 4 | 0 | 2 | 6 |
| High | 5 | 3 | 2 | 10 |
| Very high | 10 | 6 | 8 | 24 |
| Total | 20 | 10 | 14 | 44 |

Source: Analysis of data from the IMAGE Repository.

Notes: See text and notes for Table 4.2.

## Conclusions

Comparing internal migration between countries is a challenging task; making comparisons of trends over time presents additional complexities in terms of spatial and temporal harmonisation. This chapter has brought together data for some 66 countries with representatives from all continental regions, with the aim of assessing the trend in migration intensities—that is in the overall propensity, or level of migration. We have focused particularly on the last decade, broadly corresponding the 2000–2010 period, but have also reported trends for earlier decennial intervals where data are available. Recognising the importance of spatial scale in any analysis of migration, we have sought to differentiate between long-distance and intermediate/short-distance migration, distinguished by reference to movements between Major Regions and between Minor Regions where such data are available. By updating and extending recent work from the IMAGE project, we also sought to demonstrate the wide variation in migration intensities that exists between countries around the world.

The results presented here reveal that falling migration intensities are not confined to countries in one particular part of the world, nor are they a feature of a particular stage of national development. Declining intensities appear to be universal across North America and Oceania, but they are also prevalent in Latin America and Asia and even in Africa. Surprisingly, perhaps, it is Europe which reveals the greatest diversity in migration trends.

Here, while a significant number of countries registered stability or decline, rising intensities were found in a large number of countries, scattered widely across the continent. In a similar way, there is no evidence to suggest that falling intensities are a product of an advanced stage of economic development: we found no association between the trend in migration intensity and the level of development as measured by the HDI.

These trends in migration intensities appear to be largely consistent irrespective of spatial scale. The overall decline in migration intensities that is prevalent across our sample of countries is evident both for long-distance and for intermediate- and short- distance moves. Where countries register declining migration at one level of scale, this is generally replicated at other scales, including for 'all moves', as denoted by trends in the ACMI. In a similar manner, countries experiencing rising intensities generally do so at multiple scales, too. The forces operating to drive migration intensities therefore appear to act broadly upon mobility across the geographic spectrum, rather than being confined to one particular spatial level. Nevertheless, there is some evidence of differential trends, and generally it appears to be migration over longer distances (between Major Regions) that has registered the most significant declines.

Fragmentation of geographic zones over successive censuses will inflate the apparent level of mobility, and this may underpin the rises observed in a small number of countries where data harmonisation was out of reach. We also pointed to the more widespread effect of population ageing, suggesting that this accounts for around one fifth to one quarter of the apparent decline in intensities, but noted equally that ageing served to constrain the rise in countries experiencing an upward migration trend. Beyond these broad forces, a more thorough understanding of migration trends calls for careful analysis of the factors at play in individual countries.

## Acknowledgements

The work reported in this chapter forms part of the IMAGE project supported by the Australian Research Council under ARC Discovery Project (DP110101363). The chapter draws on data from several sources, including the IPUMS database maintained by the Minnesota Population Center and from CELADE (Centro Latinoamericano y Caribeño de Demografía) as well as from individual national statistical offices. We gratefully acknowledge the support from these agencies. Details of the IMAGE project are available at www.imageproject.com.au.

## Disclaimer

The views expressed in this paper are those of the authors and do not necessarily reflect the views of the United Nations.

## References

Abel, G. and Sander, N. 2014. Quantifying global international migration flows. *Science*, 343(6178), 1520–1522.

Bell, M., Blake, M., Boyle, P., Duke-Williams, O., Rees, P., Stillwell, J. and Hugo, G. 2002. Cross-national comparison of internal migration: Issues and measures. *Journal of the Royal Statistical Society A*, 165(3), 435–464.

Bell, M. and Rees, P. 2006. Comparing migration in Britain and Australia: Harmonisation through use of age-time plans. *Environment and Planning A*, 38(5), 959–988.

Bell, M. and Charles-Edwards, E. 2013. *Cross-national Comparisons of Internal Migration: An Update of Global Patterns and Trends*. Technical Paper 2013/1. New York: Population Division, United Nations Department of Economic and Social Affairs. www.un.org/en/development/desa/population/publications/pdf/technical/TP2013-1.pdf

Bell, M. and Charles-Edwards, E. 2014. *Measuring Internal Migration around the Globe: A Comparative Analysis*. KNOMAD Working Paper 3/2014. Washington, D.C.: The World Bank. www.knomad.org/docs/internal_migration/NOMAD%20Working%20Paper%203_BellCharles-Edwards_12-19-2014.pdf

Bell, M., Charles-Edwards, E., Stillwell, J., Daras, K., Kupiszewski, M., Kupiszewska, D. and Zhu, Y. 2013. *Comparing Internal Migration Around the GlobE (IMAGE): The Effects of Scale and Pattern. Keynote presentation to the International Conference on Population Geographies, Groningen, The Netherlands, 25–28 June, 2013.* https://imageproject.com.au/publications/comparing_internal_migration_around_the_globe/

Bell, M., Bernard, A., Ueffing, P. and Charles-Edwards, E. 2014. *The IMAGE Repository: A User Guide*. QCPR Working Paper No 2014/01. Brisbane: University of Queensland. https://imageproject.com.au/framework/repositoryuserguide/

Bell, M., Charles-Edwards, E., Kupiszewska, D., Kupiszewski, M., Stillwell, J. and Zhu, Y. 2015a. Internal migration data around the world: Assessing contemporary practice. *Population, Space and Place*, 21(1), 1–17.

Bell, M., Charles-Edwards, E., Ueffing, P., Stillwell, J., Kupiszewski, M. and Kupiszewska, D. 2015b. Internal migration and development: Comparing migration intensities around the world. *Population and Development Review*, 41(1), 33–58.

Bell, M. and Ward, G.J. 2000. Comparing permanent migration with temporary mobility. *Tourism Geographies*, 2(1), 97–107.

Bernard, A. and Bell. M. 2015. Smoothing internal migration age profiles for comparative research. *Demographic Research*, 32 (33), 915–948.

Bernard, A., Bell, M. and Charles-Edwards, E. 2014a. Improved measures for the cross-national comparison of age profiles of internal migration. *Population Studies*, 68(2), 179–195.

Bernard, A., Bell, M. and Charles-Edwards, E. 2014b. Explaining cross-national differences in the age profile of internal migration: the role of life-course transitions. *Population and Development Review*, 40(2), 213–239.

Blake, M., Bell, M. and Rees, P. 2000. Creating a temporally consistent spatial framework for the analysis of interregional migration in Australia. *International Journal of Population Geography*, 6, 155–174.

Castles, S. and Miller, M.J. 2009. *The Age of Migration*. 4th edition. Basingstoke: Palgrave Macmillan.

Charles-Edwards, E., Muhidin, S., Bell, M. and Zhu, Y. 2016. Regional perspectives: migration in Asia. In White, M. (ed) *International Handbook of Migration and Population Distribution*. Dordrecht: Springer, 269–284.

Coleman, D. 2013. The twilight of the census. *Population and Development Review*, 38, 334–351.

Courgeau, D., Muhidin, S. and Bell, M. 2012: Estimating changes of residence for cross-national comparison. *Population-E*, 67(4), 631–652 (Also published as: Estimer les changements de résidence pour permettre les comparaisons internationals. *Population-F*, 67(4), 747–770).

Daras, K. 2014. *IMAGE Studio 1.4.2 User Manual*. www.imageproject.com.au/docs/qcpr/IMAGE_studio.pdf

Fan, C.C. 2005. Interprovincial migration, population redistribution, and regional development in China: 1990 and 2000 census comparisons. *The Professional Geographer*, 57(2), 295–311.

Rees, P., Bell, M., Kupiszewski, M., Kupiszewska, D., Ueffing, P., Bernard, A., Charles-Edwards, E. and Stillwell, J. 2016. The impact of internal migration on population redistribution: An international comparison. *Population, Space and Place*, published online in Wiley Online Library. doi:10.1002/psp.2036

Rogers, A. and Castro, L. 1981. *Model Migration Schedules*. Laxenburg: International Institute for Applied Systems Analysis.

Skeldon, R. 1990. *Population Mobility in Developing Countries: A Reinterpretation*. London: Belhaven.

Skeldon, R. 1997. *Migration and Development: A Global Perspective*. Harlow: Longman.

Stillwell, J., Daras, K., Bell, M. and Lomax, N. 2014. The IMAGE Studio: A tool for internal migration analysis and modelling. *Applied Spatial Analysis and Policy*, 7(1), 5–23.

Stillwell, J., Bell, M., Ueffing, P., Daras, K., Charles-Edwards, E., Kupiszewski, M. and Kupiszewska, D. 2016. Internal migration around the world: comparing distance travelled and its frictional effect. *Environment and Planning A*, 48(8), 1657–1675.

United Nations. 2009. Overcoming Barriers: Human Mobility and Development. *Human Development Report 2009*. New York, NY: United Nations Development Program.

United Nations. 2014. *Human Development Report 2014*. New York, NY: United Nations Development Program. http://hdr.undp.org/sites/default/files/hdr14_technical_notes.pdf

World Bank. 2009. *World Development Report 2009. Reshaping Economic Geography*. Washington, D.C.: The World Bank.

Wrigley, N., Holt, T., Steel, D. and Tranmer, M. 1996. Analysing, modelling, and resolving the ecological fallacy. In Longley, P. and Batty, M. (eds) *Spatial Analysis, Modelling in a GIS Environment*. Cambridge: GeoInformation International, 23–40.

Zelinsky, W. 1971. The hypothesis of the mobility transition. *Geographical Review*, 61, 219–249.

# Part II
# In-depth country analyses

# 5   United States

## Cohort effects on the long-term decline in migration rates

*Thomas Cooke*

The USA forms the first of the seven case study chapters in this volume. This is entirely appropriate because, as set out in Chapter 1, this is the country which first alerted the research community to the fact that over time the world may not be becoming progressively more mobile in every way and that internal migration rates can fall as well as rise beyond the short-term effects of the business cycle. This experience is all the more notable given the long-standing perception that the USA has always been a highly mobile society (see Long, 1991). As a consequence, for a long while there was considerable resistance in the USA to the idea that the pace of internal migration could be slowing (Fischer, 2002; Wolf and Longino, 2005), so one of the roles of this chapter is to demonstrate that the evidence is incontrovertible. A second objective is to review the existing literature on migration decline in order to see how much agreement there now exists on the factors that are driving this phenomenon. Thirdly, the chapter presents the results of a new analysis that tests the explanatory power of a set of factors that has been neglected in recent research on the US case. Before tackling these three tasks, however, the chapter begins with an introduction to the data sources available for the study of internal migration in the USA and an assessment of the value of those used in the research reported in the following sections.

### Data sources

There are three main sources of data on US internal migration: the Current Population Survey (CPS), the decennial Census and panel surveys. Our understanding of long-term trends is primarily based upon the first of these. A joint project of the US Bureau of Labor Statistics and the US Census Bureau, the CPS dates from the 1930s and has a focus on providing monthly labour force statistics. Starting in 1947, the CPS added an annual supplement to the March survey (now known as the Annual Social and Economic Supplement) which includes individual-level data for change of address relative to 12 months ago, type of move and reason for move. This data is available on an annual basis since 1948, with the exception of 1972–1975 and 1977–1980. CPS microdata, most easily accessed through the IPUMS-CPS project of the Minnesota Population Center (see Flood *et al.*, 2015), allow for the breakdown of migration behaviour by subgroup and for the identification of key covariates (see Cooke, 2011).

Though valuable in spanning such a long period of time and recording all changes of address including the most local, the CPS also suffers some shortcomings. One is that the dearth of information on people's characteristics prior to the move largely precludes causal analyses of migration behaviour. Secondly, while sample sizes are relatively large, with approximately 31,000 households containing 72,000 persons in 1962 and growing to approximately 94,000 households containing 185,000 persons in 2016, the ability to calculate estimates of migration for small geographic units is limited due to confidentiality constraints. Thirdly, as detailed by Kaplan and Schulhofer-Wohl (2012), an undocumented change in the Census Bureau imputation procedure appears to have artificially inflated the inter-state migration rate for the years 1999–2005.

But perhaps the most significant limitation of the CPS is the absence of information on distance moved. Rather, it reports whether an individual remained at the same house relative to 12 months ago, changed residence but stayed within the same county, moved from one county to another but within the same state (henceforth termed an 'intra-state' migration) or moved from one state to another (henceforth an 'inter-state' migration). However, the area of these geographical units varies markedly, especially by region with a general increase in size from the North-East toward the South and West, leading to regional differences in the measurement of migration. For example, an individual living in a small north-eastern state can move 50 miles, cross a state line in the process, and be labelled an inter-state migrant. In contrast, an individual living in one of the larger California counties, for example, can move 50 miles without crossing a county line and thus not even be classified as an intra-state migrant as just defined (see Niedomysl and Fransson, 2014, for a more general discussion of issues regarding the use of administrative boundaries for defining migration).

Migration data based on the decennial US Census are available in a variety of formats. The most widely used are microdata data drawn from the original Public Use Microdata Samples (PUMS) from the 1940 through 2000 decennial censuses and the annual American Community Survey (ACS) from 2001 to present. All of these are most easily accessed, along with pre-1940 decennial census microdata, through the IPUMS project of the Minnesota Population Center (see Ruggles *et al.*, 2015). Migration data are available in each year of these series but migration definitions are inconsistent. Most notably, the PUMS reports place of residence five years ago, while the ACS reports place of residence 12 months ago. However, sample sizes are very large and so in many cases both the current and previous place of residence can be identified for smaller geographic units such as counties, larger municipalities, and areas within the largest cities. Like the CPS, both the PUMS and the ACS provide very little information prior to the move which similarly limits causal analyses of migration behaviour.

There are three main panel data surveys. The Panel Study of Income Dynamics (PSID) is a longitudinal sample of individuals and their family members collected annually since 1968 and biannually since 1997 (see Panel Study of Income Dynamics, 2016). The National Longitudinal Surveys (NLS) are a series of five panel studies which focus on specific birth cohorts differentiated, in some cases, by gender (see U.S. Bureau of Labor Statistics, 2014). The Survey of Income and Program Participation (SIPP) consists of a series of multi-year rotating panels with data stretching back to 1983 (see U.S. Census Bureau, 2017a). While sample sizes are relatively small in all three cases, these data are important for migration research because they support the examination of causal effects. Each of these projects also has procedures for gaining access to restricted data on the details of place of residence, allowing for the identification of distance-based measures of migration.

Finally, while the USA lags far behind many other (especially West European) countries in the use of administrative data for migration research, significant headway is being made through its system of Research Data Centers (see U.S. Census Bureau, 2017b). These provide access to confidential individual-level data from sources such as the Social Security Administration, the Census and the Internal Revenue Service (with some data on the geographic mobility of individual tax filers between counties and states). These data can be linked by individuals over time providing new opportunities for research on the causes and consequences of migration in the USA. As well, some administrative data are geocoded, allowing for the identification of distance-based measures of migration.

The findings presented in the rest of this chapter are drawn primarily from the CPS. This is because the CPS provides the most consistent historical series of aggregate migration data. Also, since the main research emphasis has been on identifying the demographic correlates of declining migration rates, the limited utility of the CPS for causal analyses has not been seen as an issue. However, as the emphasis switches to testing the causes of the migration decline, it is likely that future research will lean more heavily on the panel and administrative data sources (e.g. Cooke and Shuttleworth, 2017).

## Long-term trends

Figure 5.1 reports CPS-based data on the rate of annual moves across county boundaries within states and across state boundaries, respectively 'intra-state' and 'inter-state' migration as defined above, for 1948 through 2016. Up to around 1970, the incidence of both intra- and inter-state migration was generally in the range of 3.0–3.5% per year with fluctuations most likely associated with short-term business cycles. Beginning in the 1970s, rates of intra- and inter-state migration diverged. While intra-state migration rates remained stable at post-World War II levels well into the 1990s, inter-state migration rates declined consistently from the early 1970s into the 1990s. By the mid-1990s intra-state migration rates started to decline in concert with inter-state migration rates. The two best-fit curves, which are quadratics of time designed to smooth out the jumpiness of the rates between individual years and also ignore the inter-state rates for 1999–2005 (because of the imputation issue mentioned in the previous section), emphasise both the considerable extent of the fall in the two rates and the later onset of the decline in the intra-state rate.

The overall degree of fall in these migration rates is clearly very considerable. Between 1948 and 2016 the inter-state migration rate declined by 52% (from 3.1 to 1.5%) and the intra-state migration rate declined by 27% (from 3.3 to 2.4%). But as seen in Figure 5.1, the majority of these reductions have taken place since the early 1980s. The respective declines from 1984 through 2016 were by 46% for inter-state migration and by 34% for intra-state migration. Finally, while Figure 5.1 does not report values for residential change within the same county, it is important to note that it, too, has declined dramatically, falling by 34% between 1984 and 2016. This is the same degree of drop as for the intra-state rate, suggesting that these two shorter-distance types of migration are operating in concert with each other and are likely subject to the same forces of change.

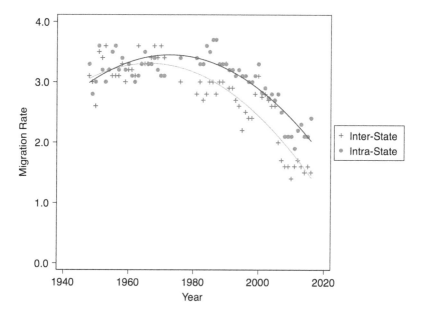

*Figure 5.1* United States: annual inter-county migration, 1948–2016, by whether intra-state or inter-state.

Source: Current Population Survey; see text for details of curve fitting.

While it is correct to make the general point that migration rates have declined in the USA, the divergent trajectories in the decline of inter-state migration and that of shorter-distance moves merit discussion. One possibility is that both intra- and inter-state migration have been drawn downwards by similar processes, but that there are additional dynamics unique to intra-state migration which may have propped up those rates for a longer period of time. One plausible explanation is that, since residential relocation from contiguous urban and suburban counties are included in the calculation of intra-state migration, the tail-end of the post-WWII suburbanisation may have masked a decline in longer-distance migration within states which was similar to that observed for inter-state migration.

A second possibility for the divergence in the decline in intra- and inter-state migration may be due to a type of measurement error as a consequence of the varying geographic sizes of states. As previously discussed, the areas of states vary dramatically by region such that there is a general increase in state size from the North East toward the South and West. Importantly, over the last 70 years there has also been a significant relocation of population from the smaller-area states of the North-east toward larger-area states of the South and especially the West. Thus, it is plausible that 70 years ago there was a greater probability that a randomly selected migrant would move across a state border than now because 70 years ago a greater proportion of people lived in smaller states. However, an examination of changes in migration rates by state precludes this explanation for the long-term decline in aggregate inter-state migration rates. For example, between 1982 and 2015 inter-state migration rates declined in every state except for only Louisiana and North Dakota.

## Explaining the migration decline

Apart from a slightly different body of research in geography that questions the focus on mobility over that of rootedness (e.g. Clark and Whiteman, 1983; Hanson, 2005), research on declining migration in the USA remains quite limited and mostly focused on the role of demographic processes. Fischer (2002), Wolf and Longino (2005), Pingle (2007), Molloy *et al.* (2011) and Cooke (2011) consider in varying ways whether the migration decline can be explained by either changes in subgroup migration behaviour (i.e. rate effects) or changes in the composition of the population (i.e. composition effects). However, none of these studies finds much of a role for either. For example, Cooke (2011) concludes that only about 17% of the decline in migration rates between 1999 and 2009 is linked to demographic changes such as the ageing of the population, while Molloy *et al.* (2011, p. 11) find that the

> ... decrease in migration does not seem to be driven by demographic or socioeconomic trends, because migration rates have fallen for nearly every subpopulation and the composition of the population has not shifted in a way that would affect aggregate migration appreciably.

As a consequence, all of these studies conclude that more structural social, cultural, political or economic changes lie at the root of the migration decline: 'The social forces that have encouraged stability…must be deep and pervasive' (Fischer, 2002, p. 193).

However, the identification of these structural processes remains highly speculative. While not focused on the migration decline, Zelinsky (1971) and Long (1991) offer some contextual guidance. Zelinsky (1971) proposes a mobility transition, paralleling the demographic transition, whereby historical changes in geographic mobility are linked to the historical process of development and modernisation. He identifies the USA at the time of his writing as an 'advanced' society marked by both high rates of migration, especially between urban regions, and high levels of both economic and non-economic circulation. Importantly, he foresees the emergence of a 'future superadvanced' society with lower migration rates due to three factors: a levelling of inter-regional differences in both the supply and demand for labour; the cancelling-out of migration '… by better communications, as travel is rendered redundant by more efficient transmission of messages for business, social, and educational purposes'; and '… strict political control over internal as well as international movements may be imposed' (Zelinsky, 1971, p. 230). Some of these same ideas are reflected in Long's (1991) cross-national analysis of internal migration. He explains high rates of migration among the 'frontier' countries of the United States, Australia and Canada as being due to the cultural traditions that emerged from less regulated housing, labour and financial markets, as well as the traditions of mobility among immigrant societies and a surplus of largely undeveloped land.

In so doing, Zelinsky (1971) and Long (1991) point toward three structural possibilities for the migration decline. First, decades of inter-regional migration have caused a levelling in regional labour and housing markets such that wages and the cost of living more fully compensate for regional differences in physical and cultural amenities (also see Ferrie, 2005). The resulting spatial equilibrium in housing and labour markets would lead to a decline in inter-regional migration by largely eliminating any gains in utility to migration. Partridge *et al.* (2012) explicitly test this idea, hypothesizing that spatial equilibrium in regional housing and labour markets should reduce the effect of amenities on in-migration because those amenities would be fully capitalised in the markets. However, they find little evidence in support of this hypothesis. Rather, their results suggest that the relationship between migration and exogenous shocks to regional labour markets has become weaker.

Second, the development and widespread adoption of advanced information and communications technologies (ICTs) has contributed to the emergence of new forms of communication that offer alternatives to migration. The common presumption is that ICTs are associated with an increase in geographic mobility in at least three ways: by improving access to information on distant locales; by allowing migrants to communicate more effectively with people and places left behind; and by allowing more freedom to choose a place of residence less on the basis of proximity to a fixed place of work and more on the basis of other geographic character-istics such as natural and cultural amenities (see Cooke and Shuttleworth, 2017). By contrast, Cooke (2013) finds that the widespread adoption of ICTs over the last several decades is strongly correlated with a reduction in migration rather than with an increase. Hence, Cooke and Shuttleworth (2017) counter that ICTs may simultaneously lower rates of migration by increasing the efficiency of migration, notably with higher-quality infor-mation meaning that fewer moves lead to disappointment and a return move, and by reducing the cost of staying put, such as through their effects on reducing isolation from kith and kin who have migrated, on providing access to higher education without having to move and on supporting em-ployees who work from a remote location. Indeed, using causal methods applied to a longitudinal sample of residents in Northern Ireland, Cooke and Shuttleworth (2017) find that, on balance, ICTs actually reduce geo-graphic mobility.

Third, there is growing interest in the relationship between a decline in migration and increasing regulation, especially in the labour market. Most importantly, individual states are now increasingly requiring a state-sanctioned licence to practise an occupation (Kleiner and Krueger, 2013). There is some evidence that this restricts inter-state migration (Holen, 1965; Kleiner et al., 1982; Peterson et al., 2014; Kleiner, 2015). However, the potential for state policies to act as a barrier to inter-state migration extends far beyond labour markets. For example, the evolution of child-custody arrangements over the last several decades is associated with lower rates of inter-state migration (Cooke et al., 2016). Tradition-ally, it was common for children of separated parents to reside primarily with one (usually the mother), with more limited visits to the other (usu-ally the father), but the contemporary trend is towards having children spend considerable residential time with both parents (Cancian et al., 2014). Since jurisdiction over associated custody agreements falls to state courts within the USA, parents are effectively constrained to living in the same state (and sometimes even the same locality) after divorce or risk having the custody arrangement nullified (Nazir, 2009). This particular issue is probably amplified for the growing number of complex blended families.

While not implied by Zelinsky (1971) and Long (1991), a general theme running through these discussions is that growing economic insecurity may also play a role in the migration decline. Potential migrants have fewer re-sources to draw from in order to move, local social capital may be seen as an increasingly important resource for buffering against economic insecurity, and the more contingent nature of employment means that the risk to mov-ing has increased. For example, and quite apart from a narrower concern for how recent housing bubbles may restrict mobility among homeowners with negative home equity (e.g. Ferreira *et al.*, 2010), Cooke (2013) finds that increasing levels of household debt and the rise of dual-income households—both of which are responses to a long-term decline in real income—are strongly correlated with the migration decline. As well, Partridge *et al.* (2012) interpret their finding of a weaker relationship between migration and exogenous shocks to regional labour markets as reflecting an increase in the risk attributed to migration. Likewise, to the degree that the adoption of advanced ICTs reduces the risk to staying (see Cooke and Shuttleworth, 2017), their use during a time of economic insecurity may be enhanced as a means of avoiding the added risk of migration. Indeed, Morrison and Clark (2016) suggest that during times of crisis the utility of possessions (such as a home) is enhanced, increasing the perceived risk of moving.

## A cohort perspective on the migration decline

Largely overlooked in these discussions of recent migration decline is a body of earlier research by Rogerson (1987), Plane and Rogerson (1991) and Pandit (1997a). Indeed, Rogerson (1987) was perhaps the first to take detailed note of the phenomenon of migration decline. He argues that the migration decline is linked to cohort effects similar to those proposed by Easterlin (1980), who hypothesised that many of the social and economic changes of the last half-century are related to the relative sizes of different birth cohorts. In particular, Easterlin argues that the size of a birth cohort determines relative labour and housing market opportunities which, in turn, shape the timing of family formation, fertility and female labour force participation, among other things. Large birth cohorts face greater competition in housing and labour markets and, as a consequence, their members delay family formation, reduce fertility and experience higher rates of female labour force participation in order to supplement household income. However, this effect is not symmetrical about the peak of a baby boom or a baby bust. Rather, those on the trailing edge of a baby boom face the greatest competition in housing and labour markets from older, and larger, age cohorts. Importantly, these effects are particularly strong for those in their 20s since they have relatively fewer resources with which to compete in housing and labour markets. Finally, since events in young adulthood shape expectations for the rest of the life course, these effects tend to persist as a cohort ages.

Rogerson (1987) and Plane and Rogerson (1991) extended this approach by applying it to migration, as too recently have Sander and Bell (2016) in the Australian context. On this basis, members of large cohorts face greater competition in housing and labour markets and so migrate less often (a cohort-size effect). Since younger adults have fewer resources to compete in housing and labour markets, the negative effect of cohort size is greatest for these (an age-specific cohort-size effect). Secondly, large numbers of older adults reduce aggregate migration rates since they crowd out opportunities in housing and labour markets for everyone (the Easterlin effect). Thirdly, since younger adults lack the resources to compete in housing and labour markets, the negative effect of a relatively large number of people in older age cohorts on age-specific migration rates is greatest among younger age groups (an age-specific Easterlin effect). Finally, since migration is a learned behaviour and the risk of (not) moving is greater for those who have (not) already moved (Bailey, 1989), the migration experience of a cohort while its members are in their 20s should persist across a cohort as it ages (a cohort effect). Note, however, that while Plane and Rogerson (1991) find that those in the baby boom had lower rates of migration in their early 20s (consistent with an age-specific cohort size effect), their migration rates recovered in their late 20s. They interpret this as a delayed migration effect (see also Pandit, 1997b). However, this finding coincided with 1985, which was also a period of economic expansion. This last point emphasises the importance of period effects associated with the business cycle: migration intensity rises during economic recovery and falls in recession, and particularly so for young adults.

From this perspective, the decline in migration rates beginning in the late 1960s and lasting through the early 1980s was partly due to the impact of the large baby-boom cohort of those born between 1946 and 1964, with a median birth year of 1955 (Carlson, 2009), passing through their 20s over this time period (an age-specific cohort-size effect) which crowded out opportunities both for themselves and for younger cohorts (Rogerson, 1987). However, migration rates have continued to decline. Rogerson's (1987) perspective suggests that this has occurred for several reasons: (1) the relatively lower rates of migration among the baby boom cohort (a cohort-size effect), (2) the ageing of the baby boom cohort into older, and less migratory, age categories (an age effect), (3) the relative size of the increasingly immobile baby boom cohort (a composition effect) and (4) the crowding of opportunities for those in the relatively younger baby bust cohort by the relatively large number of older adults (an Easterlin effect).

In the new research results that are presented below, Pandit's (1997a) approach to estimating the impact of cohort size on changing yearly migration rates is updated using data from the IPUMS version of the CPS. Data limitations restrict the analysis to the period 1966–2015 (and omitting 1972–1975, 1977–1980 and 1985) and the sample is further limited to adults aged 20 through 64. Specifically, the effect of a set of variables reflecting age-, period-, and cohort-effects on age-specific annual migration rates are estimated as an OLS linear probability model that treats the migration rate as a continuous variable, with the parameter estimates indicating the effect of a one-unit change in an independent variable on the probability of migrating (see Angrist and Pischke, 2009). The analysis also integrates the Cochrane-Orcutt procedure to address concerns for the occurrence of spurious correlation (see Gujarati, 1995; McCallum, 2010). More intuitively, the basic approach is to estimate the annual inter-county migration rate (which is the combined intra- and inter-state migration rate) for each five-year age group in the sample (i.e. 20–24, 25–29, etc.) as a function of a set of five independent variables: (1) *Age* (as indexed by each of the five-year age categories), (2) the annual *Cohort Size* of each of these age groups as defined by the population of the age group divided by the total US population, (3) the rolling three-year average State *Unemployment Rate*, (4) the annual *Easterlin Ratio* as defined by Plane and Rogerson (1991) as the number of males aged 35–64 divided by the number of males aged 15–34 (see Figure 5.2) and (5) a dummy variable indicating whether the observation is between 1999 through 2005 which captures the effect of the temporary change in imputation techniques on migration rates mentioned above.

The role of *Age* in this model is of particular importance. Reflecting standard life course and human capital theory, it is expected that younger age groups migrate more frequently than the older ones. But more importantly, the impacts of *Cohort Size*, the *Unemployment Rate* and the *Easterlin Ratio* on annual age-specific migration rates are themselves a function of age. Hence, each of these is interacted with *Age* to identify the age-specific effects of *Cohort Size,* the *Unemployment Rate* and the *Easterlin Ratio* on annual age-specific migration rates. Thus, while it is expected that *Cohort Size* has a general negative effect on annual migration rates due to greater intra-cohort competition, this negative effect is expected to be enhanced among younger age groups because younger age groups are more sensitive to

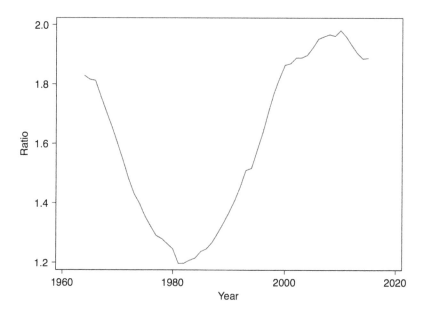

*Figure 5.2* Computed Easterlin ratio for the USA, 1966–2015.
Source: Author's calculation, see text.

intra-cohort competition. Similarly, while migration rates generally decline when unemployment increases, this effect likely enhanced among younger age cohorts. Finally, the *Easterlin Ratio* is expected to have a negative effect on the age-specific migration rates of younger age cohorts, given that younger age cohorts who enter into housing and labour markets dominated by older age cohorts will have fewer opportunities and will migrate less as a result.

The results of the linear probability model of age-specific migration rates are now presented in several ways. First, Table 5.1 shows the parameter estimates of the linear probability model of the age-specific migration rates for 1966–2015. Focusing on the main effects, it can be seen that age-specific migration rates decline with age, the unemployment rate and the Easterlin ratio, but cohort size is not statistically significant. Note also that the parameter estimate associated with observations between 1999 and 2005 are positive, reflecting the upward bias in migration rates during those years (see above). The result with respect to the main effect of the Easterlin ratio is of particular importance. When there are large numbers of 35–64 year old men relative to 15–34 year old men, migration rates for all age groups are reduced, presumably because a large number of 35–64 year olds crowd out opportunities in housing and labour markets for all age groups.

Next, the predicted values from the model are compared to observed values to confirm how the model explains the long-term trend in aggregate migration rates. Figure 5.3 shows that this model does produce an accurate estimate of the long-term decline in aggregate migration rates since the early 1980s. The only apparent deviation from the long-term trend is that the model under-predicts aggregate migration rates prior to 1970 and around 1980. Note, however, that these are periods of significant social and economic change. The 1960s saw rapid migration not only among young adults associated with the expansion of higher education opportunities and with the mandatory military draft, but migration was also a common feature of corporate employment at that time. In turn, the early 1980s were also a turbulent moment, but in this case it was associated with the peak of the disruption associated with deindustrialisation.

The third step looks at the combined main and age-specific effects for each of the three focal variables (cohort size, Easterlin ratio, and unemployment rate) on migration rates between 1986 and 2015, with 1986 selected as the base because of the steady downward decline in migration rates since then. These combined effects are presented in terms of average marginal effects, which indicate the combined effects of a one unit change in the independent variable of a particular subgroup—as estimated through the interaction effects—on annual age-specific migration rates (see Graubard and Korn, 2004). These are graphed as a function of age in Figure 5.4, which can be interpreted as follows. For example, the value of approximately −0.6 for 20–24 year olds in Figure 5.4a means that, independent of all other variables, a one-unit change in the unemployment rate is associated with a 0.6% point

*Table 5.1* United States: linear probability model of inter-county migration, 1966–2015

| Variables | Parameter estimate | p-value |
|---|---|---|
| 25–29 | −0.089 | 0.044 |
| 30–34 | −0.176 | 0.000 |
| 35–39 | −0.195 | 0.000 |
| 40–44 | −0.230 | 0.000 |
| 45–49 | −0.247 | 0.000 |
| 50–54 | −0.260 | 0.000 |
| 55–59 | −0.272 | 0.000 |
| 60–64 | −0.264 | 0.000 |
| Unemployment rate | −0.473 | 0.000 |
| Cohort size | −0.076 | 0.554 |
| Easterlin ratio | −0.094 | 0.000 |
| *Interaction of age with unemployment rate* | | |
| 25–29 | 0.167 | 0.388 |
| 30–34 | 0.302 | 0.069 |
| 35–39 | 0.329 | 0.044 |
| 40–44 | 0.403 | 0.006 |
| 45–49 | 0.439 | 0.002 |
| 50–54 | 0.510 | 0.001 |
| 55–59 | 0.445 | 0.003 |
| 60–64 | 0.372 | 0.018 |
| *Interaction of age with cohort size* | | |
| 25–29 | −0.062 | 0.682 |
| 30–34 | 0.121 | 0.354 |
| 35–39 | 0.014 | 0.917 |
| 40–44 | 0.023 | 0.861 |
| 45–49 | 0.027 | 0.836 |
| 50–54 | 0.033 | 0.796 |
| 55–59 | 0.079 | 0.538 |
| 60–64 | 0.091 | 0.479 |
| *Interaction of age with Easterlin ratio* | | |
| 25–29 | 0.039 | 0.075 |
| 30–34 | 0.059 | 0.003 |
| 35–39 | 0.064 | 0.003 |
| 40–44 | 0.074 | 0.000 |
| 45–49 | 0.077 | 0.000 |
| 50–54 | 0.079 | 0.000 |
| 55–59 | 0.086 | 0.000 |
| 60–64 | 0.083 | 0.000 |
| Observation between 1999–2005, inclusive | 0.008 | 0.000 |
| Constant | 0.313 | 0.000 |
| $N$ | 333 | |
| $R^2$ | 0.713 | |

Source: calculated by author.

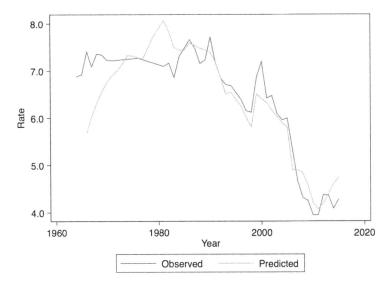

*Figure 5.3* United States: observed and predicted annual inter-county migration
rate, 1966–2015.

Source: Observed values from Current Population Survey; predicted values calculated by the
author, see text.

decline in the inter-county migration rate of 20–24 year olds. Indeed, Figure
5.4a suggests that the effect of the unemployment rate on age-specific migration rates is only statistically significant for 20–34 year olds with the largest
effects among 20–24 year olds. This is consistent with expectations, because
while the unemployment rate is generally associated with a decline in migration this is particularly true for younger age groups who have fewer resources to draw upon. However, the overall contribution of this effect to the
decline in aggregate migration rates is small: for example, of the 3.4% point
decline in inter-county migration rates among the sample from 1986 to 2015,
the change in the unemployment rate over this time period caused migration
rates to increase by only 0.2 points (see Figure 5.4d). Likewise, Figure 5.4b
shows the marginal effect of cohort size on age-specific migration rates. This
value is statistically significant only for those aged 50–54 years old. Indeed,
the overall contribution of cohort size to the decline in aggregate migration
rates is also small (see Figure 5.4d): Of the 3.4% point decline in inter-county
migration rates among the sample from 1986 to 2015, for example, changes
in cohort sizes over this time period only caused migration rates to decrease
0.4 points.

Most importantly, Figure 5.4c shows the marginal effect of the Easterlin
ratio on age-specific migration rates. The general trend is that it is significantly negative between the ages of 20 and 49. Moreover, the effect is greatest among those in younger age cohorts. Indeed, the overall contribution

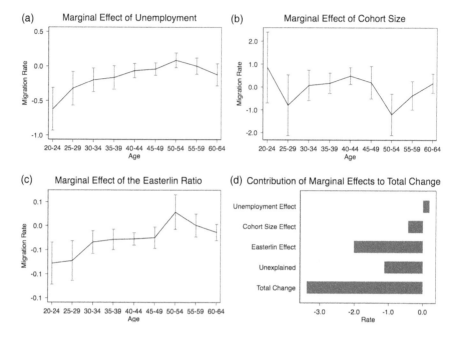

*Figure 5.4* United States: marginal effects on inter-county migration 1986–2015, by age, of (a) unemployment; (b) cohort size; (c) Easterlin ratio; and (d) contributions to total change.
Source: Author's calculations.

of cohort size to the decline in aggregate migration rates is also very large. Of the 3.4% point decline in inter-county migration rates among the sample from 1986 to 2015, for example, changes in the Easterlin ratio over this time period caused migration rates to decline by 2.0 points (see Figure 5.4d). Stated differently, approximately 60% of the decline in migration rates among this sample between 1986 and 2015 can be attributed to the Easterlin effect. The source of this effect is suggested by Figure 5.2 which shows how the Easterlin ratio has changed over the previous half-century. Since 1980 this ratio has increased as the baby boom has aged, in particular, and as the population has aged more generally. Thus, a key driver in the decline in aggregate migration rates may be the crowding out of housing and labour markets opportunities for younger age cohorts. However, with the ageing of the baby boom generation, this value has started to decline, suggesting that younger age cohorts may begin to find more opportunities. Migration rates could begin to recover in the near future as a consequence. Alternatively, the trend toward working past traditional retirement ages may also delay the recovery in migration rates.

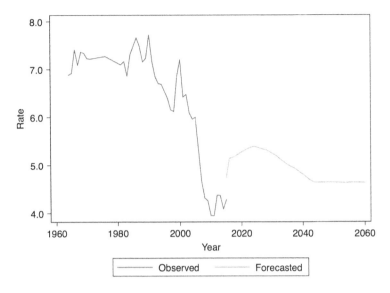

*Figure 5.5* United States: observed and forecasted annual inter-county migration rates, 1966–2060.

Source: Observed values from Current Population Survey; forecasted values calculated by the author, see text.

Finally, US Census Bureau age-specific population projections are used to predict future aggregate migration rates based upon the model estimates (see Colby and Ortman, 2014). Figure 5.5 shows the observed long-term trend in aggregate migration rates along with a forecast of aggregate migration rates based upon the parameters of the estimated model, using age-specific population projections from the US Census Bureau (the unemployment rate is held at 5% and the value for the bias variable is held at 0). This shows that, while there may be a slight rebound in aggregate migration rates over the next decade or so, long-term migration rates will remain low relative to their position prior to 1980. While the slight upswing in forecasted migration rates is a reflection of the ageing of the baby boom generation acting against the factors pushing rates up, the general ageing of the population means that the Easterlin ratio will also remain relatively high for the foreseeable future, and aggregate migration rates may consequently stabilise near historically low levels.

## Conclusion

This chapter has documented the decline in US internal migration rates, reviewed the main lines of explanation put forward by the extant literature and tested an alternative hypothesis based on the changing size of birth cohorts as originally used by Easterlin (1980). It has shown that the drop in migration rate since the 1980s has been very considerable, almost halving in the case of inter-state migration and reducing by over one third for intra-state moves. Moreover, it has progressed steadily over time, with short-term business cycles having little apparent effect.

Contemporary explanations for this long-term decline focus on six main sets of factors: (1) the impact of changing demographic behaviour and population composition on aggregate migration rates, (2) a reduction in possible gains to inter-regional migration as a consequence of decades of inter-regional population redistribution, (3) the impact of advanced ICTs on both the risks and rewards associated with both moving and staying, (4) the potential for ICTs to offer alternatives to migration, (5) the increasing state regulation—in both direct and indirect ways—of housing and labour markets and (6) increasing rootedness as a rational response to economic uncertainty.

However, this recent body of research on the migration decline has ignored an earlier series of papers that considered how cohort effects might contribute to the migration decline (see Rogerson, 1987). An update of an earlier analysis by Pandit (1997a) has confirmed that the large post-World War II baby boom cohort has absorbed work and housing opportunities for everyone, but especially so for younger cohorts, and that this has been a significant factor in the long-term decline in US migration rates. Projecting forward on the basis of these results, it has been found that, while this crowding effect may wane in the next several decades as the baby boom ages, the forecasted long-term trend is of a continuation of the current historically low migration rates because of the growing number of older Americans—many of whom are continuing to work later in life—increasingly dominating the housing and labour markets.

While this new analysis lends support to a demographic perspective, this chapter argues that the migration decline is the consequence of four fundamental shifts in American society. First, there has been a decline in the ability to migrate in order to gain access to economic opportunity. The cohort effects documented here basically establish that migration has declined because older age cohorts crowd out opportunities for younger age cohorts. Also, decades of inter-regional population redistribution may have levelled inter-regional variations in both living costs and economic opportunities. Second, while the immediate presumption is that ICTs are associated with an increase in geographic mobility, empirical evidence suggests that the opposite may be true: ICTs are associated with a reduction in geographic mobility. Third, to the degree that ICTs offer alternatives to migration, they provide a means for avoiding the heightened risk associated with migration during a period of economic uncertainty. Fourth, migration may be hindered by the increasing regulation of state housing and labour markets. Overall, it is apparent that the long-term decline in US migration rates represents a transition toward a more rooted society. Indeed, to the degree that migration is a learned behaviour and the risk of (not) moving is greater for those who have (not) already moved (Bailey, 1989), the impact of historically low rates of migration for those entering adulthood over the last decade or so implies that they may be much less migratory over the rest of their life course than previous cohorts.

## References

Angrist, J.D. and Pischke, J.-S. 2009. *Mostly Harmless Econometrics: An Empiricist's Companion*. Princeton: Princeton University Press.

Bailey, A.J. 1989. Getting on your bike: What difference does a migration history make? *Tijdschrift voor Economische en Sociale Geografie*, 80(5), 312–317.

Cancian, M., Meyer, D., Brown, P. and Cook, S. 2014. Who gets custody now? Dramatic changes in children's living arrangements after divorce. *Demography*, 51(4), 1381–1396.

Carlson, E. 2009. *20th Century US Generations*. Washington, DC: Population Reference Bureau.

Clark, G.L. and Whiteman, J. 1983. Why poor people do not move: Job search behaviour and disequilibrium amongst local labour markets. *Environment and Planning A*, 15(1), 85–104.

Colby, S.L. and Ortman, J.M. 2014. Projections of the size and composition of the U.S. Population: 2014 to 2060. Washington, DC: U.S. Census Bureau.

Cooke, T.J. 2011. It is not just the economy: Declining migration and the rise of secular rootedness. *Population, Space and Place*, 17(3), 193–203.

Cooke, T.J. 2013. Internal migration in decline. *The Professional Geographer*, 65(4), 664–675.

Cooke, T.J., Mulder, C. and Thomas, M. 2016. Union dissolution and migration. *Demographic Research*, 34(26), 741–760.

Cooke, T.J. and Shuttleworth, I. 2017. *The Internet and Rootedness*. Storrs, CT: Department of Geography, University of Connecticut.

Easterlin, R.A. 1980. *Birth and Fortune: The Impact of Numbers on Personal Welfare*. New York, NY: Basic Books.

Ferreira, F., Gyourko, J. and Tracy, J. 2010. Housing busts and household mobility. *Journal of Urban Economics*, 68(1), 34–45.

Ferrie, J.P. 2005. History lessons: The end of American exceptionalism? Mobility in the United States since 1850. *The Journal of Economic Perspectives*, 19(3), 199–215.

Fischer, C.S. 2002. Ever-more rooted Americans. *City & Community*, 1(2), 177–198.

Flood, S., King, M., Ruggles, S. and Warren, J.R. 2015. *Integrated Public Use Microdata Series, Current Population Survey: Version 4.0*. Minneapolis, MN: University of Minnesota. doi:10.18128/D030.V4.0

Graubard, B.I. and Korn, E.L. 2004. Predictive margins with survey data. *Biometrics*, 55, 652–659.

Gujarati, D.N. 1995. *Basic Econometric*. New York, NY: McGraw-Hill.

Hanson, S. 2005. Perspectives on the geographic stability and mobility of people in cities. *Proceedings of the National Academy of Sciences*, 102(43), 15301–15306.

Holen, A.S. 1965. Effects of professional licensing arrangements on interstate labour mobility and resource allocation. *The Journal of Political Economy*, 73, 492–498.

Kaplan, G. and Schulhofer-Wohl, S. 2012. Interstate migration has fallen less than you think: Consequences of hot deck imputation in the current population survey. *Demography*, 49(3), 1061–1074.

Kleiner, M.M. 2015. Border battles: The influence of occupational licensing on interstate migration. *Employment Research Newsletter*, 22(4), 2.

Kleiner, M.M., Gay, R.S. and Greene, K. 1982. Barriers to labour migration: The case of occupational licensing. *Industrial Relations: A Journal of Economy and Society*, 21(3), 383–391.

Kleiner, M.M. and Krueger, A.B. 2013. Analyzing the extent and influence of occupational licensing on the labour market. *Journal of Labour Economics*, 31(2), S173–S202.

Long, L. 1991. Residential mobility differences among developed countries. *International Regional Science Review*, 14(2), 133–147.

McCallum, B. 2010. Is the spurious regression problem spurious? *Economics Letters*, 107(3), 321–323.

Molloy, R., Smith, C.L. and Wozniak, A. 2011. Internal migration in the United States. *Journal of Economic Perspectives*, 25(2), 1–42.

Morrison, P.S. and Clark, W.A.V. 2016. Loss aversion and duration of residence. *Demographic Research*, 35, 1079–1100.

Nazir, S. 2009. The changing path to relocation: An update on post-divorce relocation issues. *Journal of the American Academy of Matrimonial Law*, 22, 483–498.

Niedomysl, T. and Fransson, U. 2014. On distance and the spatial dimension in the definition of internal migration. *Annals of the Association of American Geographers*, 104(2), 357–372.

Pandit, K. 1997a. Cohort and period effects in US migration: How demographic and economic cycles influence the migration schedule. *Annals of the Association of American Geographers*, 87(3), 439–450.

Pandit, K. 1997b. Demographic cycle effects on migration timing and the delayed mobility phenomenon. *Geographical Analysis*, 29(3), 187–199.

Panel Study of Income Dynamics. 2016. Panel Study of Income Dynamics. Public use dataset produced and distributed by the Survey Research Center, Institute for Social Research, University of Michigan, Ann Arbor, MI. https://psidonline.isr.umich.edu/

Partridge, M.D., Rickman, D.S., Olfert, M.R. and Ali, K. 2012. Dwindling U.S. Internal migration: Evidence of spatial equilibrium or structural shifts in local labour markets? *Regional Science and Urban Economics*, 42(1), 375–388.

Peterson, B.D., Pandya, S.S. and Leblang, D. 2014. Doctors with borders: Occupational licensing as an implicit barrier to high skill migration. *Public Choice*, 160(1–2), 45–63.

Pingle, J.F. 2007. A note on measuring internal migration in the United States. *Economics Letters*, 94(1), 38–42.

Plane, D.A. and Rogerson, P.A. 1991. Tracking the baby boom, the baby bust, and the echo generations: How age composition regulates US migration. *Professional Geographer*, 43(4), 416–430.

Rogerson, P.A. 1987. Changes in U.S. national mobility levels. *The Professional Geographer*, 39(3), 344–351.

Ruggles, S., Genadek, K., Goeken, R., Grover, J. and Sobek, M. 2015. *Integrated Public Use Microdata Series: Version 6.0*. Minneapolis, MN: University of Minnesota. doi:10.18128/D010.V6.0

Sander, N. and Bell, M. 2016. Age, period, and cohort effects on migration of the baby boomers in Australia. *Population, Space and Place*, 22, 617–630.

U.S. Bureau of Labor Statistics. 2014. *National Longitudinal Survey of Youth 1979 Cohort, 1979–2012 (rounds 1–25)*. Produced and distributed by the Center for Human Resource Research, The Ohio State University. Columbus, OH.

U.S. Census Bureau. 2017a. Survey of Income and Program Participation. www.census.gov/programs-surveys/sipp/about.html

U.S. Census Bureau. 2017b. Research Data Centers Research Opportunities. www.census.gov/ces/rdcresearch/

Wolf, D. and Longino, C. 2005. Our 'increasingly mobile society?' The curious persistence of a false belief. *The Gerontologist*, 45, 5–11.

Zelinsky, W. 1971. The hypothesis of the mobility transition. *Geographical Review*, 61, 219–249.

# 6 United Kingdom

## Temporal change in internal migration

*Nik Lomax and John Stillwell*

As seen in Chapter 4 (Figure 4.1), the United Kingdom (UK) occupies an intermediate rank in terms of overall migration intensity, coming below the USA, Australia and Sweden but above Germany, Japan and Italy. The integrated results of the population censuses carried out in 2011 across its four countries indicate that 6.8 million individuals, or almost 11 in every 100 persons, were living at a different usual address from that of 12 months earlier. The censuses captured individual migrants who were alive at the start of that year as well as at the end, regardless of the distance over which they travelled. Whilst 40% of these transition flows occurred between local authority areas (the administrative units for which resource allocation decisions are made), the majority were of shorter distance and took place within local authority areas. The census provides the only reliable source of information about *total* migration in the UK, but its infrequency and problems of definitional consistency necessitate the adoption of data from other sources to answer the question at the heart of this chapter: how has the intensity of internal migration in the UK changed over time?

Our starting point for answering this question must therefore be to acknowledge that there is no source of readily available data that allows us to monitor *total* migration propensities in the UK consistently over long periods of time: many of the problems of constructing time-series migration data sets that have been fully documented in Chapter 3 are exemplified in the UK. However, there have been a number of studies of temporal change in internal migration propensities on which we can draw in this quest to establish whether intensities in the UK have followed a trajectory of decline similar to that which has occurred in migration between counties in the USA since the 1970s (Cooke, 2013; Kaplan and Schulhofer-Wohl, 2012). In particular, we make substantial reference to two recent studies by Champion and Shuttleworth (2016a, 2016b), which use, respectively, administrative data from the National Health Service Central Register (NHSCR) and microdata from the Office for National Statistics (ONS) Longitudinal Studies of linked census records to map out the trajectory of internal migration within England and Wales since the 1970s.

Previous research in the UK (e.g. Champion *et al.*, 1998) recognises that migration propensities are determined by both personal and place characteristics. Whilst the availability of data by age group reveals differences in both migration propensities and geographical patterns across the life course, spatial attributes at different scales may also play key roles in determining migration behaviour, as may the level of development and the condition of the national economy. Moreover, there are clear geographical-scale effects on the measurement of migration indicators which obscure the comparison between countries, as demonstrated by Bell *et al.* (2015a). Here we make use of an estimated annual time series of migration between local authority districts (LADs) in the UK that builds on Lomax *et al.* (2014). The time series runs from 2001/2002 until 2012/2013, covers the whole of the UK at a finer spatial scale (404 LADs) than that used by Champion and Shuttleworth (2016a) and involves a more detailed set of age groups to decompose the aggregate counts. Importantly, this set of time-series data allows us to monitor changes in intensities and patterns of migration over a period in which the UK experienced rapid population growth in parallel with the worst recession in post-war times.

The remainder of this chapter develops these points more fully in three sections. The first describes the main types of data sets that are available for studying the UK's internal migration patterns and trends and outlines their strengths and weaknesses for present purposes. The second summarises the temporal changes observed by Champion and Shuttleworth (2016a, 2016b) and goes on to provide additional insights into migration across the North-South divide for the whole UK using an effectiveness indicator that measures the impact of migration on population settlement. Finally, using Lomax *et al.* (2014)'s framework, we seek to identify what changes, if any, have occurred in migration within and between urban and rural areas for different age ranges. Particular attention is paid to London since it is the hub of the UK's national migration network, with almost one in every five migrants in 2010/2011 moving to, from or within the capital.

## Key data sources and spatial systems

The construction of consistent time-series data on migration within the UK is a difficult and time-consuming task. This is partly because the collection of internal migration data is undertaken by three separate agencies covering England and Wales, Scotland and Northern Ireland. Even more problematic are changes over time in the instruments for data capture and also in the boundaries of the geographical units that are used for recording changes of residence between areas. This explains why there are a plethora of cross-sectional studies of migration in the UK (as elsewhere) but relatively few that examine trends and processes that extend beyond a time period of one decade at most.

Our understanding of internal migration and its temporal change is captured from a number of different sources that are either census, administrative or survey based. Censuses in the UK since 1961 have contained a one-year migration question that continues to provide the most reliable and comprehensive data, some of which have been released in aggregate statistics or origin-destination tables and some as microdata in either cross-sectional (i.e. Samples of Anonymised Records) or longitudinal (i.e. Longitudinal Studies) form. In this chapter, we summarise novel analysis reported by Champion and Shuttleworth (2016b) that uses data from the ONS-LS relating not to one-year but 10-year periods for England and Wales from record linkage between censuses since 1971 and which also contains flows disaggregated by distance of move and by a range of personal characteristics.

Whilst UK census data are extremely valuable for their reliability, comprehensive coverage throughout the national territory (including all distances of address-changing, unlike many other countries) and the relatively rich detail that they provide for small areas, they are somewhat limited by a number of factors including: the infrequency of their collection (once every 10 years apart from a sample census in 1966), changes in the definition of migration flows (for example, including or excluding flows between students' parental domiciles and their term-time places of residence), changes in the statistical disclosure control mechanisms that are required to ensure confidentiality and changes in the boundaries of the geographical areas used for reporting migration statistics. It is these constraints that have prompted researchers to look for alternative sources of data on migration, one of the most popular of these being the National Health Service Central Register (NHSCR), which collates the changes of address of NHS patients recorded by their doctors. The NHSCR provides movement data on a mid-year to mid-year basis dating back to 1971 for inter-regional moves and back to 1975 for intra-regional (between health service areas) moves. The NHSCR captures each migration event, with the result that studies comparing NHSCR data with census transition data—which count migrants rather than moves—indicate higher levels of mobility for the former, although data from both sources illustrate very similar spatial patterns for equivalent geographical areas (Ogilvy, 1980; Devis and Mills, 1986; Boden *et al.*, 1992; Stillwell *et al.*, 1995).

Amongst the disadvantages of the NHSCR data are the lack of data on shorter-distance flows within health areas and the changes that have taken place to the NHSCR geography over the 36 year time series constructed by Champion and Shuttleworth. Their approach has been to identify the lowest common denominator (LCD) areas, resulting in a set of 80 consistent polygons for the period from 1976 to 2011, which aggregate into 10 former Government Office Regions (GOR) in England and Wales and form the geographical basis for the analysis reviewed in the next section.

Another source of migration data, introduced in more recent times, is the Patient Register Data Service (PRDS), an annual download of home addresses of NHS patients which is compared with the download from the

previous year to indicate where changes have occurred. These data are therefore transition data insofar as they capture migrants alive at both start and
end points of each period, with ONS providing annual estimates of flows
between local authorities in England and Wales. Lomax *et al.* (2013) report
how these data, together with administrative data on flows from National
Records for Scotland (NRS) and the Northern Ireland Statistical Research
Agency (NISRA), have been used to create a time series of flows disaggregated by age and sex between LADs for the whole of the UK, including flows
between LADs in each of the four constituent UK nations. The estimated
time series is used as the basis for spatial analysis of time series change between 2001/2002 and 2012/2013 later in the chapter.

The third source of internal migration data are surveys such as the Labour
Force Survey (LFS) and the General Household Survey (GHS) which have
been running from 1973 and 1971 respectively. As outlined in Stillwell *et al.*
(2010), these are often of great value when used for analysis at a national or
regional level, but sample size normally precludes their use for sub-regional
analysis. Additionally, in the case of the GHS, the origin of the migration
flows is either the current region or elsewhere, preventing the distinction
between an internal and an international move. We do not make use of any
government survey data in what follows, although some results from a large
consumer survey undertaken by Acxiom Ltd in the mid-2000s (Thomas
*et al.*, 2014) are reported in the next section.

## Long-term fluctuations in annual migration intensity

Amongst the earliest studies of internal migration in the UK, the most influential was Ravenstein (1885) which, using birthplace and enumeration data
from the 1871 and 1881 Censuses, set out a series of generalisations about migration that have since become reference points for later researchers. Many
studies have been conducted subsequently, but virtually all have used either
census or NHSCR data, particularly in the last 50 years since a one-year
census question was first asked in 1961, and have involved the analysis of
migration over relatively short time spans. Examples include Ogilvy (1982),
Devis (1984), Stillwell and Boden (1989), Champion (1989a), Rosenbaum
and Bailey (1991), Stillwell (1994), Stillwell *et al.* (1995, 2015, 2017), ODPM
(2002), Duke-Williams and Stillwell (2010), Fielding (2012) and Lomax *et al.*
(2013, 2014), although Stillwell *et al.* (1992) traced the fluctuation in migration intensities from 1960/1961 to 1988/1989 using a combination of census
and NHSCR data.

Most recently, Champion and Shuttleworth (2016a) have constructed a
longer time series that captures trends in intensities of migration between
the former GOR and between health areas within each GOR in England
and Wales from 1975/1976 to 2010/2011 using data supplied by ONS entirely
from the NHSCR. These two types of flow allow a distinction to be drawn
between longer-distance, inter-regional migration and shorter-distance,

'intra-region, inter-health-area' migration, but exclude the flows taking place within health areas. The time-series schedules of inter- and intra-regional migration intensities computed for each year based on start-of-period populations are shown in Figure 6.1 and demonstrate both the fluctuation year on year in the aggregate (all age) migration intensity and the variation in age-group intensities around the all age trend. In both cases, there is little evidence of a decline in the rates of migration since the 1970s equivalent to that observed for inter-county migration between and within states in the USA (see Chapter 5 of this book).

The trend for all age migration observed in both series begins in the mid-1970s with a continuation of a decline that Stillwell *et al.* (1992) suggest had been occurring since 1970/1971. This is followed by a rise after 1980/1981 as the country pulled out of recession to a peak of nearly 24 per thousand

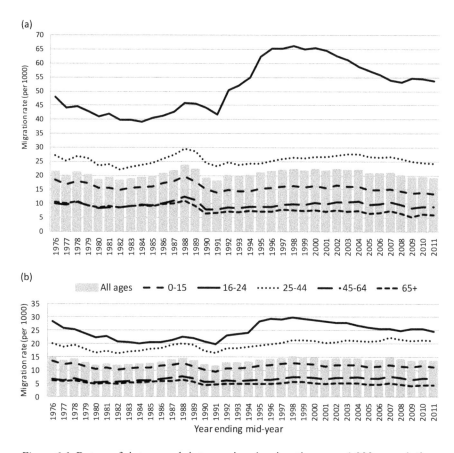

*Figure 6.1* Rates of inter- and intra-regional migration per 1,000 population, 1975/1976–2010/2011, for consistent areas in England and Wales: (a) inter-regional rates; (b) intra-regional rates.

Source: Adapted from Champion and Shuttleworth (2016a, Figures 1, 3, 5 and 6).

moves in the case of inter-regional migration at the height of the boom in 1987/1988, before dropping rapidly to its lowest point across the whole time series in 1990/1991, a time when the country was back in recession and experiencing civil unrest as a result of unemployment and social discontent. Following the end of this recession, the British economy enjoyed a record run of unbroken economic growth lasting more than 15 years, until falling back into an economic downturn that was ultimately much worse than that of the early 1990s. Migration rates gradually increased throughout most of this period. Whilst it is possible to attribute the declining rates since 2007/2008 to the onset of the Great Recession, the drop appears to have begun well beforehand. For inter-regional migration, the decline in rates began in the first half of the 2000s, perhaps due to the arrival of large numbers of migrants from the eight eastern European countries that joined the European Union in 2004 which, as Champion and Shuttleworth (2016a) suggest, may have reduced the need for longer-distance, inter-regional movements by the existing population. For migration between areas within regions, the turning point appears to have been earlier still, with the rate falling marginally after 1997/1998 but not consistently from year to year.

These aggregate rates of migration are essentially determined by the migration propensities of individuals in different demographic and socioeconomic subgroups of the population. Figure 6.1 also contains the migration intensity schedules over the same period for five age groups that reflect those in broad stages of the life course: children aged 0–15, students and young adults aged 16–24, labour force migrants aged 25–44, older workers and (pre-) retirement migrants aged 45–64 and post-retirement migrants aged 65 and over. There are clear variations in migration propensities between age groups which conform to familiar patterns observed in the UK by Champion (2005) and Dennett and Stillwell (2010), but the most significant feature of the time-series trends in age-specific migration is the dramatic increase in the intensity of migration exhibited by those aged 16–24. Rates of inter-regional migration in this age group increased from 40 per thousand in 1990/1991 to 65 per thousand in 1995/1996 before stabilising and then dropping back to around 53 per thousand by the end of the period, whereas intra-region movement followed a similar pattern but with rates of half that magnitude. The primary reason for this increase is likely to be the policy changes in the UK university sector (Wyness, 2010) that have resulted in the expansion of numbers participating in higher education, particularly during the 1990s following the introduction of the student loan system in 1990 and the Further Education and Higher Education Act in 1992 that granted university status to 48 polytechnics, making them more attractive to students beyond their own localities. Changes over time are less evident for the other age groups, with greatest stability in rates apparent for those aged 45–64 and 65 plus and decline over the last decade being observed for inter-regional moves but not for moves between the health areas within regions.

Drawing on the same time-series data used by Champion and Shuttle-worth (2016a) to establish the temporal trends in migration intensity, we have examined the spatial pattern of migration within the UK across the 'North-South' divide, a framework frequently used by politicians and social commentators concerned with highlighting spatial inequalities in social and health indicators. In this case, the West Midlands region is included in the 'North' together with the North West, Yorkshire and The Humber, the North East, Wales, Scotland and Northern Ireland. The time-series graphs presented in Figure 6.2 for persons of all ages and for those in the five age groups used earlier show a migration effectiveness ratio (MER) indicator in which net migration for the North is computed as a percentage of the gross migration flows between the North and the South—a useful measure for monitoring migration impact (Bell *et al.*, 2002; Rees *et al.*, 2016). The time series for aggregate migration is one of cyclical fluctuation, with the North having a positive net migration balance in only six years of the time series. During the late 1970s and first half of the 1980s, the MER indicator shows increasing losses from the North to the South through net migration which reached over 50,000 at its peak in 1986, before falling sharply in the recession years of the late 1980s and becoming positive in 1989 before retreating back to modest losses of less than 5,000 in the early 1990s. Net losses from the North increased to over 15,000 per year in the mid-1990s as the country pulled out of recession and, after 1997, when economic growth reached 2–3% per year, negative net migration dwindled and gains from the South were experienced in the early 2000s. By the onset of the Global Financial Crisis in 2007, the North was once again losing migrants to the South at an increasing rate, a trend which continued until the last year of the time series when a small upturn was evident. In terms of the long-term trend in migration between the North and the South, 'the net drift to the latter is now but a pale shadow of its former self' (Champion, 2016, p. 129).

Although the time series appears to be cyclical, there is no clear cut relationship between the North-South net migration divide and the trend in the rate of economic growth. Moreover, it is revealing to observe that the age groups with the most prominent fluctuations in MER as far as the North-South balance is concerned are the two oldest and the youngest age groups. This was particularly the case in the first period of net gain by the North in the late 1980s, whereas the effectiveness of net gains in the 2000s was much less apparent for those aged 65 and over. The MER time-series schedule for the 16–24 year olds, whose migration intensity is the highest of all the age groups, illustrates the relative attractiveness of the South across all years except 2004/2005 and 2005/2006.

The trends in net migration effectiveness ratio observed for the North in Figure 6.2 are a reflection of the changes in the difference between the flows in both directions, as shown for all-age migration in Figure 6.3. Migration in each direction during the time series fluctuates around 200,000 moves per year. Initially, in the first 10 years of the series, moves from the North to the

(a)

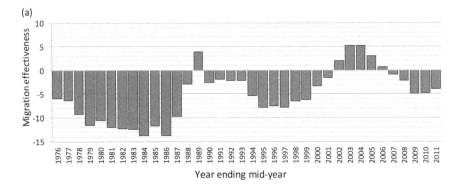

(b)

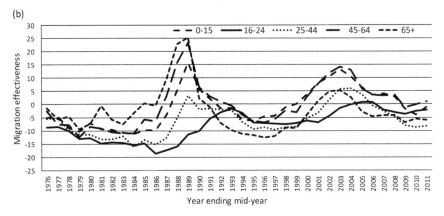

*Figure 6.2* Migration effectiveness ratio for the North of the UK, 1975/1976–2010/2011: (a) all ages; (b) broad age groups.

Source: Authors' calculations based on NHSCR data supplied by Tony Champion.

South were considerably higher than in the opposite direction, but in the late 1980s it was the migration from the South to the North that increased the more rapidly, creating the first incidence of net gain in the North in 1989. The 1990s, however, saw an increase in moves from the North which created the migration deficit throughout the decade, until migration from the South increased again in the first half of the 2000s causing the North's net balance to become positive. During and since the last recession, moves from North to the South have remained above 200,000, whereas those in the opposite direction have fallen below this threshold, bringing the North's balance back into deficit. These North/South fluctuations in MER reflect the relative rates of job creation in the two regions, higher in the South than the North. But, at the end of each boom, job growth in the South has raised housing prices so much that some commuters to London seek cheaper residences in the southern fringes of the North, leading, briefly, to net internal migration outflow from the South.

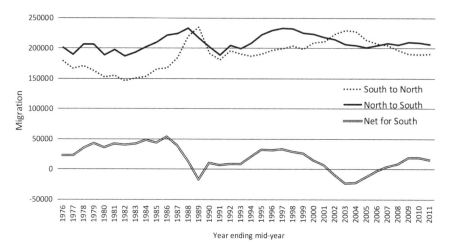

*Figure 6.3* Migration flows between the North and South, 1975/1976–2010/2011.
Source: Authors' calculations based on NHSCR data supplied by Tony Champion.

Until this point, our review and commentary has been concerned with migration over relatively long distances, but it is well known that the intensity of migration increases as the zones used for measuring migration reduce in size, a relationship measured using Courgeau's $k$ (Courgeau, 1973; Bell *et al.*, 2015b). In the UK, the crude intensity for inter-regional migration recorded by the 2011 Census was 18 per thousand in 2010/2011, whilst the crude intensity for aggregate migration reached 108 per thousand, as mentioned at the start of the chapter. The effects of scale on migration distance have been observed in recent studies by Stillwell and Thomas (2016) using origin-destination postcode data for England from a large consumer survey undertaken by Acxiom Ltd, whilst changes in migration distance over time have been documented by Champion and Shuttleworth (2016b) using categorised microdata from ONS Longitudinal Studies for England and Wales. The former study highlights the advantage of having access to detailed geographic locations of origin and destination and thereby enabling precise measures of migration distance to be computed. The consumer survey data show that in the mid-2000s, whereas the mean distance of migration in England was around 25 km, the median was just under 3 km (Stillwell and Thomas, 2016, p. 11, Table 2), revealing the extent of skew in the distribution of migration flows towards short-distance residential mobility. The study by Champion and Shuttleworth (2016b) is particularly valuable because it provides time-series data over four decades by presenting the percentages of people who were living at an address at the end of each intercensal period that was different from that at the previous census date and cross-classifies these migrants according to how far they moved. The graph in Figure 6.4 illustrates

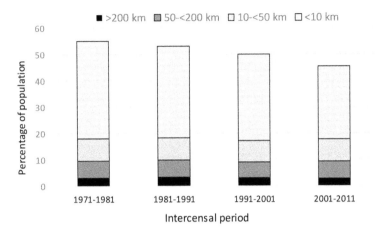

*Figure 6.4* Percentage of the England and Wales population surviving from one census to the next but having a different address, by distance band.
Source: Adapted from Champion and Shuttleworth (2016b, Figure 1).

how migration propensity declined consistently over the four decades from a rate of 55% in the 1970s to 45% in the 2000s. Moreover, the graph shows that the fall in the aggregate rate is almost entirely explained by the decline in the rate of moves taking place over less than 10 km, which dropped from 36.9% in the 1970s to 27.5% in the 2000s, with the percentages of migration in the three longer-distance bands remaining relatively unchanged.

Despite the limitations of the use of 10-year data rehearsed by Champion and Shuttleworth (2016b)—including the less than perfect linkage of individuals between censuses, the possible undercounting of short-distance moves in the 1980s and, most importantly, the change in the census definition of the usual residence of students living away from home during term-time between 1991 and 2001—the conclusion is that there has been a sustained decline in shorter-distance residential mobility over the last four decades. The authors go on to validate these findings and to identify the types of individual that experienced the highest migration intensities and the greatest declines in rates of migration over different distances between the first and last decades of the time series. Those in the older age groups (55 plus), together with those in the retired and widowed categories, for example, stand out as being the individuals whose rates have declined most in the less than 10 km distance band and, thus, for all distances of move.

The analyses of longitudinal census data undertaken by Champion and Shuttleworth are particularly valuable for exposing the longer-term temporal changes in residential mobility at the national level that are so difficult to capture from cross-sectional census or administrative data sets. However, the results reported by Champion and Shuttleworth refer only to England

and Wales rather than the UK, use only five rather broad age groups and provide no information about changes in the spatial patterns of migration over time. In the next two sections, we seek firstly to understand how recent intensities have changed within the UK as a whole, making use of estimated annual flows for 11 age groups between 404 LADs and examining changing spatial patterns for all age migration from 2001/2002 to 2012/2013, a period which has seen rapid population change and fluctuations in the economy on a scale not experienced since the inter-war years. Secondly, we recognise the importance of London's influence in the UK migration network and the key role which the capital plays at the core of south-eastern England, particularly as a social mobility escalator that attracts young adults at rates which are higher than elsewhere in the country but then experiences significant losses of migrants in all other age groups as they step off the escalator and move away from London (Fielding, 1992). We explore changes in MER for Greater London in both the short-term, by using data from the latest two censuses, and in the long term with data from the database compiled by Champion and Shuttleworth (2016a).

## Recent changes in migration intensities and spatial patterns

Our estimates show that between 2.8 and 3 million individuals, or around 5% of the UK population, migrated between districts in the UK in any one year of the 12-year time series from 2001/2002 to 2012/2013. Overall, the average intensity declined from 48.4 migrants per thousand people in the first six years of the time series to 46.3 per thousand in the last five years, a relative change of only 4.3% between two periods with very different national economic conditions. Figure 6.5 shows the year-on-year rates of migration for all ages represented as vertical bars, and the time series of 11 age-specific migration rates as individual lines. Up to mid-year 2007, the all-age migration rates remain fairly stable (between 47.7 and 48.9 migrants per thousand), while the rate in 2006/2007 represents a peak in the time series (49.4 migrants per thousand). From this high point, total migration rates drop to between 45 and 47 migrants per thousand until 2011/2012, with some recovery evident towards the end of the time series (to 46.7 per thousand in 2012/2013). When age-specific rates are compared, the 20–24 age group consistently demonstrates the highest, with a peak of 142 migrants per thousand in 2008/2009. This is the age group identified as 'leaving university for work' by Champion *et al.* (2003), with much of the mobility attributed to individuals who leave a university town or city to take up their first graduate job (or increasingly, it might be argued, return home to their parents' address while they look for work). Inter-district migration rates for those aged 15–19 are lower because only those aged 18–19 will be making their first move away from home to study in higher education and those leaving school at 16, whose move may be prompted by a first job, are more likely to move over relatively shorter distances.

Migration per 1,000 people

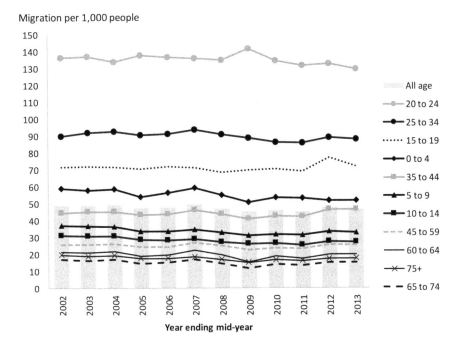

*Figure 6.5* Rates of migration between 404 consistent local districts, 2001/2002–
2012/2013, by age group.

Source: Authors' calculations based on their time series estimates of flows between districts.

The 25–34 group exhibits the second highest inter-district rate, but this
is substantially lower (peaking at 94 migrants per thousand in 2006/2007)
than those in their early 20s. This age group encompasses a variety of life
course events, including couple formation and family commencement, both
of which drive migration for cohabitation and upsizing to accommodate
children. The migration intensities of the four family age groups—children
aged 0–4, 5–9 and 10–14, together with their parents aged 35–44—are
closest to the all-age intensity time series. Parents cannot be isolated from
non-parents in the latter group but the time series schedules indicate higher
rates for infants (aged 0–4) relative to their parental age group and lower
rates for children aged 5–9 and 10–14 as education becomes increasingly
important. Migrants aged 0–4 peak in intensity in 2006/2007 at 59 per thousand, but the overall trend is for their intensities to fall during the period
whereas those migrants in the parental age group show stability with a peak
of 46 per thousand in 2012/2013. The 10–14 age group is less mobile (perhaps
a product of the need for stability in these important school years). Rates of
inter-district migration are relatively low for those aged 45–59 and appear to
remain relatively stable throughout the time series, whereas the least mobile
age groups are those in the 60 plus categories. Amongst these, 60–64 year

olds are slightly more mobile than those aged 65–74 and 75 plus. Certain older ages are associated with retirement migration (65 years for men and 60 for women during the period), moves to areas of greater amenity and a desire to downsize and, ultimately, with moves to areas which provide suitable care provision or to be closer to family.

While it is clear from the schedules presented in Figure 6.5 that migration intensities have shown some degree of fluctuation from year to year, Figure 6.6 summarises changes over the whole period by illustrating the percentage change between average rates in the first (2001–2007) and second (2007–2013) 'halves' of the period and revealing that all age groups experienced a fall in migration rates. At younger ages (0–14), the change is between 8% and 10%, while for ages 15 through 44 the change is lower at between 0.4% and 5%. Thereafter, from age group 45–49 through age group 70–74, the negative percentage change gets progressively larger, with the most pronounced being for those aged 65 to 74, where migration rates are 12.7% lower in the second half of the decade than the first. The oldest age group, those aged over 75, shows a fall of 10.5%. Many individuals in the middle age groups (25–34, 35–44 and 45–59) will be parents of those in the childhood age groups (0–4, 5–9 and 10–14), but the percentage decline in migration rates for the middle age groups is far lower than that of the childhood ages. Because not all persons in the middle age groups are parents, this difference implies that middle-aged parents have experienced a greater decline in migration rates than non-parents.

So, whilst these results suggest that there has been an overall reduction in mobility over the time period for all age groups, it is important to bear in mind that the rates reported are a product of the number of people moving between LADs *and* the size of the population in each group. The former fluctuates throughout the time series but, overall, total flows are larger in 2012/2013 than in 2001/2002 (2.99 million compared with 2.88 million),

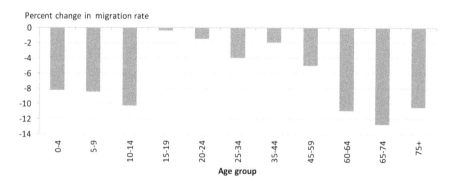

*Figure 6.6* Percentage change in inter-district migration rates between 2001/2007 and 2007/2013, by age group.

Source: Authors' calculations based on their time series estimates of flows between districts.

with all ages except 5–14 and 35–44 showing an increase in the total number of migrants. The population is larger at the end of the time series than the beginning in all age groups, except for those aged 10–14 and 35–44 (although the population in age group 5–9 only surpasses the 2001/2002 total in 2012/2013).

The aggregate migration intensity is a composite measure of migration in different age groups as indicated in Figure 6.5, but it also conceals spatial variations in migration propensity that reflect different processes taking place in various parts of the country at different spatial scales. One key feature of sub-national migration in the UK at the district scale is that of counter-urbanisation (Champion, 1989b), characterised by losses from the major metropolitan areas and gains in the smaller towns and rural areas, as illustrated by the spatial variation in MER for the 404 LADs in the UK in 2001/2002 (Figure 6.7a) and juxtaposed against the equivalent indicator for the same areas in the last year of the time series (Figure 6.7b).

*Figure 6.7* Migration effectiveness ratio for districts, all ages: (a) 2001/2002; (b) 2012/2013.

Source: Authors' calculations based on their time series estimates of flows between districts; light grey circles represent positive MER, dark grey represent negative MER, and the size of symbol represents the MER value.

When comparing the spatial patterns of MER at the start and end of the period, two striking differences are apparent. First, there is a general decline in MER across most LADs, as reflected in the general reduction in the size of the circles; the net gains and losses have tended to decrease in magnitude, suggesting a weakening of the counter-urbanisation process. Second, and related to the previous observation, the London region has undergone a substantial shift in MER pattern: while London boroughs almost uniformly had negative MER scores in 2001/2002, by 2012/2013 the losses were limited to central London, with Outer London boroughs showing a positive MER score.

In order to get a better handle on the interaction between areas over time, we use a classification which enables us to assess moves between aggregate 'metro' and 'non-metro' areas. These area categories are aggregations of LADs where metro areas comprise 13 core urban areas of the UK (Aberdeen, Belfast, Birmingham, Bristol, Cardiff, Edinburgh, Glasgow, Leeds, Liverpool, London, Manchester, Newcastle and Sheffield) and their immediate peripheries, while the non-metro areas comprise areas that are more distant from these urban centres. This city region classification was first used by Stillwell *et al.* (2000, 2001) and is further explained in Stillwell *et al.* (2015). Other city region classifications have been developed (e.g. Marvin *et al.*, 2006), but the one used here has the advantage of extension beyond England and Wales to incorporate Scotland and Northern Ireland. Moves within the 13 metro areas and moves within associated non-metro areas (i.e. all intra-area moves) are excluded, so we are dealing with approximately 27% of all UK migration (e.g. 1.8 million of the 6.8 million individuals identified in the 2011 Census) which occurs between different metro or non-metro areas.

Figure 6.8 provides an overview of the absolute number of people moving between the two area types for each year between 2001/2002 and 2012/2013. Some clear trends can be seen in these overall numbers. Moves from metro to non-metro declined overall, with 526,000 people making the move in 2001/2002 compared with 476,000 in 2012/2013. At the same time, the number of moves occurring in each year in the other direction, from non-metro to metro, increased from 405,000 in 2001/2002 to 426,000 in 2012/2013. Moves between non-metro areas declined in the period from 445,000 to 395,000, while the number of people moving between metro areas increased from 474,000 in 2001/2002 to 520,000 in 2012/2013. Thus, moves which are often categorised as counter-urbanisation declined in the first decade of the 2000s, while the volume of moves in the other direction increased.

The metro and non-metro populations also increased over time, so Figure 6.9 shows rates in terms of the number of people moving between metro or non-metro areas related to the total population at the origin. Total migration rates are shown by the solid black lines on each graph in Figure 6.9. A small increase can be seen in the rate of migration between different metro areas, from 16.3 to 16.4 per thousand in 2001/2002 and 2012/2013 respectively (Figure 6.9a); while a small decrease can be seen for

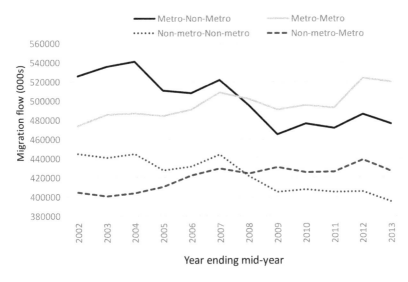

*Figure 6.8* The total number of people moving between metro and non-metro areas, 2001/2002–2012/2013.
*Source*: Authors' calculations based on their time series estimates of flows between districts.

moves from non-metro to metro areas, from 13.5 per thousand at the beginning of the time series to 13.3 per thousand at the end (Figure 6.9c). A larger decline in migration rate is evident for moves between different non-metropolitan areas (Figure 6.9d), where the rate drops from 14.8 per thousand in 2001/2002 to 12.3 per thousand in 2012/2013. The most notable change in total rates is for moves from metro to non-metro areas (Figure 6.9b), with a drop from 18.1 per thousand in 2001/2002 to 15.1 per thousand in 2012/2013.

Figure 6.9 also displays migration rates for seven broad age groups, revealing that the patterns seen for overall rates are not uniform for all origin/ destination combinations. The most consistent trend across all age groups can be seen for moves between different non-metro areas (Figure 6.9d) with a notable decline at age 0–14, down from 11.6 per thousand in 2001/02 to 8.0 in 2012/2013, a fall of 32%. At age 25–29, the fall in rate is around 19%, from 30.9 to 25.0 per thousand. The slight increase seen in the overall rate of migration for moves between metro areas (Figure 6.9a) is mirrored at ages 45–59, 30–44 and notably 15–19, for whom the rate increases from 24.8 per thousand at the beginning of the time series to 28.1 at the end. It is moves by those aged 60 and over which most notably buck the trend, with their rate falling from 4.3 to 3.9 per thousand. The slight decrease in rate for moves from non-metro to metro areas (Figure 6.9c) is most apparent at age 0–14 (7.1 per thousand drops to 6.2) and at age 20–24 (71.3 per thousand to 62.5). Those at ages 30–44 and 45–59 counter the general downward trajectory, where migration rates rise from 11.6 to 12.7 and 4.4 to 5.0 per thousand

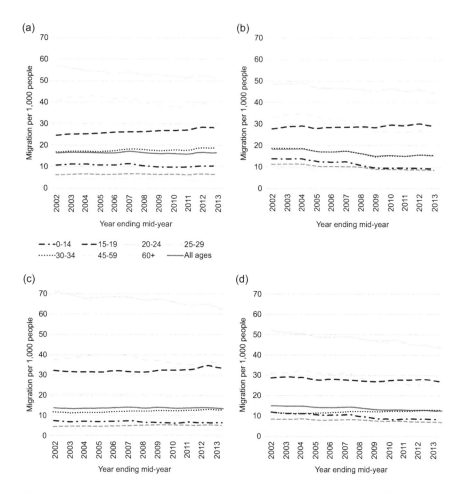

*Figure 6.9* Rates of migration between metro and non-metro areas, 2001/2002–2012/2013, by age group: (a) metro to metro; (b) metro to non-metro; (c) non-metro to metro; (d) non-metro to non-metro.

Source: Authors' calculations based on their time series estimates of flows between districts.

respectively. The drop in migration rate for moves from metro to non-metro areas (Figure 6.9b) is evident at all ages, except 15–19 where it rises from 27.7 to 29.2 per thousand at the end of the time series. The largest declines are at ages 0–14 (from 13.7 to 9.1 per thousand), 25–29 (from 33.4 to 26.4 per thousand), 45–59 (from 11.1 to 8.5 per thousand) and 60 plus (from 8.9 to 7.1 per thousand).

Overall, there is variation by age but the most compelling patterns can be seen in the declining rates of migration between non-metro areas and the fall in migration rates for moves from metro to non-metro areas. Urban renaissance is a term that has been used to reflect the demographic fortunes of

UK cities in the last decade (ODPM, 2006). Champion (2015) has used ONS mid-year population estimates which confirm the extent to which counter-ubanisation has diminished as major cities and large towns across England have grown at significantly higher rates than in the previous two decades. Whilst international migration has been a key component of urban growth during the 2000s and increasing rates of natural change have bolstered the uplift, the reduction in the numbers of people leaving cities for rural areas has also been a critical driver of change, together with the increase in those moving into cities, particularly more central areas where 'city living' has become a prominent feature of most provincial capitals. The larger-scale trends reported by Champion (2015) are supported by the inter-LAD and metro/non-metro analyses reported here, but moves at a finer spatial scale are not available from the above dataset, as mentioned earlier.

## London's migration trends and patterns

The capital city, when defined as equivalent to the former London GOR which is identical to the area administered by the Greater London Authority, is the hub of the UK migration system, thereby exercising a great influence over the intensity of migration across the whole country. According to the 2011 Census Special Migration Statistics (SMS), flows to and from London in 2010/2011 represented nearly 15% of the total migration flows between LADs in the UK, with London losing around 36,000 in this exchange. The regional MER scores shown in Figure 6.10 indicate London's role as a net exporter of migrants alongside Northern Ireland, the North West and the West Midlands in both one-year periods, although the impact of net migration is less in 2010/2011 than it was in 2000/2001. Net migration is negative for all age groups apart from those in the 20s. The fall in negative effectiveness is apparent across all age groups except those aged 15–19 and 85–89, and is most emphatic in the ages from 55 to 74 (Figure 6.11). The impact of net gains of those aged 20–24 dropped between the two periods, whereas the MER value increased for those aged 25–29.

In terms of the longer year-on-year NHSCR time series, London's negative migration with the rest of the UK can be seen to have ebbed and flowed over the last four decades with the greatest losses tending to occur during periods of greatest national prosperity and lowest net outflows to the rest of the UK evident when economic conditions are less buoyant. The time-series MER schedule shown in Figure 6.12a for London *vis a vis* the rest of the UK indicates, in particular, the extent to which the impact of all age net migration losses increased in the first half of the 2000s and declined rapidly between 2004 and 2009 before dropping back in the last two years. All the constituent age groups follow the same trend more or less but there are distinct differences in the levels of effectiveness by age group, as shown in Figure 6.12b. At one end of the spectrum, migration associated with the oldest age group has the greatest negative MER with net migration loss

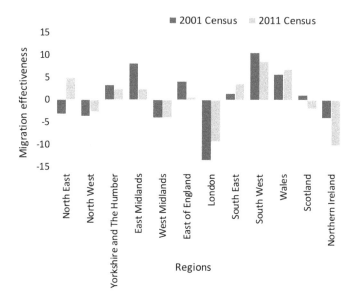

*Figure 6.10* Migration effectiveness ratio for the regions, 2000/2001 and 2010/2011.
Source: 2001 and 2011 Census Special Migration Statistics; 2001 Census data have not been adjusted to include count of those with origin not stated.

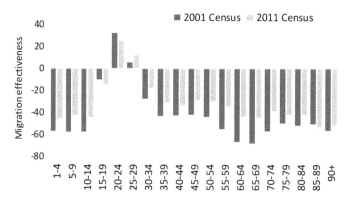

*Figure 6.11* Migration effectiveness ratio by 5-year age group, London, 2000/2001 and 2010/2011.
Source: 2001 and 2011 Census Special Migration Statistics; 2001 Census data have not been adjusted to include count of those with origin not stated.

from Greater London reaching a peak in 1988 at over 60% of gross turn-over and reducing to its lowest level, under 50%, by the end of the time series. The MER schedules for those aged 45–64 and 0–15 appear to have converged during the period, having changed in parallel in the 1970s and

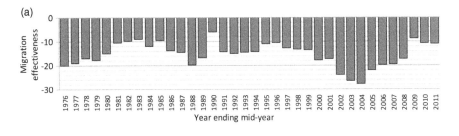

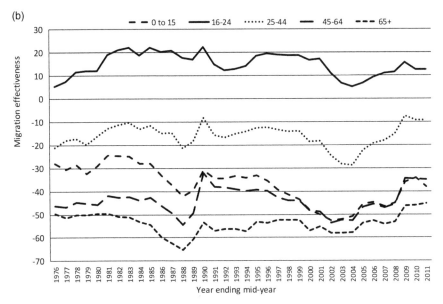

*Figure 6.12* Migration effectiveness ratio for London with respect to the rest of the
UK, 1975/1976–2010/2011: (a) all ages; (b) broad age groups.

Source: Authors' calculations based on NHSCR data supplied by Tony Champion.

1980s. Whilst those in the 25–44 age group experience net migration losses
throughout the time series with an impact that is lower than for the other
age groups, the MER for the 16–24 year olds remains positive in each year
and shows greater stability than all the other broad age groups, reflecting
how the London escalator (Fielding, 1992) has continued to attract young
adults from across the rest of the country.

An important conclusion from the analysis reported for London and its relationship with the rest of the country is that the 2000s were characterised by a
slowdown in net losses driven by fewer out-migrants to the rest of the UK and
more incoming migrants, as confirmed by Champion (2015) who uses mid-
year population estimates to indicate that growth in Inner London outpaced
the rest of the London primary urban area in both 2001–2008 and 2008–2014.

## Concluding discussion

The two headline conclusions that can be drawn from recent literature about the changing intensity of migration in the UK can be summarised as follows. First, there is little evidence to support a long-term decline in the relatively long-distance migration either between regions or between health areas within regions in England and Wales. There is evidence of fluctuations in the time series of migration at both scales, but these are accounted for partly by cyclical changes in national economic prosperity as well as varying conditions in labour and housing markets and changing locational preferences amongst certain groups rather than any pronounced fall in the underlying propensity to move home. Second, analysis of longitudinal data over five censuses reported by Champion and Shuttleworth has indicated that shorter-distance migration (under 10 km) has declined in each decade since the 1970s. This decline has been particularly evident in the most recent decade, with those sub-groups for whom the fall has been most apparent between the 1970s and the 2000s including older people (especially those in their 60s), the widowed and the retired. When we consider an intermediate geography of local authority districts, there is some evidence of a decline in migration rates (and effectiveness) over the 12 years since 2001/2002, especially when we contrast our estimated intensities in the first half of this period with those in the second half.

The scale at which migration is taking place is therefore a critical dimension when assessing changing migration intensity in the UK because different sub-groups of the population will be involved in varying proportions at different scales and the range of motivations that determine migration behaviour will vary accordingly. We have observed how longer-distance internal migration intensities vary by age across the life course and, using the migration effectiveness ratio, we have demonstrated how the changes over time translate into spatial impacts on the population, with particular reference to movements across the North-South divide and between London and the rest of the UK.

Given that the majority of migration is residential mobility taking place over short distances (i.e. mainly within LADs) and that moves of less than 10 km have declined progressively whereas longer distance migration has moved up and down on a cyclical basis, it would be logical to surmise that the overall crude migration intensity in the UK has been in decline. However, with censuses in the UK only recording migration for one year per decade and NHS data only recording migration between health areas (in the long term), there are no data available for tracking aggregate migration intensities and provide reliable authentication of this trend. One of the key questions is why, at a time when the conventional paradigm suggests that mobility is increasing, residential mobility should have been in long-term decline in the UK, at least over relatively short distances.

We are conscious of the range of determinants of migration intensities and patterns in the UK (Champion *et al.*, 1998) at different spatial and temporal scales, together with the underlying theoretical perspectives outlined by Fielding (2012), but we have not reported any explanatory model-based analysis of the intensities or patterns in this chapter. Neither have we considered in any detail the conventional arguments about living in an 'age of migration' (Castles and Miller, 2009) and about the emergence of hypermobility, as featured in the 'new mobilities' paradigm (Sheller and Urry, 2006). At the same time, technological change, in spearheading the death of distance (Cairncross, 1998), may be having a negative effect on migration rates. For example, as people have become more mobile as commuters, they may well be less concerned to move short distances to be nearer their workplace. Added to this, the idea of a 'job for life' no longer exists, with people changing employer (and potentially employment location) more frequently, while the costs of buying and selling homes have increased substantially over the last 40 years. This resultant shift in attitudes towards commuting may have dampened the enthusiasm to migrate unless it is really 'necessary' to do so (i.e. over longer distances). In some localities, the availability of properties to move to in the locality has also reduced, which further intensifies the disincentive to move over shorter distances.

In addition to reviewing trends in the intensity of migration at different spatial scales, the chapter has also provided insights into the spatial imbalance in flows between the North and the South and between London and the rest of the UK. We have seen how the historic North-South drift has been reversed temporarily on a couple of occasions over the last 40 years and that the cyclical pattern of net migration has been a function of changes in the gross migration taking place in both directions. The evidence in the case of London is of a slowdown in the capital's net migration outflow to the rest of the UK, particularly during 2008/2009. The analysis of time series data at LAD scale reveals a rise in the number of people moving between different metro areas between 2001/2002 and 2012/2013, coupled with an increase in the number of moves from non-metro to metro areas. The number of moves in the other direction, from metro to non-metro areas, has declined, as has the number of moves occurring between different non-metro areas. The most consistent patterns across different age groups are the declines in migration rates for moves from metro to non-metro areas and for moves between non-metro areas.

The slowdown in the urban exodus and general decline in migration intensity in the latter part of the 2001/2002–2012/2013 period are likely to be partly due to the impact of economic recession, but they can also be linked to demographic changes, primarily an increase in urban populations driven by international migrants, predominantly from Eastern Europe since 2004, who live and work in the UK's core cities. This hypothesis is supported by declining rates at key ages: for example, 25–29 year olds (who move for employment) have experienced decline in migration rate when moves from metro to non-metro areas are assessed. At the same time, it is likely that urban redevelopment is having the effect of retaining populations within cities. Other researchers have referred to a process of re-urbanisation taking place in parts of inner London (e.g. Docklands) and in central areas of other British cities (e.g. Butler, 2007).

These conclusions beg a series of questions about what the future may hold. Will short-distance migration intensities continue to fall? What will be the spatial impacts of changing migration intensities at different scales and what will be the cumulative impact on the aggregate migration intensity? Will migration across the north-south divide continue its dampening cyclical pattern? Will varying intensities of movement into and out of cities result in net internal migration balances that, when combined with the dynamics of natural change and international migration, produce a slowdown in city growth, as implied by Champion (2015), and mimic what Frey (2015) suggests is happening in the USA? What exactly is the relationship between the economic cycle (including fluctuations in house prices) and the national migration intensity and are the cyclical effects lagged according to geographical location? Can we contemplate a new regime of migration intensity born out of increasing housing shortages, new household-formation behaviour and new working practices, or will aggregate migration rates over longer distances continue to exhibit the same stability as they have done over the past 40 years?

These are all critical questions but are difficult to answer because of the uncertainty surrounding the multitude of factors that impact on the components of change at national and local levels as well as the relative paucity of data on short-distance migration for the years between censuses. Aggregate migration intensities in the UK in the long term do not fluctuate widely over time, despite fluctuating trends for different age groups such as the 16–24 year olds. The evidence of the past decade suggests that the impact of the Great Recession on the national internal migration intensity has been less than might have been anticipated; it was children and the elderly that experienced the largest falls in migration rates. Unless there is an upturn in the supply of appropriate housing, it is likely that migration intensities for the elderly, whose numbers are growing rapidly, will decline further.

Recognition of the shortcomings of census data in the UK has encouraged ONS, as part of its Census Transformation Programme (Teague *et al.*, 2017), to explore the potential of administrative data to produce more frequent population statistics. Since internal migration is a key component of population change at small area scales, one area of research might involve the estimation of flows within local authorities from the range of sources available. Moreover, whilst migration intensity is an important social indicator (and is the focus of this book), it is the impact of migration on population settlement which is critical as far as planning and policy making is concerned. The primary indicator of migration impact is the aggregate net migration rate, which is determined as a function of the crude migration intensity multiplied by the migration effectiveness index (Rees *et al.*, 2016). Despite the difficulties observed in creating consistent historical time series of migration data, further work on quantifying the relationship between these three variables (net migration rate, intensity and effectiveness) at the national level would further enhance our understanding of migration within the UK. Similarly, further work is required to clarify the relationship between internal and international migration intensities and impacts over time both nationally and sub-nationally.

## References

Bell, M., Blake, M., Boyle, P., Duke-Williams, O, Rees, P., Stillwell, J. and Hugo, G. 2002. Cross-national comparison of internal migration: Issues and measures. *Journal of the Royal Statistical Society A*, 165(3), 435–464.

Bell, M., Charles-Edwards, E., Kupiszewska, D, Kupiszewski, M., Stillwell, J. and Zhu, Y. 2015a. Internal migration data around the world: Assessing contemporary practice. *Population, Space and Place*, 21(1), 1–17.

Bell, M., Charles-Edwards, E., Ueffing, P., Stillwell, J., Kupiszewski, M. and Kupiszewska, D. 2015b. Internal migration and development: Comparing migration intensities around the world. *Population and Development Review*, 41(1), 33–58.

Boden, P., Stillwell, J. and Rees, P. 1992. How good are the NHSCR data? In Stillwell, J., Rees, P. and Boden, P. (eds) *Migration Processes & Patterns Volume 2 Population Redistribution in the United Kingdom*. London: Belhaven Press, 13–27.

Butler, T. 2007. Re-urbanising London Docklands: Gentrification, suburbanisation or new urbanism? *International Journal of Urban and Regional Research*. 31(4), 759–781.

Cairncross, F. 1998. *The Death of Distance: Communications Revolution and its Implications*. London: Orion Business.

Castles, S. and Miller, M.J. 2009. *The Age of Migration*. Fourth edition. Basingstoke: Palgrave Macmillan.

Champion, A.G. 1989a. Internal migration and the spatial distribution of the population. In Joshi, H. (ed) *The Changing Population of Britain*. Oxford: Blackwell, 110–132.

Champion, A.G. (ed) 1989b. *Counterurbanisation: The Changing Pace and Nature of Population Deconcentration*. London: Edward Arnold.

Champion, A.G. 2005. Population movement within the UK. In Chappell, R. (ed) *Focus on People and Migration 2005 Edition*. Basingstoke: Palgrave Macmillan, 91–113.

Champion, A.G. 2015. Urban population—can recovery last? *Town and Country Planning*, 84, 338–343.

Champion, A.G. 2016. Internal migration and the spatial distribution of population. In Champion, A.G. and Falkingham, J. (eds) *Population Change in the United Kingdom*. London: Rowman and Littlefield, 125–142.

Champion, A.G., Fotheringham, A.S., Rees, P., Boyle, P. and Stillwell, J. 1998. *The Determinants of Migration Flows in England: A Review of Existing Data and*

*Evidence.* Report for the Department of the Environment, Transport and the Regions. Newcastle upon Tyne: Newcastle University.

Champion, A.G., Bramley, G., Fotheringham, A.S., Macgill, J. and Rees, P. 2003. A migration modelling system to support government decision-making. In Geertman, S. and Stillwell, J. (eds) *Planning Support Systems in Practice.* Berlin: Springer, 269–290.

Champion, A.G. and Shuttleworth, I. 2016a. Is longer-distance migration slowing? An analysis of the annual record for England and Wales. *Population, Space and Place,* published online in Wiley Online Library. doi:10.1002/psp.2024

Champion, A.G. and Shuttleworth, I. 2016b. Are people moving address less? An analysis of migration within England and Wales, 1971–2011, by distance of move. *Population, Space and Place,* published online in Wiley Online Library. doi:10.1002/psp.2026

Cooke, T.J. 2013. Internal migration in decline. *The Professional Geographer,* 65(4), 664–675.

Courgeau, D. 1973. Migrants and migrations, *Population,* 28, 95–128.

Dennett, A. and Stillwell, J. 2010. Age-sex migration patterns at the start of the new millennium. In Stillwell, J., Duke-Williams, O. and Dennett, A. (eds) *Technologies for Migration and Commuting Analysis: Spatial Interaction Data Applications.* Hershey: IGI Global, 153–174.

Devis, T. 1984. Population movements measured by the NHSCR. *Population Trends,* 36, 18–24.

Devis, T. and Mills, I. 1986. A comparison of migration data from the National Health Central Register and the 1981 Census. *OPCS Occasional Paper* 35. London: OPCS.

Duke-Williams, O. and Stillwell, J. 2010. Temporal and spatial consistency. In Stillwell, J., Duke-Williams, O. and Dennett, A. (eds) *Technologies for Migration and Commuting Analysis: Spatial Interaction Data Applications.* Hershey: IGI Global, 89–110.

Fielding, A.J. 1992. Migration and social mobility: South East England as an escalator region. *Regional Studies,* 26(1), 1–15.

Fielding, A.J. 2012. *Migration in Britain: Paradoxes of the Present, Prospects for the Future.* Cheltenham: Edward Elgar.

Frey, W. 2015. *Diversity Explosion: How New Racial Demographics are Remaking America.* Washington, DC: Brookings Institution Press.

Kaplan, G. and Schulhofer-Wohl, S. 2012. *Understanding the long-term decline in interstate migration.* Working Paper 18507. Cambridge, MA: National Bureau of Economic Research.

Lomax, N., Norman, P., Rees, P. and Stillwell, J. 2013. Subnational migration in the United Kingdom: Producing a consistent time series using a combination of available data and estimates. *Journal of Population Research,* 30, 265–288.

Lomax, N., Stillwell, J., Norman, P. and Rees, P. 2014. Internal migration in the United Kingdom: Analysis of an estimated inter-district time series, 2001–2011. *Applied Spatial Analysis and Policy,* 7(1), 25–45.

Marvin, S., Harding, A. and Robson, B. 2006. *A Framework for City-regions.* London: Office of the Deputy Prime Minister.

ODPM. 2002. *Development of a Migration Model.* London: Office of the Deputy Prime Minister.

ODPM. 2006. *State of the English Cities Volumes 1 and 2*. London: Office of the Deputy Prime Minister.

Ogilvy, A. 1980. Inter-regional migration since 1971: An appraisal of data from the National Health Service Central Register and Labour Force Surveys. *OPCS Occasional Paper* 16. London: OPCS.

Ogilvy, A. 1982. Population migration between the regions of Great Britain, 1971–9. *Regional Studies*, 16(1), 65–73.

Ravenstein, E. 1885. The laws of migration. *Journal of the Statistical Society of London*, 48(2), 167–235.

Rees, P., Bell, M., Kupiszewski, M., Kupiszewska, D, Ueffing, P., Bernard, A., Charles-Edwards, E. and Stillwell, J. 2016. The impact of internal migration on population redistribution: an international comparison. *Population, Space and Place,* published online in Wiley Online Library. doi:10.1002/psp.2036

Rosenbaum, M. and Bailey, J. 1991. Movement within England and Wales during the 1980s, as measured by the NHS Central Register. *Population Trends*, 65, 24–34.

Sheller, M. and Urry, J. 2006. The new mobilities paradigm. *Environment and Planning A*, 38(2), 207–226.

Stillwell, J. 1994. Monitoring intercensal migration in the United Kingdom. *Environment and Planning A*, 26, 1711–1730.

Stillwell, J., Bell, M., Blake, M., Duke-Williams, O. and Rees, P. 2000. A comparison of net migration flows and migration effectiveness in Australia and Britain: Part 1, Total migration patterns. *Journal of Population Research*, 17(1), 17–41.

Stillwell, J., Bell, M., Blake, M., Duke-Williams, O. and Rees, P. 2001. A comparison of net migration flows and migration effectiveness in Australia and Britain: Part 2, Age-related migration patterns. *Journal of Population Research*, 18(1), 19–39.

Stillwell, J. and Boden, P. 1989. Internal migration: The United Kingdom. In Stillwell, J. and Scholten, H. (eds) *Contemporary Research in Population Geography: A Comparison of the United Kingdom and the Netherlands*. Dordrecht: Kluwer, 64–75.

Stillwell, J., Dennett, A. and Duke-Williams, O. 2010. Definitions, concepts and interaction data sources. In Stillwell, J., Duke-Williams, O. and Dennett, A. (eds) *Technologies for Migration and Commuting Analysis: Spatial Interaction Data Applications*. Hershey: IGI Global, 1–30.

Stillwell, J., Duke-Williams, O. and Rees, P. 1995. Time series migration in Britain: the context for 1991 Census analysis. *Papers in Regional Science*, 74(4), 341–359.

Stillwell, J., Rees, P. and Boden, P. 1992. Internal migration trends: An overview. In Stillwell, J., Rees, P. and Boden, P. (eds) *Migration Processes and Patterns Volume 2: Population Distribution in the United Kingdom*. London: Belhaven Press, 28–55.

Stillwell, J., Lomax, N. and Chatagnier, S. 2017. Changing intensities and spatial patterns of internal migration in the United Kingdom. In Stillwell, J. (ed) *The Routledge Handbook of Census Resources, Methods and Applications: Unlocking the UK 2011 Census*. London: Routledge, 362–376.

Stillwell, J., Lomax, N. and Sander, N. 2015. Monitoring and visualising subnational migration trends in the United Kingdom. In Geertman, S., Ferreira, J. Godspeed, R. and Stillwell, J. (eds) *Planning Support Systems Smart Cities*. Dordrecht: Springer, 427–445.

Stillwell, J. and Thomas, M. 2016. How far to internal migrants really move? Demonstrating a new method for the estimation of intra-zonal distance. *Regional Studies, Regional Science*, 3(1), 28–47.

Teague, A., Phelan, L. Elkin, M. and Compton, G. 2017. Towards 2021 and beyond. In Stillwell, J. (ed) *The Routledge Handbook of Census Resources, Methods and Applications: Unlocking the UK 2011 Census.* London: Routledge, 453–468.

Thomas, M., Stillwell, J. and Gould, M. 2014. Exploring and validating a commercial lifestyle survey for its use in the analysis of population migration. *Applied Spatial Analysis and Policy*, 7(1), 71–95.

Wyness, G. 2010. Policy changes in UK higher education funding, 1963–2009. *Working Paper* 10–15. London: Department of Quantitative Social Science, Institute of Education.

# 7   Australia

## The long-run decline in internal migration intensities

*Martin Bell, Tom Wilson, Elin Charles-Edwards and Philipp Ueffing*

Australians are among the most mobile people in the world (Bell *et al.*, 2015b; see also Chapter 4 of this book) with one person in seven changing address in a single year, and around two-fifths of the population over a five-year period. High mobility is a long-standing feature of the Australian landscape. It is integral to the lives of Indigenous Australians and was essential to the process of European settlement (Taylor and Bell, 1996, 2004; Blainey, 1963). Levels of mobility fluctuated during the late nineteenth and early twentieth centuries in response to phases of drought, recession and bursts of mineral exploration (Rowland, 1979), but by the 1960s, when the first comprehensive data became available from the 1971 Census, migration seemed to have become intrinsic to the Australian way of life, a feature it shared with New Zealand, Canada and the USA. In a seminal contribution on cross-national comparisons, Long (1991) attributed this high mobility among these four countries to their institutional frameworks, open housing and labour markets and peripatetic traditions inherited from immigrant forbears. By the turn of the twenty-first century, Australia was still ranked among the most mobile of nations but, as in the mentioned counterparts, a growing body of evidence has begun to show that the intensity of migration is on a downward trend.

Australian internal migration has been the subject of a series of comprehensive analyses over the past three decades, based around successive censuses (Rowland, 1979; Maher and McKay, 1986; Bell, 1992, 1995; Bell and Hugo, 2000; Hugo and Harris, 2011) and edited collections (see, e.g., Newton and Bell, 1998). This work has focused particularly on the spatial patterns of migration, urbanisation and the redistribution of particular population groups, but it has also documented the sequence of changing migration intensities at a range of spatial scales and the demographic composition of this population movement. In this chapter, we update these earlier analyses to include data from the 2011 Census. We also draw upon a temporally harmonised spatial framework that provides migration flows between 69 regions of Australia (Blake *et al.*, 2000), and a similar framework of movement between 1,182 statistical local areas. Our primary focus is trends in migration intensities, but in order to better understand these we also examine the composition of internal migration and its spatial patterns.

The chapter comprises five main sections. The first one sets out the nature of the internal migration data collected in Australia and the spatial frameworks we employ for analysis. We then utilise these data to establish trends in the intensity of migration at a range of spatial scales over the past 40 years. Age is a key determinant of the propensity to move, so in the following section we examine the effects of population ageing on the crude migration intensity and explore the way that the age profile of migration itself has shifted over time. Attention then turns to the spatial patterns of population movement and the way these have changed over the period under review. Finally, drawing on the ideas outlined in Chapter 2, we review a range of possible explanations for the long-term decline in Australian mobility, before making our concluding comments.

## Australian data on internal migration

Compared with many other countries, Australia is well served with data on internal migration. While limited data are available from administrative collections and household surveys, the census is the principal source and has a number of distinctive features in terms of spatial and temporal detail. Internal migration data were first collected at the 1971 Census and have been sought in much the same form at each successive census since 1976. Allied to this, the census is unusual in measuring data over both a one-year and a five-year interval, one of only 10 countries worldwide to do so (Bell *et al.*, 2015a). The census therefore opens two separate windows into internal migration. While the 1-year data provide a concise snapshot comparable to that available in many other countries, the five-year data coupled with the quinquennial census interval offer a continuous picture of changing spatial patterns spanning more than 40 years.

In the spatial domain, a key feature of the Australian Census is that it collects information on all changes of address, thereby providing a direct measure of aggregate population movement that incorporates both local residential mobility and longer-distance migration. As shown in Chapter 4, it is this aggregate measure that places Australia towards the top of the league table of internal migration. In addition, census coding of current and previous place of residence provides a fine spatial matrix of inter-regional migration flows and a hierarchy of administrative and statistical geographies enables construction of migration flow matrices at a number of spatial levels. It is therefore possible to calculate a set of migration intensities, distinguishing between local, regional and long-distance moves.

The Australian Census records migration in the form of transitions between two discrete points in time. It therefore measures migrants rather than migrations, since it fails to capture return and repeat moves (Long and Boertlein, 1990). One-year data therefore provide a better measure of the underlying propensity to move, but are more susceptible to the influence of periodic events. Five-year data offer a more reliable basis from which to assess temporal trends. Despite generally high quality, Australian Census-based migration data have a number of shortcomings. At the 1976 Census only 50% of returns were processed in an effort to limit census costs, thereby undermining the overall reliability of the data (Bell, 1992). For similar reasons, the 1991 Census omitted the one-year question on all changes of address and only captured place of residence one year earlier at State level, thereby interrupting the time series (Bell and Maher, 1995). More significant still was a change in the coding and processing of migration data from the 1996 Census which generated a number of anomalies (Bell and Stratton, 1998). As will be shown below, one result appears to have been a significant inflation in the apparent level of local residential mobility.

A more general issue affecting the comparability of migration intensities over time arises from changes in the geographic boundaries of administrative and statistical districts against which migration is recorded. As noted in Chapters 3 and 4, differences between countries in the number of zones and in their spatial configuration directly affect the measured intensity of migration, an issue widely recognised as the Modifiable Areal Unit Problem (Wrigley *et al.*, 1996). In a similar manner, boundary changes over time therefore compromise the comparability of migration intensities recorded at different spatial scales in Australia from one census to the next. At the state and territory level, geographic boundaries in Australia have been stable since Federation in 1901, although a ninth jurisdiction 'Other Territories' was formally added to the statistical framework in the 1990s. By contrast, statistical local areas—the most finely grained zonal units against which migration has traditionally been recorded—rose progressively in number from 1,336 in 1996 to 1,390 in 2011 (ABS, 1996, 2011a). Prior to the introduction of a new Australian Statistical Geographical Standard in 2011, there were also repeated changes to the boundaries of several intermediate regional geographies.

We circumvent these boundary-change issues by drawing on the Australian Internal Migration database which contains inter-regional migration flows, disaggregated by five-year age groups and sex and organised around a bespoke geographical classification of 69 Temporal Statistical Divisions (TSDs) that are spatially harmonised over the period 1976–1981 to 2006–2011. TSDs were originally designed to provide a temporally consistent set of regions for the period 1981 through 1996 (see Bell *et al.*, 1999; Bell and Rees, 2006; Blake *et al.*, 2000; Rees *et al.*, 2000) and were later extended to include data from the 2001, 2006 and 2011 Censuses. TSDs essentially represent functional regions, based largely around the Statistical Divisions formerly defined by the Australian Bureau of Statistics (ABS 2011a). They primarily distinguish the coast from the hinterland and outback, but also divide the major metropolitan centres into inner, middle and outer zones. Following Stillwell *et al.* (2000), TSDs can be further aggregated into City Regions (38 zones) which reflect Australia's functional geography of metropolitan cores and hinterlands. These in turn provide the basis for a formal classification into eight distinctive types of regions which reflect a spatial hierarchy that has been found valuable in cross-national comparisons of migration flows and spatial impacts (see Stillwell *et al.*, 2000, 2001).

Temporal Statistical Local Areas (TSLAs) are a purposive geography created for the specific purpose of the time-series analysis of Australian internal migration presented in this chapter. These were derived by overlaying the digital boundaries of Statistical Local Areas (SLAs) at the 1996, 2001, 2006 and 2011 Censuses and then by aggregating mismatched units to a new geography that was consistent over time. TSLAs are the smallest spatial unit used in this analysis of census data with 1,182 zones across Australia. These units range in size from 0.33 km$^2$ to 670,000 km$^2$ and had a median population of 5,487 in 2011. They are most numerous in metropolitan regions, making them well suited for capturing short-distance intra-urban mobility. Outside the major cities, TSLAs can cover very large areas. This spatial heterogeneity is a product of Australia's highly concentrated settlement system, with two-thirds of Australians living in eight Australian capital cities and 80% living within 50 km of the coast (ABS, 2004).

While the census provides the most comprehensive picture of internal migration, data are only collected every five years. An alternative source which provides annual statistics on changes of address is the Medicare data, drawn from Australia's national health care system. The system relies on individuals notifying Medicare of changes of address, but this usually occurs only at the time of the next visit to a GP, so coverage of address-changes among healthy individuals, such as highly mobile young adults, is poor. Medicare data measure migration events rather than transitions so they are not directly comparable to data from the Census (see, e.g., Bell, 2002). Nevertheless, they provide a second, independent source of information on migration trends, with the advantage of quarterly statistics. Medicare data form the basis for post-census estimates of interstate migration published annually by the Australian Bureau of Statistics (ABS, 2015a) and have recently been extended to generate estimates at the regional level (ABS, 2011b). For the purposes of this chapter, we draw on published Medicare-based estimates of internal migration between the states and territories (1981–1982 to 2014–2015), and unpublished estimates between postcodes, covering annual intervals from 1985–1986 to 2011–2012.

Table 7.1 summarises the various levels of geography used to analyse Australian internal migration in this chapter. Because of changes in statistical geography and variable data availability, the length of the time series varies for different levels of geography. For the reasons outlined above, we focus primarily on migration over a five-year interval, but single-year data are also reported where these are available, including for the Medicare statistics. Following Rees *et al.* (2000), we adopt Crude Migration Intensity (CMI) as the key metric for comparing migration rates, computed using the population at the start of the interval, excluding immigrants, emigrants, children born and persons who died during the interval, as the population

Table 7.1 Australia: summary of geographies used in the analysis, 1971–2011

| Name of zone | Description | Migration interval (years) | Number of zones | Census years |
|---|---|---|---|---|
| Census data | | | | |
| All moves | Captures all changes of address | 5 and 1 | Na | 1971, 1976, 1981, 1986, 1991, 1996, 2001, 2006, 2011 |
| Statistical local areas (SLA) | Small areas | 5 and 1 | Varies | 1976, 1981, 1986, 1991 (5), 1996, 2001, 2006, 2011 |
| Temporal statistical local areas (TSLA) | Small areas based on SLA geographies and harmonised using GIS overlays | 5 | 1,182 | 1996, 2001, 2006, 2011 |
| Temporal statistical divisions (TSDs) | Regions based on 1996 ABS Statistical Divisions and harmonised using GIS overlays. AIM data base contains inter-regional flows disaggregated by age and sex. | 5 | 69 | 1981, 1986, 1991, 1996, 2001, 2006, 2011 |
| City regions | Derived from TSDs and further classified into functional regions | 5 | 38 | 1981, 1986, 1991, 1996, 2001, 2006, 2011 |
| City region types | Identifies six discrete types of city region | 5 | 6 | 1981, 1986, 1991, 1996, 2001, 2006, 2011 |
| States and territories | ABS standard geography | 5 and 1 | 8 | 1971, 1976, 1981, 1986, 1991, 1996, 2001, 2006, 2011 |
| Medicare data | | | | |
| Postcodes | Australia Post postcodes as used by Medicare and reported by ABS | 1 | Approx. 1400 | 1985–1986 to 2011–2012 |
| States and territories | ABS standard geography | 1 | 8 | 1981–1982 to 2014–2015 |

Source: Compiled by the authors, see text; 'na' not applicable.

at risk. For a given level of geography, the CMI is calculated by dividing the total number of migrants ($M$) by the population at risk of moving ($P$) and expressed as a percentage, so that:

$$CMI = M / P * 100 \qquad (7.1)$$

Age-specific rates are calculated on a similar basis.

## Trends in migration intensity

Australia is unusual on the global stage in collecting information on all changes of address. As argued in Chapters 3 and 4, this is the mobility statistic that is most readily comparable between countries and is termed the Aggregate Crude Migration Intensity (ACMI) to differentiate it from intensities observed at other spatial scales (Bell *et al.*, 2015b). Table 7.2 sets out the total number of internal migrants and the associated ACMI for one and five year intervals for each Census since 1971. The results confirm Australia's high level of mobility but also reveal significant fluctuations over the 40-year period. Five-year intensities fell in the early 1970s, probably reflecting the OPEC oil crisis and economic recession, then rose to around 40% and remained stable through the 1980s. The 1996 Census shows a sharp rise of almost 3% points in the five-year intensity and this is mirrored in the data for the single year 1995–1996. Both series then register a steady decline, such that migration intensities fell by some 12–20% (five-year and one-year data respectively) over the 15-year period between the 1996 and 2011 Censuses. Indeed, such was the scale of the decline that the absolute number of internal migrants counted by the 2011 Census was well below that recorded 10 years earlier, despite growth of more than 13% in the national population.

*Table 7.2* Australia: aggregate crude migration intensities, 1971–2011

| Census year | 5-year interval | | 1-year interval | |
|---|---|---|---|---|
| | *Migrants* | *CMI* | *Migrants* | *CMI* |
| 1971 | 3,928,567 | 39.41 | nc | Nc |
| 1976 | 4,157,579 | 36.49 | 2,071,232 | 16.14 |
| 1981 | 5,120,100 | 40.77 | 2,276,532 | 16.44 |
| 1986 | 5,537,475 | 41.09 | 2,465,596 | 16.57 |
| 1991 | 5,734,245 | 40.38 | nc | nc |
| 1996 | 6,568,414 | 43.13 | 3,079,390 | 18.34 |
| 2001 | 6,805,790 | 42.38 | 3,072,386 | 17.55 |
| 2006 | 6,598,190 | 40.20 | 2,812,386 | 15.49 |
| 2011 | 6,670,260 | 37.73 | 2,891,684 | 14.58 |

Sources: Compiled by authors from Rowland (1979, Table 32) and ABS Censuses (unpublished data); nc—not collected.

As can be seen from Figure 7.1, this downward trend in the overall propensity to move has been echoed at other geographic scales. Migration between the 69 TSDs and between the 38 City Regions has fallen at an accelerating rate in successive inter-census periods since the 1970s. Movement between SLAs displays the same general trajectory, though it started a little later and was arrested by a short-lived plateau in the early 1990s. This fluctuating trajectory can be traced in part to the changing geography of SLAs noted earlier. When SLA boundaries are harmonised on a consistent spatial geography of 1,182 zones (TSLAs), the graph reveals a sustained decline since 1981–1986, again with an accelerating fall in migration intensity. What is also apparent from Figure 7.1 is that the sharp rise in the ACMI recorded in the 1996 Census was entirely a product of changes of address within the same SLA. The difference between the curve for 'all moves' and

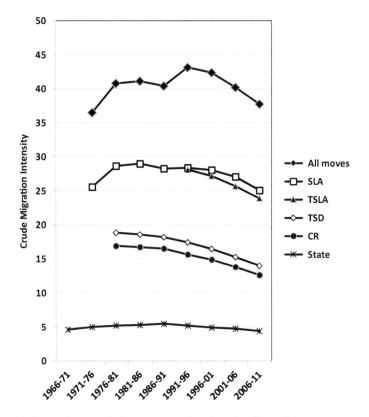

*Figure 7.1* Australia: trends in 5-year crude migration intensity, at various spatial scales, 1966–1971 to 2006–2011.

Source and notes: Rowland (1979, Table 32); ABS Censuses (unpublished data). SLA—statistical local area; TSLA—temporally consistent SLAs (1,182); TSD—temporally consistent Statistical Divisions (69); CR—city regions (38); State—states and territories (8).

for 'moves between SLAs' is comprised solely of moves *within* SLAs and the hump in the trajectory for aggregate mobility is conspicuously absent from that for any other spatial scale. As noted earlier, this appears to be the product of a change in coding procedures at the 1996 Census and resulted in a one-off, upwards shift in the measured level of aggregate migration intensity (Bell and Stratton, 1998). Discounting this brief hiatus, the ACMI too shows a persistent downward trend. The same is true of the profile for migration between the eight states and territories. For these long-distance moves, however, the peak was delayed until the late 1980s, with migration rates continuing to rise for a full decade after they had begun to fall over short and intermediate distances. From the 1990s the five-year interstate data show a sustained fall in migration intensities and this same decline is confirmed in the one-year interstate migration data (not shown).

Notwithstanding these variations in timing, it appears to be longer-distance migration that has borne the brunt of the decline in Australian migration. Table 7.3 sets out the individual components of internal migration, distinguishing local residential mobility from that over longer distances. Migration intensities declined at all spatial scales, but the fall was greatest between TSDs and between City Regions, and a little more attenuated between states and territories. Movement within TSLAs, and between TSLAs within TSDs, also dropped but at a more modest rate. The consequence of these changes was that local mobility (within and between TSLAs) accounted for an increasing share of all internal migration, rising from 59.5% over the 1991–1996 interval to 63% for 2006–2011.

SLAs vary widely in area, from relatively small suburbs in the major metropolitan centres to extensive, though sparsely inhabited, zones in the Australian outback. Movements between SLAs therefore encompass a

*Table 7.3* Australia: change in intensities by type of migration, 1991–1996 to 2006–2011

| Type of migration | CMI | | | Share of migration | | |
|---|---|---|---|---|---|---|
| | 1991–1996 | 2006–2011 | % change | 1991–1996 | 2006–2011 | Change in share |
| Within SLAs | 15.01 | 13.87 | −7.6 | 34.8 | 36.8 | 2.0 |
| Between SLAs within TSDs | 10.66 | 9.87 | −7.4 | 24.7 | 26.2 | 1.4 |
| Between TSDs within city regions | 1.79 | 1.38 | −22.6 | 4.1 | 3.7 | −0.5 |
| Between CRs within states | 10.45 | 8.21 | −21.4 | 24.2 | 21.8 | −2.5 |
| Interstate | 5.23 | 4.40 | −15.9 | 12.1 | 11.7 | −0.5 |
| Total | 43.13 | 37.73 | −12.5 | 100.0 | 100.0 | 0.0 |

Source: Calculated from ABS Censuses (unpublished data).

range of migration distances. To better understand the spatial dimensions of recent mobility decline, the CMI between TSLAs was decomposed into discrete bands based on the distances between their geographic centroids (Table 7.4). The results confirm and amplify the analysis classified by type of move. Migration intensities declined across all distance bands. Long-distance moves, involving displacements of 100 km or more, accounted for only 37% of all moves between TSLAs between 1991 and 1996, reflecting a CMI of just 10.4%, but it was over these longer distances that the sharpest falls were registered over the 15-year interval. Migration intensities over distances of 100 km or more fell by more than a quarter, compared with less than 10% for moves of 50 km or less. As a result, the decline in long distance moves accounted for almost two-thirds (63.5%) of the overall decline in the CMI after 1996. Nevertheless, it is clear that local mobility and migration over intermediate distances, however measured, also fell over the period.

This profile suggests that the overall decline in Australian migration intensity cannot be attributed simply to forces such as the housing or labour market, operating in a specific locality or at a particular level of spatial scale. Instead, a range of forces, including demographic structure and transformation of the national space economy, would appear to be implicated. Allied to this, it is also clear that reduced migration cannot be traced to a single time-bound event. Migration has fallen steadily over a 15- to 25-year period, and the decline shows no sign of abating. Indeed, the rate of decline has accelerated since the first half of the 1990s, and this has occurred across all types of move and distance bands. From the first to the second half of the 1990s, the ACMI fell by just 1.7%, but the following census saw a reduction of 5.1%, while for 2006–2011 it declined by a further 6.2%.

Results from the 2016 Census will not become available until 2018, but migration data from the Medicare system provide the basis for a broad comparison of trends against results from the census (Figure 7.2). Postcodes are more numerous than SLAs, and Medicare data report migration events,

*Table 7.4* Australia: migration intensities between TSLAs by distance, 1991–1996 to 2006–2011

| Distance band (km) | CMI | | | Change (%) | Share of decline (%) |
|---|---|---|---|---|---|
| | *1991–1996* | *2006–2011* | *Change* | | |
| 0–10 | 7.04 | 6.51 | −0.53 | −7.5 | 12.5 |
| 10–25 | 6.21 | 5.73 | −0.48 | −7.7 | 11.3 |
| 25–50 | 2.50 | 2.25 | −0.25 | −10.2 | 6.0 |
| 50–100 | 1.97 | 1.68 | −0.29 | −14.6 | 6.7 |
| 100–250 | 2.89 | 2.19 | −0.70 | −24.3 | 16.5 |
| 250–1,000 | 4.36 | 3.08 | −1.28 | −29.3 | 30.1 |
| 1,000+ | 3.14 | 2.42 | −0.72 | −23.0 | 17.0 |
| Total | 28.12 | 23.86 | −4.26 | −15.1 | 100.0 |

Source: Calculated from ABS Censuses (unpublished data).

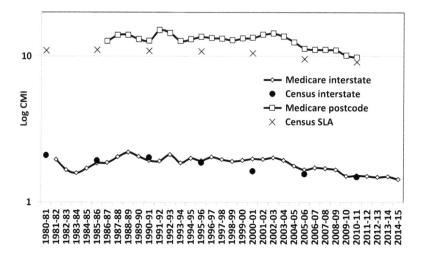

*Figure 7.2* Australia: Census and Medicare based CMI, at various spatial scales, 1980–1981 to 2014–2015.

Source and notes: ABS Censuses and Medicare estimates (unpublished data). SLA—statistical local area.

rather than transitions, so there are underlying differences between the two sources, even at the level of interstate migration. Nevertheless, it is clear from Figure 7.2 that the Medicare-based data show a downward trend in migration intensities similar to that recorded by the census, both for long- and short-distance migration. The Medicare data underscore the volatility inherent in single year measures of migration, with short-term fluctuations in migration intensities observed over this period, but a sustained decline in migration intensity at both spatial scales has been observed since 2002–2003.

## The changing age profile of migration

Shifts in migration intensity can arise from a number of sources, but one important component comprises structural effects which result from changes in the underlying composition of the population. Changes in dimensions such as occupational mix, labour force status, household composition or ethnic structure can all impact on aggregate migration intensities, but perhaps the greatest effect is from population ageing. Migration is an age selective process, peaking among young adults and declining in the later adult years and among children (Rogers and Castro, 1981). As fertility falls and populations age, larger cohorts transition into the lower-mobility years, placing downward pressure on overall migration intensities. Australia, like other advanced economies, has undergone significant ageing in recent decades, with the median age rising from 27.5 in 1971 to 37.3 in 2011 (ABS,

2011c), so the declines described above are likely due, at least in part, to age structure effects.

To assess the effect of ageing on the change in crude migration intensities, five-year CMIs for TSDs for each Census year since 1981 were age-standardised using direct standardisation procedures with the 1976 population at risk as the base (see, e.g., Shryock *et al.*, 1973). As illustrated in Figure 7.1, the raw data reveal a decline in inter-TSD migration from 18.9% in the 1976–1981 interval to 14.0% for 2006–2011. However, if the age profile of the population had remained unchanged, the 2006–2011 intensity would have been 15.3%, a fall of 3.6% points instead of 4.9%. Thus, changes in age composition accounted for fully one quarter of the 26% decline in inter-TSD migration intensity over the 30-year period. These shifts in age composition will have affected all forms of internal migration, but the precise contribution in each case depends on the extent of the overall decline and on the interaction between ageing and changes in age-specific migration intensities. Age-specific migration data are not available for all levels of geography, but age-standardisation of one-year intensities reveals that population ageing contributed a similar share (22.5%) of the decline at the level of 'all moves' between the 1996 and 2011 Censuses.

In parallel with the ageing of Australia's population over the past four decades has been a steady shift in age-specific migration intensities. Figure 7.3 shows the age profile of all changes of address from each census since 1996. Data are reported for one-year migration intervals because these more accurately show the ages at which migration occurs. The results follow the classic profile described by Rogers and Castro (1981), peaking among young adults, but with a clear and persistent decline, especially from the late teens to the early 30s, but also among young children and at older ages. At age 18, for example, the age-specific migration intensity fell by one third over the 15-year period. Proportionately, the decline was similar among people of retirement age, although the absolute reduction in migration rates was much smaller. For ages 25–45, the graph suggests little change in migration intensities, but closer analysis (not shown) reveals that this is a product of compensating trends between the sexes: male intensities fell in these prime working-age groups over the 15-year period, whereas migration intensities among females actually rose. This differential almost certainly reflects the increased professionalisation of women and their greater participation in the labour force, where mobility is high. Changing outcomes following separation and divorce may also be implicated. Beyond age 60, on the other hand, it was women who recorded the sharpest reduction in migration intensities, probably reflecting growing policy emphasis on remaining in the family home and delayed widowhood as a result of increasing male life expectancy.

Also perceptible from Figure 7.3 is a modest rightwards shift, or ageing, of the migration profile itself, continuing a trend identified previously by Bell and Hugo (2000). In fact, the age at peak migration has increased steadily

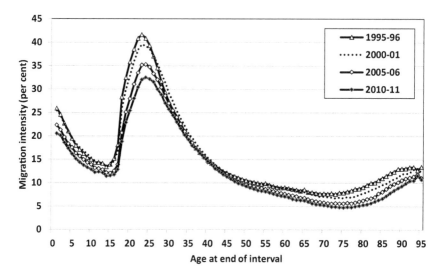

*Figure 7.3* Australia: intensities of all address changing by age, 1995–1996 to 2010–2011.

Source: ABS Censuses (unpublished data).

since 1981, from 23 to 25 for men and from 22 to 24 among women. At the same time, there is evidence that migrations have become more widely dispersed around the peak. These shifts in the age profile of migration are amplified in the data for migration between Australia's 69 TSDs, measured here by five-year age groups over five-year intervals for successive decades since 1981 (Figure 7.4). As with the profiles for 'all moves', the reductions have been sharpest over the last decade and are concentrated especially at ages below 30 and above 50. Declines first occurred among young adults in the 1980s and were followed by falling migration intensities among children and adolescents under 20 (the only difference in the 1986–1991 and 1996–2001 age profiles). The rightwards shift in the age profile is clearly apparent in the last period (2006–2011), as is a decline in mobility at older ages.

Bernard *et al.* (2014a, 2014b) showed that variations between countries in the age profile of migration among young adults could be traced to differences in the timing, spread and prevalence of key life course events. Computed across 27 countries, the intensity of peak migration and the age at which it occurred were closely correlated with four life course transitions— completion of education, entry into the labour force, formation of a union and birth of the first child. The later the age at which men and women completed these transitions, and the more concentrated the transition within a narrow age range, the higher the peak in migration and the older the age at the peak. Shifts over time in the age profile of migration within countries might therefore be expected to reveal a similar association with changes

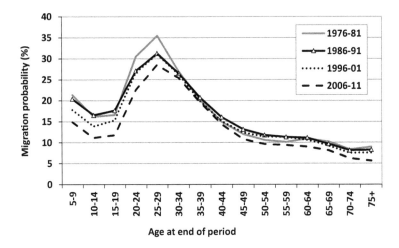

*Figure 7.4* Australia: inter-TSD migration intensities by age, 1976–1981 to 2006–2011.
Source: ABS Censuses (unpublished data).

in the timing and spread of life course events. Table 7.5 sets out the timing of key transitions for Australian men and women at selected censuses, together with the associated shift in the age and intensity of migration derived from the one-year transition data. Time series data are not available to allow computation of the mean age and interquartile range used by Bernard *et al.* (2014a) to track the timing of transitions, but the data in Table 7.5 nevertheless show a clear link to the rising age at peak migration. Median age at first marriage rose by around five years for both men and women over the 30-year period, and median age at first birth increased by four years. The timing of transitions in the economic domain is less clear but the proportions aged 15–24 still in tertiary education rose rapidly for both sexes after 2001. There was also a sharp fall in the proportion of young men in the labour force from 1981 to 1996, again suggesting a later transition from education into employment.

These shifts in the age profile of migration are of significance themselves in understanding the nature of population mobility but they also entrain broader implications for the overall level of migration intensity. It is well established that the age at first birth shapes completed family size for women, and similar effects may occur with migration, as later ages at leaving home and the transition to adulthood delay the onset of mobility. On the other hand, a longer period unencumbered by children, a marital union and home ownership raises the opportunity for migration. Moreover, the propensity to move is significantly higher among those with advanced educational qualifications and those who work in professional occupations (Bell, 2002).

*Table 7.5* Australia: age at migration, peak intensity and life course transitions, 1981–2011

| Sex and year | Age at peak migration | Peak migration intensity | Per cent aged 15–24 in full-time tertiary education | Per cent aged 15–24 in the labour force | Median age at first marriage | Median age at first birth |
|---|---|---|---|---|---|---|
| *Males* | | | | | | |
| 1981 | 23 | 38.2 | na | 75.4 | 24.4 | – |
| 1986 | 23 | 36.7 | na | 71.9 | 25.4 | – |
| 1996 | 24 | 40.2 | na | 67.2 | 26.7 | – |
| 2001 | 25 | 37.9 | 11.7 | 66.2 | 27.6 | – |
| 2006 | 25 | 33.7 | 12.6 | 67.2 | 28.7 | – |
| 2011 | 25 | 31.0 | 15.3 | 64.9 | 29.6 | – |
| *Females* | | | | | | |
| 1981 | 22 | 41.1 | na | 63.1 | 22.1 | 25.3 |
| 1986 | 22 | 40.4 | na | 63.6 | 23.2 | 26.5 |
| 1996 | 22 | 43.5 | na | 63.1 | 24.5 | 27.8 |
| 2001 | 23 | 42.0 | 15.8 | 64.1 | 25.7 | 28.8 |
| 2006 | 23 | 37.5 | 16.0 | 65.9 | 26.9 | 28.8 |
| 2011 | 23 | 34.2 | 18.6 | 64.2 | 27.6 | 29.3 |

Sources: ABS, 2008, 2015b, 2016; Censuses (unpublished data).

## The spatial dimensions of mobility decline

Migration is fundamentally a spatial process which operates to redistribute population across the settlement system. People move because regions differ and fulfil varying roles in the national space economy. It follows that shifts in the pattern of movement between regions should also aid understanding of changes in the level of mobility itself. Following Stillwell *et al.* (2000, 2001), we examine inter-regional flows by assembling the 69 TSDs into a hierarchy of six types of regions, oriented around the eight state and territory capitals. This classification sets the metropolitan cores at the apex of the spatial hierarchy, consisting largely of the older parts of the urban fabric in Sydney, Melbourne, Brisbane, Adelaide and Perth, together with Hobart, Darwin and the Australian Capital Territory. Surrounding each of these cores lies a 'Metro Rest' region, encompassing the outlying suburbs and peri-urban periphery. Beyond this, TSDs are classified into four regional types, based on proximity and functional links to the core, and designated as 'Metro Near', 'Metro Far', 'Coast' and 'Remote', the last of these dominating the centre and north-west of the continent (Figure 7.5).

Table 7.6 extends the analysis presented by Stillwell *et al.* (2000) to include estimates of net migration and migration effectiveness for the 2006–2011 interval. The data reveal a distinctive pattern of net losses from the metropolitan cores and from the remote inland, with compensating gains in the Metro Near and Coastal regions. Underlying these patterns are four longstanding streams of inter-regional migration in Australia: from inland towns and rural areas to the major cities, from the cities themselves to the metropolitan hinterland and the coast, from the inner cities to their urban peripheries and (at the national scale) from the south and east of the continent to the north and west (Bell, 1992). As Table 7.6 makes clear, however, substantial changes have taken place over the past 30 years. Net losses from

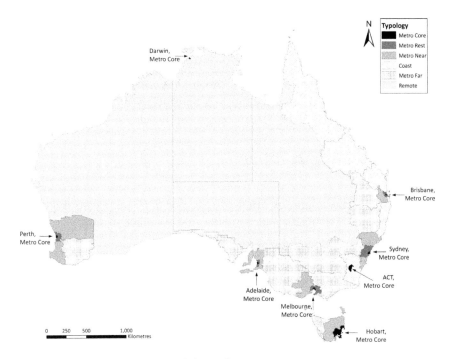

*Figure 7.5* Australia: typology of city regions.

*Table 7.6* Australia: net migration and effectiveness, 1976–1981, 1991–1996 and 2006–2011, by type of region

| Region type | Net migration (000) | | | Migration effectiveness (%) | | |
|---|---|---|---|---|---|---|
| | 1976–1981 | 1991–1996 | 2006–2011 | 1976–1981 | 1991–1996 | 2006–2011 |
| Metro core | −283 | −135 | −113 | −19.5 | −9.2 | −10.5 |
| Metro rest | 188 | 56 | −8 | 14.3 | 4.0 | −0.6 |
| Metro near | 37 | 35 | 68 | 7.5 | 5.7 | 13.3 |
| Coast | 109 | 150 | 67 | 21.1 | 20.3 | 13.3 |
| Metro far | −27 | −60 | −5 | −5.5 | −12.2 | −1.3 |
| Metro remote | −26 | −46 | −9 | −11.2 | −18.8 | −5.3 |
| Net redistribution | 334 | 241 | 135 | 16.5 | 10.6 | 7.0 |

Sources: Stillwell *et al.* (2000), Table 1; 2011 Census (unpublished data).

Notes: The MER or Migration Effectiveness Ratio is calculated as net migration divided by gross flows, expressed as a percentage. For the system as a whole, the MEI or Migration Effectiveness Index is calculated as half of the sum over all zones of the absolute values of net migration divided by the total flows between all zones, expressed as a percentage.

the Metro cores have fallen by 60% with a corresponding reduction in areas designated as Metro Rest. Part of this difference has been picked up by Metro Near regions and the Coast, though the latter too registered a substantial fall in net migration gains after 1996. At the same time Metro Far and Remote regions recorded a reduction in net outflows after rising losses between the late 1970s and the early 1990s.

Together, these changing flows have generated a marked reduction in the net redistribution of population between the six levels of the spatial hierarchy, dropping from 334,000 in the 1976–1981 interval to just 135,000 over the period 2006–2011. The effect of this reduction is most readily captured in the Migration Effectiveness Index, which is the net gain (or loss) summed across all gaining (or losing) regions as a percentage of the gross flows between the six regional types.

$$\mathrm{MEI} = 100 \times 0.5 \sum_i |D_i - O_i| / M \tag{7.2}$$

where $D$ represents migration inflows to a region ($i$), $O$ represents migration outflows from a region ($i$), and $M$ is the total number of migrants in the system.

Since the total volume of migration flows between regions remained relatively stable over the 30-year period, at around 2 million, the net effect was a sharp decline in the MEI, falling from 16.5% in 1976–1981 to just 7% in 2006–2011. Thus, flows between the six levels of the spatial hierarchy became progressively more balanced, with significantly less redistribution of population for the given volume of migration. Similar trends are apparent in the Migration Effectiveness Ratios which summarise the balance between net and gross flows for individual region types (Table 7.6).

Figure 7.6 illustrates the direction and magnitude of net flows between the six levels of the spatial hierarchy, this time comparing data for the 2006–2011 interval with that of two decades earlier. The broad picture remains unchanged between the two periods. With only minor exceptions, the direction of net flows between levels of the hierarchy is the same, being dominated by flows from the Remote and Metro Far regions to the Coast and Metro Centre, and by a cascade of movements back down the spatial hierarchy from the Metro Core and Metro Rest to Metro Near and the Coast. By 2006–2011, however, the scale of redistribution was greatly diminished. The rate of net outflow from the Metro Core to Metro Rest fell by half, from the Core to Metro Near by one third, and from the Core to the Coast by three-quarters. Outflows from Metro Near fell almost as sharply: the rate of net loss to the Coast fell by half, while losses to Metro Near and Metro Far were down by 20%. At the same time, migration rates from Remote, Metro Far and Metro Near to the Coast all fell by 70% or more. Only in the case of the Remote and Metro Rest did the direction of the net flow actually reverse between the two periods, resulting in a small net gain to the former.

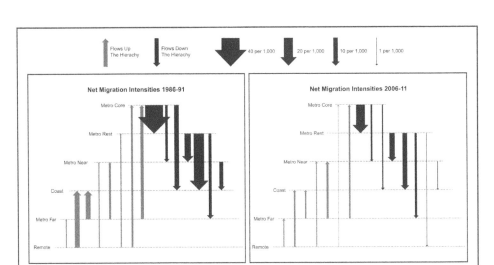

*Figure 7.6* Australia: net migration through the hierarchy of city-regions, 1986–1991 and 2006–2011.

Source: Analysis of the AIM Database drawn from ABS Censuses (unpublished data).

These shifts can be traced in part to the nature of the regional boundaries, particularly those between Metro Cores and Metro Rests, as the larger cities have expanded and suburban development has spilled into the peri-metropolitan region. However, losses from the Metro Cores have also been contained by the rising popularity of inner city living which has drawn population back into new apartment complexes developed in and around transit centres and the central business districts of Australian cities. Other phase shifts are also apparent, for example in the sharply reduced net flows to the Coast and rising gains by Metro Near regions. Sander (2011) ascribes this shift to the drop in affordability of coastal living, especially for those of retirement age who played a significant role in the counter-urban migration that triggered this trend in the 1970s and 1980s (see, e.g., Hugo and Smailes, 1985; Champion, 1989). As coastal towns have become more crowded and housing costs have risen, retirees have begun to favour alternative destinations in Metro Near regions. In a similar way, the sharp fall in net losses from Metro Far and Metro Remote regions signals a historic shift in the almost century-long outflow from inland Australia, as rural populations approach minimum levels. While young adults continue to leave the countryside, attracted by educational and employment opportunities in the cities (Argent and Walmsley, 2008), indigenous Australians have increasingly engaged in a 'return to country' and make up a rising proportion of Australia's outback population (Taylor, 2010). At the same time, the resources boom in the first decade of the twenty-first century triggered considerable development activity in Remote regions of the country, attracting significant streams of labour. While much of this was based around fly-in-fly-out arrangements, it also attracted permanent migration to mining regions.

These changing patterns of migration are the product of nuanced shifts in the nature and utilisation of regional space but the underlying framework of the Australian settlement system remains essentially stable. It follows that declining migration effectiveness might be interpreted as reflecting an overall trend towards maturation of the space-economy. As argued by Rowland (1979), in this situation migration serves to maintain dynamic equilibrium whereby population exchanges serve to rejuvenate and renew population structures rather than to redistribute population across the settlement system. In Australia, as in many other parts of the world (Rees *et al.*, 2016), falling levels of population redistribution are the product of declines both in migration intensity and in migration effectiveness. These effects are most readily captured in the Aggregate Net Migration Rate, which is measured as the sum of net migration gains across all gaining regions as a percentage of the total population at risk:

$$\text{ANMR} = 100 \times 0.5 \sum_i |D_i - O_i| / P \tag{7.3}$$

Combining Equations 7.1–7.3, it can be seen that the ANMR is a product of the CMI and the MEI (Bell *et al.*, 2002), such that:

$$\text{ANMR} = \text{CMI} \times \text{MEI} / 100 \tag{7.4}$$

Table 7.7 sets out the constituent elements of the ANMR for TSDs for five-year migration intervals from each Census since 1981. As the results clearly show, there has been a fall in aggregate redistribution between TSDs over the 40-year period, but this represents the joint product of falls in both migration intensity and migration effectiveness. As noted in discussion of

*Table 7.7* Australia: components of migration change, TSDs, 1976–1981 to 2006–2011

| Migration interval | Crude migration intensity (%) | Migration effectiveness index (%) | Aggregate net migration rate (%) |
|---|---|---|---|
| 1976–1981 | 18.9 | 14.1 | 2.7 |
| 1981–1986 | 18.6 | 10.5 | 2.0 |
| 1986–1991 | 18.2 | 11.6 | 2.1 |
| 1991–1996 | 17.5 | 10.1 | 1.8 |
| 1996–2001 | 16.5 | 7.7 | 1.3 |
| 2001–2006 | 15.3 | 9.6 | 1.5 |
| 2006–2011 | 14.0 | 8.6 | 1.2 |
| Decline 1976–1981 to 2006–2011(%) | 25.8 | 38.7 | 54.5 |

Source: Analysis of AIM database, derived from ABS Censuses (various years).

Figure 7.1, the CMI for movement between TSDs fell by more than a quarter over the period, but migration effectiveness dropped even more sharply over the period as a whole. Small increases in effectiveness were observed in 1986–1991 and 2001–2006, corresponding to periods of rapid growth in the peri-urban fringe of major cities and selected high amenity coastal communities which produced more unidirectional flows. Notwithstanding these short-term increases in effectiveness, the overall level of population redistribution between TSDs fell by half between 1981 and 2011, due both to lower rates of movement and greater equilibrium between inter-regional flows.

## Understanding declining mobility

Structural effects such as population ageing, together with greater stability in the settlement system, undoubtedly contributed to the decline in migration intensities across Australia after the 1970s, but other factors also influenced the level of population mobility. Table 7.8 sets out a suite of national economic and social indicators synchronised with each quinquennial Census since 1971, and Table 7.9 reports the association between selected indicators, for which extended annual time series were available, and the annual CMIs for interstate movement (derived from Medicare-based estimates) and for movement between SLAs (interpolated from the single year interval Census data). SLAs were used instead of TSLAs to provide the longest possible time series. Changes in the SLA geography over time means that results should be treated with some caution.

Economic growth creates employment opportunities and an environment conducive to migration while recession is likely to limit mobility. Australia's economic growth has fluctuated over the past 40 years but the range from high to low is modest and the trends set out in Table 7.8 provide no rationale for a secular decline in migration. Notably, Australia was the one OECD member country in which policy settings avoided recession during the global financial crisis of 2008–2009. Nevertheless, annual growth in total GDP dipped to 1.4% in 2008 and there was a weak positive correlation between migration and growth in GDP over the period 1981–2011 (Table 7.9): as growth in GDP slowed, migration intensities also fell. Unemployment has also fluctuated, peaking at around 10% in the early 1980s and again in the early 1990s before falling to just 5% in 2010–2011. The unemployed display mobility rates above those for people within the labour force, even when standardised for age (Bell, 2002), so declining unemployment serves to reduce migration propensities and this again is reflected in a positive association with the fall in Australian migration intensities over the 30 years 1981–2011 (Table 7.9). There was a weak positive correlation between unemployment and the CMI for interstate migration ($r = 0.35$), but a much stronger association with movement between SLAs ($r = 0.75$). Female labour force participation, on the other hand, has risen progressively and is negatively associated with the decline in migration intensities. In this instance it seems likely that greater

*Table 7.8* Australia: key social and economic indicators, 1971–2011

| Indicator | 1971 | 1981 | 1986 | 1991 | 1996 | 2001 | 2006 | 2011 |
|---|---|---|---|---|---|---|---|---|
| Annual economic growth rate (%)[1] | 5.8 | 2.9 | 2.9 | 2.6 | 4.0 | 3.8 | 3.5 | 2.6 |
| Unemployment rate (%)[2] | 1.7 | 5.8 | 8.1 | 9.6 | 8.5 | 6.8 | 4.8 | 5.1 |
| Female labour force participation (%)[3] | Na | 44.8 | 48.3 | 52.0 | 53.8 | 55.1 | 57.6 | 59.0 |
| Mortgage rate (%)[4] | 7.5 | 13.8 | 16.8 | 10.5 | 7.2 | 5.1 | 5.8 | 4.7 |
| Average household size[5] | 3.3 | 3.0 | 2.9 | 2.8 | 2.6 | 2.6 | 2.6 | 2.6 |
| Dwellings owned/ being purchased (%)[5] | 72.3 | 73.2 | 72.9 | 75.6 | na | 70.6 | 70.5 | 69.1 |
| House price index[6] | | | 100 | 167 | 184 | 266 | 437 | 574 |
| Percentage with tertiary education[7] | 2.0 | 4.1 | 5.0 | 8.6 | na | 12.9 | 15.6 | 18.8 |
| Median age at first marriage (brides)[8] | 21.1 | 22.1 | 23.2 | 24.5 | 25.7 | 26.9 | 27.6 | 27.7 |
| Persons living with parents[5] | | | | | | | | |
| Aged 20–24 | Na | 34.1 | na | 40.0 | na | na | 45.2 | 45.6 |
| Aged 25–29 | Na | 9.4 | na | 13.1 | na | na | 17.7 | 17.1 |
| Percentage women divorced or separated, 40–44[5] | 4.8 | 11.8 | 12.2 | 13.7 | 16.0 | 17.5 | 17.2 | 17.0 |
| Net overseas migration (%)[9] | 0.79 | 0.68 | 0.69 | 0.29 | 0.39 | 0.50 | Na | na |

*Notes*: na—not available; 1 Calculated for the prior five year interval with GDP at 2005 prices; 2 Average monthly unemployment rate (%) for the calendar year; 3 Average monthly female labour force participation rate (%) for the calendar year; 4 Average monthly interbank cash-rate (%) for the calendar year; 5 ABS Census data, various years; 6 Weighted Average of Capital City House Price Indexes represented as a ratio to the 1986 index value, ABS House Price Indexes: Eight Capital Cities, Various editions; 7 ABS Census data, various years, Persons aged 15+ with a Bachelor degree or higher; 8 Australian Institute of Family Studies. 9 ABS annual estimates of net overseas migration expressed as a percentage of the midyear population. Data are not shown for 2006 and 2011 due to a break in the series.

attachment to the workforce has reduced the opportunities for migration, particularly among the growing proportion of dual income households. Net overseas migration was not found to be significantly associated with internal migration intensities at the interstate ($r = 0.31$) nor the SLA level ($r = -0.02$). The former is unsurprising as any displacement and housing adjustment arising from internal migration is likely to be relatively local. The lack of association at the local level suggests that international migration is not an important trigger of internal migration in Australia.

Home owners are less mobile than renters so increases in home ownership are likely to reduce mobility (Bell, 2002). In practice, however, home ownership rates in Australia peaked in the early 1990s and have fallen over

*Table 7.9* Australia: bivariate correlates of change in crude migration intensity (Pearson *r*), all years 1981–2011

| Indicator | Interstate migration (Medicare) | Inter-SLA migration (Census interpolated) |
|---|---|---|
| Annual economic growth (%)[1] | 0.28 | 0.12 |
| Unemployment rate (%)[2] | 0.35 | 0.75** |
| Female labour force participation (%)[2] | −0.19 | −0.84** |
| Mortgage rate (%)[2] | 0.16 | 0.65** |
| House price index[3] | −0.84** | −0.99** |
| Annual net overseas migration (% total population)[4] | 0.31 | 0.02 |
| N | 30 | 30 |

*Notes*: **$p < 0.01$; Correlations were run pooling data for all years, 1981 to 2011. 1 Economic growth (%) is calculated for annual intervals with GDP at 2005 prices; 2 Annual variables as defined in Table 7.8; 3 Weighted average of Capital City House Price Indexes, 2003–2004 = 100, ABS House Price Indexes: Eight Capital Cities, Various editions; 4 $n = 25$ due to a break in the time series in 2006.

successive censuses. Home mortgage rates (the cost of loans to home buyers) have also fallen since they peaked at nearly 20% in the late 1980s: *ceteris paribus*, this should facilitate entry to home ownership, and hence migration, so the positive association with falling inter-SLA migration rates (Table 7.9) is counter-intuitive. Reduced constraints from the lower cost of money have likely been offset by the rapid escalation in housing costs. The house-price index shows a rapid rise since the 1980s and there is a strong negative association with housing costs. NATSEM (2011) demonstrate that disposable income kept pace with housing prices between 1991 and 2001, but over the next decade growth in prices outpaced income growth by more than two to one. As a result, housing affordability fell sharply and the house price to after-tax income ratio climbed from five to one in 2001 to more than seven to one in 2008–2009, with further rises only halted by the global financial crisis. Rising costs almost certainly constrained new household formation among young adults, but may well have impeded upward mobility in the housing market, and hence migration, among existing home owners.

Allied to these changing conditions in the economy and housing market are secular shifts in key social indicators that shape migration. As noted earlier, there has been a progressive rightwards shift in the age profile of migration associated with the later transition to adulthood. Table 7.8 confirms the steady rise in median age at first marriage among brides and the rising share of the population with tertiary education, again contributing to a later transition into the labour force. Rising educational participation is particularly significant in Australia because, unlike the UK and USA, a majority of young people remain in the parental home during their tertiary studies (Bornholt *et al.*, 2004; Holdsworth, 2009; Bernard *et al.*, 2016). Together with the rising cost of housing, this underpins the steady rise in the proportion of young adults living with their parents (ABS, 2009). As shown in Table 7.8, by 2011 almost half of 20–24 year olds and one in six 25–29 year olds were still living in the parental home. While this increase in the age at leaving home has exerted a depressing effect on mobility, at least one family transition later in life has exerted the opposite effect. Rates of divorce among women aged 40–44 rose steadily until 2001, signalling a rise in household fission which is generally associated with residential mobility by at least one partner. On the other hand, migration intensities among older Australians have also fallen sharply, as Figure 7.3 makes clear. Lower rates of migration at retirement age are linked to a progressive increase in retirement age (ABS, 2006; Sander, 2011), but at older ages it is housing policy that largely accounts for the fall in mobility. The Commonwealth Government Housing and Community Care programme, introduced in 1984, reduced the need for older people to move into nursing home accommodation by providing a range of support services that enabled them to remain in their own homes, reducing the incidence of dependency-driven migration (Rowland, 1991).

## Conclusion

Australia is firmly established on a long term trajectory of internal migration decline. We have drawn on a lengthy time series of data from successive five yearly censuses to trace migration intensities since the 1970s at a range of spatial scales, harmonised for changes in geographical boundaries. We also analysed data from Medicare registers which track annual mobility between States and between Postcodes, thereby providing a window on mobility since the 1980s at a finer temporal resolution. The results are unequivocal, documenting a consistent and accelerating decline in migration intensities at all spatial scales. The onset of falling mobility varies, dating from the early 1980s in the case of local mobility but delayed until the early 1990s in the case of interstate migration. Between 1991–1996 and 2006–2011, however, it was long-distance mobility that recorded the sharpest decline. Migration intensities between Australia's 69 TSDs fell by more than one-fifth, and between the eight states and territories by 16%, compared with just 7–8% for local moves. Migration over distances of 250 km or more was the most strongly affected, with intensities falling by a quarter. Estimates from the Medicare data trace the same general pattern, but with the decline in intensities not commencing until 2003.

We found that age-structure effects accounted for between one-fifth and a quarter of the overall decline in migration intensities, but there was also a progressive rightwards shift in the age profile of migration, characterised by both a fall in the height and an increase in the age at the peak. Later migration could be traced to increases in the average age at key life course transitions among young adults. It also suggested a potential reduction in lifetime migration expectancy as age at leaving the parental home increased and the scope for future mobility narrowed. Migration intensities also fell at older ages, particularly for those aged 50 and over, constrained both by opportunities for migration upon retirement and by public policies aimed at maintaining individual autonomy in old age.

Conditions in labour and housing markets have placed further downward pressure on migration. Australia weathered the global financial crisis of 2007–2008 better than most nations, but declining mobility has been positively correlated with falling growth in GDP per capita as well as a decline in unemployment (since employed workers are less mobile). Equally important has been a sharp fall in housing affordability since 2001, constraining new household formation and limiting opportunities for mobility within the housing market. Allied to these factors, we also identified a trend towards greater stability in the Australian space economy, with a systematic decline in the degree of redistribution between regions. Thus, declining migration intensities have been accompanied by parallel falls in migration effectiveness. While this suggests a continuing downward trajectory in the pace and impact of migration, population movement remains central to the effective functioning of the economy, the development of human capital and the achievement of individual goals and aspirations, and is integral to Australian culture.

## Disclaimer

The views expressed in this paper are those of the authors and do not necessarily reflect the views of the United Nations.

## References

ABS. 1996. *Statistical Geography Volume, Australian Standard Geographical Classification (ASGC), 1996 Edition, Cat. No. 1216.0.* Canberra: Australian Bureau of Statistics.

ABS. 2004. *Year Book Australia, 2004, Cat. No. 1301.0.* Canberra: Australian Bureau of Statistics.

ABS. 2006. *Retirement and Retirement Intentions, ABS Catalogue 6238.0.* Canberra: Australian Bureau of Statistics.

ABS. 2008. *Marriages, Australia, 2007, Cat. No. 3306.0.55.001.* Canberra: Australian Bureau of Statistics.

ABS. 2009. *Australian Social Trends, June 2009, Cat. No. 4102.0.* Canberra: Australian Bureau of Statistics.

ABS. 2011a. *Australian Standard Geographical Classification (ASGC), 2011, Cat. No. 1216.0.* Canberra: Australian Bureau of Statistics.

ABS. 2011b. *Discussion Paper: Assessment of Methods for Developing Experimental Historical Estimates for Regional Internal Migration, Cat. No. 3405.0.55.001.* Canberra: Australian Bureau of Statistics.

ABS. 2011c. *Population by Age and Sex, Regions of Australia, 2011, Cat. No. 3235.0.* Canberra: Australian Bureau of Statistics.

ABS. 2015a. *Migration Australia, 2013–2014, Cat. No. 3412.0.* Canberra: Australian Bureau of Statistics.

ABS. 2015b. *Marriages and Divorces, Australia, 2014, Cat. No. 3310.0.* Canberra: Australian Bureau of Statistics.

ABS. 2016. *Australian Historical Population Statistics, Cat. no. 3105.0.65.001.* Canberra: Australian Bureau of Statistics.

Argent, N. and Walmsley, J. 2008. Rural youth migration trends in Australia: An overview of recent trends and two inland case studies. *Geographical Research*, 46, 135–244.

Bell, M. 1992. *Internal Migration in Australia, 1981–86.* Canberra: AGPS.

Bell, M. 1995. *Internal Migration in Australia, 1986–91: Overview Report.* Canberra: AGPS.

Bell, M. 2002. Comparing population mobility in Australia and New Zealand. In Carmichael, G.A. and Dharmalingam, A. (eds) Populations of New Zealand and Australia at the Millennium. Joint Special Issue of the *Journal of Population Research* and *New Zealand Population Review*, Canberra and Wellington: Australian Population Association and Population Association of New Zealand, 169–193.

Bell, M., Charles-Edwards, E., Kupiszewska, D., Kupiszewski, M., Stillwell, J. and Zhu, Y. 2015a. Internal migration around the world: Assessing contemporary practice. *Population, Space and Place*, 21(1), 1–17.

Bell, M, Charles-Edwards, E, Ueffing, P., Stillwell, J., Kupiszewski, M. and Kupiszewska, D. 2015b. Internal migration and development: Comparing migration intensities around the world. *Population and Development Review*, 41(1), 33–58.

Bell, M. and Hugo, G.J. 2000. *Internal Migration in Australia 1991–96: Overview and the Overseas-born.* Canberra: Department of Immigration and Multicultural Affairs.

Bell, M. & Maher, C.A. 1995. *Internal Migration in Australia 1986–91: The Labour Force.* Canberra: AGPS.

Bell, M. and Rees, P. 2006. Comparing migration in Britain and Australia: Harmonisation through use of age-time plans. *Environment and Planning A*, 38(5), 959–988.

Bell, M., Rees, P., Blake, M. and Duke-Williams, O. 1999. *An age-period-cohort database of inter-regional migration in Australia and Britain, 1976–96.* Working Paper 99/2. Leeds: School of Geography, University of Leeds.

Bell, M. and Stratton, M. 1998. Understanding the 1996 Census migration data. *Journal of the Australian Population Association*, 15(2), 155–170.

Bernard, A., Bell, M. and Charles-Edwards, E. 2014a. Life-course transitions and the age profile of internal migration. *Population and Development Review*, 40, 213–239.

Bernard, A., Bell, M. and Charles-Edwards, E. 2014b. Improved measures for the cross-national comparison of age profiles of internal migration. *Population Studies*, 68(2), 179–195.

Bernard, A., Bell, M. and Charles-Edwards, E. 2016. Internal migration age patterns and the transition to adulthood: Australia and Great Britain compared. *Journal of Population Research*, 33(2), 123–146.

Blainey, G. 1963. *The Rush that Never Ended: A History of Australian Mining.* Melbourne, VIC: Melbourne University Press.

Blake, M., Bell, M. and Rees, P. 2000. Creating a temporally consistent spatial framework for the analysis of interregional migration in Australia. *International Journal of Population Geography*, 6, 155–174.

Bornholt, L., Gientzotis, J. and Cooney, G. 2004. Understanding choice behaviours: Pathways from school to university with changing aspirations and opportunities. *Social Psychology of Education*, 7(2), 211–228.

Champion, A.G. 1989. *Counterurbanisation.* London: Edward Arnold.

Holdsworth, C. 2009. Going away to uni': Mobility, modernity, and independence of English higher education students. *Environment and Planning, A,* 41(8), 1849–1864.

Hugo, G. and Harris, K. 2011. *Population Distribution Effects of Migration in Australia.* Canberra: Department of Immigration and Citizenship.

Hugo, G. and Smailes, P. 1985. Urban-rural migration in Australia: A process view of the turnaround. *Journal of Rural Studies*, 1(1), 11–30.

Long, J. and Boertlein, C. 1990. Using Migration Methods Having Different Intervals: Current Population Reports, Series P-23. Washington, D.C.: U.S. Government Printing Office.

Long, L. 1991. Residential mobility differences among developed countries. *International Regional Science Review*, 14(2), 133–147.

Maher, C. and McKay, J. 1986. *Final Report: Internal Migration in Australia, 1981 Internal Migration Study.* Canberra: Department of Immigration and Ethnic Affairs.

NATSEM. 2011. *AMP-NATSEM Income and Wealth Report 29.* Canberra: National Centre for Social and Economic Modelling, The University of Canberra. www.natsem.canberra.edu.au/publications/?publication=the-great-australian-dream-just-a-dream

Newton, P. and Bell, M. (eds) 1996. *Population Shift: Mobility and Change in Australia.* Canberra: AGPS.

Rees, P., Bell, M., Blake, M. and Duke-Williams, O. 2000. *Harmonising Databases for the Cross National Study of Internal Migration: Lessons from Australia and Britain. Working Paper 00–05.* Leeds: School of Geography, University of Leeds.

Rees, P., Bell, M., Kupiszewski, M., Kupiszewska, D., Ueffing, P., Bernard, A., Charles-Edwards, E. and Stillwell, J. 2016. The impact of internal migration on population redistribution: An international comparison. *Population, Space and Place*, published online in Wiley Online Library. doi:10.1002/psp.2036

Rogers, A. and Castro, L.J. 1981. *Model Migration Schedules. Research Report RR-81–30.* Laxenburg, Austria: International Institute for Applied Systems Analysis.

Rowland, D. 1979. *Internal Migration in Australia.* Canberra: Australian Bureau of Statistics.

Rowland, D. 1991. *Ageing in Australia.* Melbourne, VIC: Longman Cheshire.

Sander, N. 2011. *Retirement Migration of the Baby Boomers in Australia: Beach, Bush or Busted?* Unpublished PhD Thesis. Brisbane: The University of Queensland.

Shryock, H.S., Siegel, J.S. and Associates 1973. *The Methods and Materials of Demography.* Washington, DC: US Bureau of the Census.

Stillwell, J., Bell, M. Blake, M., Duke-Williams, O. and Rees, P. 2000. A comparison of net migration flows and migration effectiveness in Australia and Britain: Part 1, total migration patterns. *Journal of Population Research*, 17(1), 17–38.

Stillwell, J., Bell, M., Blake, M., Duke-Williams, O. and Rees, P. 2001. A comparison of net migration flows and migration effectiveness in Australia and Britain: Part 2, age-related migration patterns. *Journal of Population Research*, 18(1), 19–39.

Taylor, J. 2010. Population futures in the Australian desert, 2001–2016. *Australian Geographer*, 34(3), 355–370.

Taylor, J. and Bell, M. 1996. Population mobility and indigenous peoples: The view from Australia. *International Journal of Population Geography*, 2, 153–169.

Taylor, J. and Bell, M. (eds) 2004. *Population Mobility and Indigenous Peoples in Australasia and North America*. London: Routledge.

Wrigley, N., Holt, T., Steel, D. and Tranmer, M. 1996. Analysing, modelling, and resolving the ecological fallacy. In Longley, P. and Batty, M. (eds) *Spatial Analysis, Modelling in a GIS Environment*. Cambridge: GeoInformation International, 23–40.

# 8 Japan

## Internal migration trends and processes since the 1950s

*Tony Fielding*

Japan's demography, economy and society have been truly transformed since the late 1950s. What was then a youthful, high population growth country has become the aged, population-decline country that we see to-day. Its booming economy and export success turned into the 'lost decades' of the period since 1990, accompanied by a debt crisis for the state. The confident society of secure family life backed up by lifetime employment has morphed into an apprehensive society which is fearful of natural disasters, full of uncertainty about the future, held back by mistrust of government and big corporations (for example, over the Fukushima disaster) and by a general pessimism which seems somehow to extinguish political activism and corrode commitment to personal relationships, marriage and family formation.

How might these changes have been expected to impact on the volume and characteristics of internal migration? Other things being equal, one might expect the ageing of the population to result in a lower intensity of migration. This is because young adults are almost universally more geographically mobile than elderly people. So, as the top of the age pyramid becomes wider due to greater longevity, and the base of the pyramid becomes narrower due to lower fertility, a higher proportion of the population would be located in the older low-mobility age groups and a smaller proportion in the younger high-mobility ones. Under two conditions (i) a steady downward curve in in-ternal migration rates matching the upward curve in median age of the pop-ulation and (ii) no change in age-specific migration rates over the period, one might be tempted to assume that it is solely, or very predominantly, the changing age composition of the population that is responsible for changes in the volumes and rates of internal migration.

The problem is that neither of these conditions is met. Figure 8.1 shows the volume of inter- and intra-prefectural migration in Japan from 1959 to 2014 and the rates of inter-prefectural and intra-prefectural migrations over the same period (intra-prefectural migrants are those who in changing ad-dress crossed a municipal boundary but not a prefectural one. Information on the total number of people changing addresses is not available). The data used are based upon the fairly reliable household registration data set,

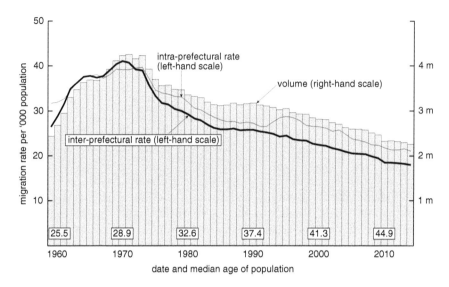

*Figure 8.1* Japan: annual inter-prefectural migration, 1959–2014.
Source: Author's calculation from annual household registration data.

but clearly do not include the internal displacement consequent upon the Tohoku earthquake and tsunami in March 2011—such displacement would be regarded as temporary migration, not a change in permanent residence (for the migration effects of disasters in Japan, see Mizutani, 1989). The Figure also shows the median age of the Japanese population.

The first thing to notice is that the dominant trend is for the annual volumes and rates to decline over time. There are, however, two major qualifications that need to be made to this generalisation. One is that during the 1960s, despite the increase in the median age of the Japanese population from 25.5 to 28.9, there was a rapid rise in inter- and intra-prefectural migration. It is true, of course, that the 'baby-boom' of 1945–1950 had produced an increase in the number of young people coming onto the labour market in the 1960s (Itoh, 1984; for an analysis of cohort effects on migration, see Inoue, 2002; Kawabe, 1985; Oe, 1995). But, as we shall see below, the main cause of this rapid rise in migration was rural depopulation—the massive transfer of mostly young adults into the three main metropolitan regions (Tokyo, Osaka and Nagoya). The second qualification is that the decline after 1970 was not regular. Instead, the decrease in both volume and rates was very rapid in the early mid-1970s, moderately rapid from the mid-1970s to the mid-1980s and then hardly rapid at all until 2014.

So, in a nutshell, much more seems to be going on than the migration effects of population ageing. Indeed, we can make a rough calculation of the effects of the changing age structure of the population on migration levels by first multiplying the proportions of the population in each group at two dates by the age-specific migration rates applying at that time. The difference in the sum of these figures gives us a measure of the total decline in migration. Then we calculate the volume of migration that would result if the age-specific migration rates of the first date were applied to the population age structure at the second date, and if the age-specific migration rates of the second date were applied to the population age structure at the first date. The result (that is, the mean of the two values) is very clear—only about 45% of the decline in migration between 1959–1960 and 2005–2010 (see Figure 8.2) is due to the change in age composition.

Figure 8.2 shows something that is crucial to the argument (that the decline in migration rates is simply a product of the ageing of the population). This diagram has been constructed in the following way. First, the one-year inter-prefectural migration rates from the 1960 Census have been multiplied by 3.16 to scale them up to five-year rates (US Bureau of the Census, 1990), thereby making them comparable with the rates for 2005–2010 from the 2010 Census (the quality of data from the Censuses is universally regarded as being very high). Please note that had the multiplier been 3.00 or 3.50 it would have made no difference to the shapes of the curves, and only a slight difference to their positions relative to one another. Please note also that, as in Figure 8.1, the 1960 data do not include migration flows to and from Okinawa (only after 1972 were they included). Finally, the 1960 data exclude those aged less than 1 year; so, to ensure comparability, the aged 0–15 migration rate had to be raised by 7%.

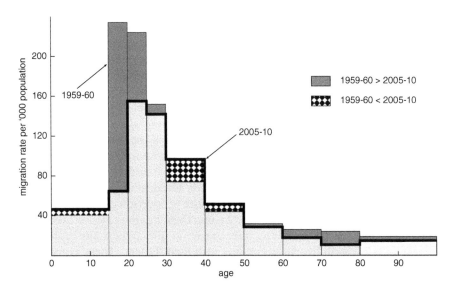

*Figure 8.2* Japan: age-specific inter-prefectural migration rates in 1959–1960 and 2005–2010.

Source: Author's calculation from 1960 and 2010 Censuses.

For the age composition explanation to work, it is necessary for the age-specific migration rates to remain constant, or at least approximately so, over time. This is because only then would an increase in the proportion of low-mobility elderly and a decline in the high-mobility young adult population be directly, and without mediation, translated into a decline in migration rates. Figure 8.2 shows that this is manifestly not the case. It compares the age-specific migration rates for the late 2000s with those for the late 1950s. The picture is as dramatic as it is perhaps surprising. First, the overall inter-prefectural migration rate, which stood at 89 per thousand population in 1959–1960, dropped sharply to just 54 per thousand in 2005–2010. Secondly, while there is little change in the rates of those aged under 15, those aged 15–19 in 1959–1960 were more than three times as mobile as those in the same age group in 2005–2010. The difference is smaller for those aged 20–24, but the 1959–1960 figure is still about one third larger than that for 2005–2010. A reasonable interpretation of this is that many of the major migration generating processes (such as leaving the parental home, entry into full-time permanent employment, becoming married, buying a house, or moving house to accommodate a growing family) were being delayed in the recent period compared with 1959–1960. This view is supported by the rates for older age groups; whereas the figures for those aged 25–29 are not very different, those for men and women aged 30–39 and 40–49 show that migration was rather more frequent for these groups in the recent period than was the case in 1959–1960. For those aged 50 and over, the rates for the earlier period once again become higher than those for the recent period. One cannot escape the general conclusion that, far from being stable over time, the whole curve has become flattened and extended to the right in the recent period. What this clearly does not mean is an overall reversal of the tendency for those who are in the younger adult age groups to migrate more than those who are in the older ones. What it does mean, however, is that, at the individual level, in line with many other demographic variables, migration has not just become less, it has become later.

Were these the only issues, it might still be possible to hold on to a demographically determinist view of the trends in migration rates depicted in Figure 8.1. But what about the migration effects of the other economic and social changes listed above? The problem is that almost all of these changes reinforce our expectations that migration rates should have *increased* over the last 50–60 years, not *decreased*. This is because in Japan marriage is almost universally desired and only a minute proportion of children (about 3%) are born to unmarried women. As marriage has become less and later, and as childlessness and having only one child in marriage has grown, the constraints on mobility have become significantly less. These constraints relate to the many forms of embeddedness of married couples in their communities (such as strongly felt obligations to their families and to their neighbours), but above all to their extreme reluctance to do anything, such as migrate to another area, that would disrupt their children's education.

In a 'one-chance society', parents (but especially mothers—the *kyouiku mamas*) do everything they can, notably by avoiding instability, to ensure exam success and a problem-free transition for their offspring from school to university and/or into work.

Changes in the world of work should have had a similar effect on migration. As the security and stability of full (indeed, very full) employment and of life-time employment have become partially replaced by 'flexible' labour markets with high levels of short-term contracts and part-time work, so the constraints on migration arising from very low levels of inter-organisational and inter-occupational mobility, and from early entry into permanent full-time employment, have been lifted. Flexibility for the employer, of course, implies insecurity for the employee, and insecurity of employment generally means greater mobility and more migration. 'Freeters' (young adults in temporary and part-time employment) cannot be expected to buy houses— hence many are 'parasite singles' still living in their parental homes—but they *can* migrate.

These arguments favouring higher migration rates are backed up by broader social changes. In 1960, Japan had a very large blue collar (manual worker) labour force. These workers had company-specific skills and were locationally tied to the factories and workshops in which they worked. Very rarely did they migrate, and then usually only if the company was restructuring and thereby relocating its operations. Thus, manual worker immobility was at least as much a feature of Japanese society as it was of other high-income industrial capitalist economies. With the 'hollowing out' of the Japanese economy as major Japan-based multinational companies shifted their investments to countries which are major consumers of their goods, or to low-wage and low-regulation developing countries, employment in manufacturing (despite some return of manufacturing investment to Japan) has come to form a much smaller proportion (less than 20%) of total employment. It has been replaced by financial and business services and, more generally, by an increase in traditionally high-mobility white-collar employment, especially in professional and managerial occupations—hence the expectation that, overall, migration rates would increase (as described in Chapter 2). Other changes in employment point in the same direction; the many small farmers and family owned service sector businesses, locked into their locations by historic land ownership and a highly localised customer base, have similarly lost ground to the fewer, larger agri-businesses and service sector chain stores (such as supermarkets and shopping malls). In this way, jobs that tied people to localised labour markets in specific places have decreased while those involving generic skills linking people to delocalised labour markets set in spatially extended regions have increased.

To summarise, the problem is this: we cannot fully explain the trends in inter-prefectural (or for that matter intra-prefectural) migration rates shown in Figure 8.1 as a product of population ageing, itself an outcome of the combination of higher longevity and lower fertility. But, in addition, the national socio-economic changes since the late 1950s seem weighted on the side of causing greater volumes of internal migration rather than smaller ones. So what then is going on?

## Internal migration and regional economic development in Japan

To answer this question it is necessary to delve into two groups of relationships and processes that are contextual to migration: the first is politico-economic and the second is socio-demographic.

Japan's space-economy can be conceived as a system undergoing changes that are occurring at different rates: (i) those that can be characterised as 'conjuncture' (see Fielding, 2012) are rapid—they are the ups and downs of the business cycle, (ii) those that involve the restructuring of urban and regional economies—such as the decline of agricultural areas and of 'old industrial regions'—are much slower and (iii) those that relate to the underlying geography of opportunity—such as the attractiveness of the capital city region compared to the rest of the country—change almost imperceptibly slowly. Each of these bundles of processes has migration effects.

The business cycle tends to produce high rates of internal migration during the boom phase and low rates during recessions. This can be seen very clearly in Figure 8.3 where the deep recessions of the early mid-1970s (first oil shock), the early mid-1990s (burst of the 'bubble economy') and of the post-2008 global financial crisis coincide with very low levels of net migration to the major cities (particularly Tokyo). It is crucially important to note, however, that these recessions are not sufficient in their migration effects to dominate the overall schedule of migration rates shown in Figure 8.1 (for example, notice the relatively minor effect of the economic crisis after 2008). It is clear that the migration changes associated with the business cycle represent only a small part of the complex reality that we are trying to understand. They tend to affect the direction of flow rather than the overall level of mobility (Higuchi, 2008).

The much slower changes arising from the second bundle of processes— the restructuring of urban and regional economies—are central to an explanation of the trends in Japan's internal migration flows (for a useful overview, see Matsuhashi *et al.*, 2013). In the early period, from the early 1950s until about 1970, the dominant flow was from rural and peripheral regions to the main metropolitan cities, often displaying semi-separate migration fields (Saito and Higashi, 1978) and taking the form of chain migration (Miyazaki, 1998; Tsutsumi, 1987). In Figure 8.3 we can see that all three of the main metropolitan regions were major gainers by migration with the largest gains going to the largest cities, and especially to Tokyo (defined as the Tokyo Metropolitan Region so as to include suburban areas and nearby cities such as Yokohama). This rural-urban migration can best be explained as resulting from the spatial division of labour that is known

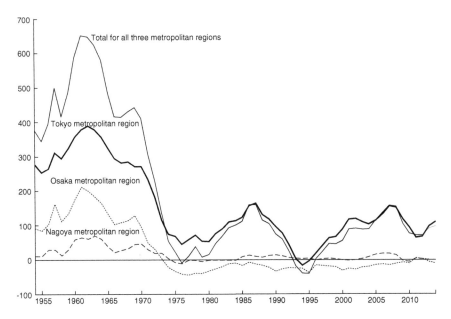

*Figure 8.3* Japan: net internal migration, 1954–2014, for the three main metropoli-
    tan regions.
Source: Author's calculation from annual household registration data; data in thousands.
Data for 1954–1972 do not include migrants to or from Okinawa. Composition of metropol-
itan regions by prefecture: Tokyo—Tokyo, Kanagawa, Saitama and Chiba; Osaka—Osaka,
Hyogo, Kyoto and Nara; Nagoya—Aichi, Gifu and Mie.

as 'regional sectoral specialisation'. Employment is divided up amongst
the regions and cities of a country in such a way as to reflect the special
properties of those places. Places that have reserves of useful resources
(for example, coal in the early post-war period) will have those branches
of employment that depend on that resource (such as iron and steel, heavy
engineering and shipbuilding). Places that have fertile soils located on
workable land (given that much of Japan is mountainous) will have rice
and vegetable/fruit crop agriculture. Coastal locations with small bays and
inlets will tend to have small-scale fisheries industries. Places that have unusual centrality and/or a physical geography favourable to major harbour
construction will tend to become centres of commerce and trade. Places
that have inherited particular work traditions (such as those associated
with the spinning of yarn and the making of cloth) will tend to become the
centres of modern textile manufacture. Such patterns of regional sectoral
specialisation are maintained by the exchange of goods and services between the regions such that each survives by selling its products to the others, thereby earning the means to buy the products it wants and needs from
the others. In this way, regional sectoral specialisation reflects one-to-one

the social division of labour in society—a division of labour brought about by market exchange.

This point is important for explaining the internal migration flows of this early post-war period because, as economies and societies develop, their demands for goods and services change. In those regions in which the goods and services produced were in low demand, unemployment and out-migration would result; in regions of high demand, wages would rise and in-migration (some of it seasonal; see Oshiro, 1976) would result (Abe, 1997; Gauthier *et al.*, 1992; Ito, 1998; Kondo and Okubo, 2015; Tabuchi, 1986; Watanabe, 1989). More specifically, at this time incomes were rising fast and, as they did so, those products that had a low income elasticity of demand (that is, those on which very little more was spent as incomes went up), such as food products, lost out to those products which were 'must buys' for the modern household such as mass-produced consumer electrical and electronic goods. The former were produced in rural and peripheral regions. The latter were produced in or close to major cities—this was both because this is where the large workforces needed for large-scale factory production were located, and because it was these large cities and metropolitan regions, enjoying rapid service sector employment growth, that had the vast numbers of well-paid consumers which made them profitable markets for these products. The result was the mass migration of young adults (predominantly men) from rural and remote areas to the main metropolitan city-regions (Inoue and Liaw, 2004; Kakiuchi and Hasegawa, 1979; Kawabe, 1980; Liaw, 2003; Ogawa, 1986; Sakuno, 1994; Uchino, 1990).

Regional sectoral specialisation did not completely disappear after the early post-war period (indeed traditional local products remain one of the mainstays of the domestic tourist industry). But as the post-war years unfolded, regional sectoral specialisation came to be overlain and then progressively replaced by a 'new spatial division of labour'. This new geography of production was characterised by a spatial separation of the roles that different kinds of labour played in the overall operations of a large national corporation (Fukurai, 1991). Head offices would gravitate to Tokyo, research and development establishments would locate in those places where the highly educated staffs on which they depended could be recruited and retained (often environmentally attractive places within the capital city-region, see Nakazawa, 2001) and routine production would be sent to the provinces—old industrial regions if specialist manual skills were called for, otherwise to rural, small town and peripheral regions to take advantage of the remaining, now scarce, reserves of unskilled and semi-skilled labour (Kamo, 1999). The problem for major Japanese employers was that the early post-war rate of capital accumulation and investment in production in the major cities had been so rapid and so prolonged that both the indigenous urban and the rural in-migrant labour reserves had been largely used up. To find cheap and manageable labour (and low-cost sites large enough to house modern forms of production), firms began to locate away from the metropolitan cities and deep into provincial Japan (Kajita, 1998; Koganezawa, 1987; Matsuda, 1986; Okahashi, 1996).

The migration effects of the changeover to this new spatial division of labour were dramatic (Arai *et al.*, 2002; Ishikawa, 1978, 2001; Otomo, 1990). Coinciding with a shift in national policy favouring non-metropolitan regional development (Mera, 1988), the 'era of the localities' witnessed three major transformations in Japanese inter-prefectural migration rates. The first of these was the sudden decline in overall spatial mobility shown in Figure 8.1 (see also Yamaguchi, 2002). The second was the so-called 'U-turn' in the pattern of flows whereby those places that had been major attractions for internal migration (notably the three main metropolitan regions) became locations of net migration loss or at least of no net migration gain, whereas those places that had previously been suffering depopulation now experienced either net migration gain or at least very much lower rates of loss (shown in Figure 8.3; see also Imai, 1989; Ito *et al.*, 1979; Liaw, 1992; Murayama, 1990; Traphagan, 2000; Tsuya and Kuroda, 1989; Wang, 1991; Wiltshire, 1980). Much of the centripetal flow of migrants was fairly local— this was, after all, the high point for suburbanisation and peri-urbanisation (Ishiguro, 1976; Nakagawa, 1990; Otomo, 1996; Ozeki, 1997; Tachi, 1971; Yamagami, 2003; Yano, 1989). In other cases, the longer-distance return migration that formed much of the counter-urban flow to rural and peripheral regions was not so much to the villages and small towns from which the migrants had originally migrated, but to the prefectural or regional capital cities—this is sometimes called 'J-turn' migration (Yahata, 1997).

The third major transformation in inter-prefectural migration rates was in their social composition (Nakagawa, 1996, 2006). The larger part of early post-war migration consisted of very young adults and late adolescents from farm and workshop family backgrounds migrating to the proletarianised manufacturing and (often manual) service sector jobs now being offered in abundance in the major metropolitan cities. Now that the 'work was going to the workers' rather than the other way round, manual working class migration fell sharply (Tanaka and Nakano, 1998; Yamaguchi *et al.*, 2000), to be replaced by, or throwing into relief, forms of middle class migration associated with the new spatial division of labour (Inoki and Suruga, 1981; Nishihara and Saito, 2002). In particular, to coordinate the operations in many parts of the country, large firms and other large organisations, through intra-organisational transfers, posted their employees to new locations such as regional cities to manage affairs or provide technical or professional support for their operations at those locations (Ishikawa, 1994; Uchino, 1982; Wiltshire, 1995). Most of these relocations were of families (Arai and Ohki, 1999). However, since many of these intra-organisational transfers resulted in a threat of disruption to family life and obligations (for example, towards a child's education, mentioned above), it became quite normal for the male breadwinner to move away to the new posting on his own (*tanshin funin*) leaving behind his wife and family (and maybe elderly parents as well) (Iwao *et al.*, 1991; Tanaka, 1991) thereby enhancing the male sex bias of migration streams (Sakai, 1991).

Yet, no sooner had the new spatial division of labour become fairly established in Japan (by about the early 1980s) than another spatial restructuring of production began. The 'spatial fix' represented by the decentralisation of industrial investment into provincial Japan, even sometimes to its more rural and peripheral regions, turned out to be relatively short-lived (i.e. about 20 years or so) for two main reasons: firstly, because the reserves of young adult and older female labour locked up in these peripheral regions were limited (Ishikawa, 1992) and secondly because new sources of plentiful cheap labour became available in other countries of East and Southeast Asia. Thus a new international division of labour came to replace the intranational 'new spatial division of labour'. Large manufacturing companies such as Toyota became 'global' not just in the sense that they sold their products made in Japan to consumers located all over the world, which they had done for some time, but also in that they now became global producers. Toyota came to produce more than half of its cars outside Japan. In doing so, it forged links between its operations in Japan, concentrated in the region around Nagoya and its operations in many countries lying very distant from Japan. This process, repeated thousands and thousands of times, resulted in a spatial division of labour which we shall call 'regional functional disconnection' (Dunford and Fielding, 1997) because, despite the partial continuation of, and the legacies from, earlier divisions of labour, the places within the country were becoming increasingly disconnected with one another as they became increasingly connected with places located outside the country. Nor was it confined to manufacturing industry; the financial and business service industries of the Tokyo Metropolitan Region (Ishikawa and Fielding, 1998) became more and more integrated into the global financial and business service system located in New York and London (and also in Shanghai, Hong Kong and Singapore) at the same time as the domestic market for these services expanded more slowly and then, after the bursting of the bubble economy around 1990, went, for a time, into reverse.

The migration effects of regional functional disconnection have proved very significant for Japan. Largely gone are the massive rural-to-urban and inter-city-region flows of the 1950s and 1960s. Largely gone too are the U-turn, J-turn and suburban and peri-urban migrations of the era of the localities of the 1970s and 1980s (Hama, 1995) when, apart from relocations associated with infrastructure and new factory investments, manual worker migration for many people became almost unnecessary, and white-collar migration (some of it intra-organisational transfers) came to the forefront. From the 1990s, low internal migration rates have become the norm. Rural depopulation continues, especially from the remotest and most peripheral regions—except perhaps where tourism success reverses the trend (Elis, 2011). Provincial cities remain attractive for short-distance migrations (Isoda, 1993). The Tokyo Metropolitan Region as the 'one-point concentration' of the Japanese space-economy continues to attract migrants from all over Japan, but especially from its northern regions such as Tohoku (Isoda, 1995). Suburbanisation and migration to peri-urban areas continues, though now at a very low level (Kawaguchi, 2002). Despite all this, the principal feature of the Japanese urban and regional demographic system has become its remarkable geographical stability—its immobility (Inaba and Mita, 1995; Yano, 2007).

Yet, as Figure 8.1 has shown, there is still a lot of migration—about two and a quarter million people moved between prefectures in 2014. This encourages us to turn to the third bundle of processes—those that change only very slowly over time. In addition to the socio-demographic factors listed below, what is it about the political economy of Japan that maintains from one decade to the next a groundswell of internal migration? A significant part of the answer is the formation of middle-class careers, and upward social mobility more generally, in the context of a massive concentration of wealth and power in the Tokyo Metropolitan Region (Gedik, 1995; Hirai, 2006; Kawada, 1993; Lutzeler, 2008). Tokyo is the location of the 'iron triangle' of power. In the first place, it is the seat of the influential bureaucracy—Japan is often characterised as a bureaucratically authoritarian country. Its government ministries in Tokyo are seen as semi-independent 'empires' of power (*METI, Gaimushou, Mombushou* etc). Secondly, so powerful is the bureaucracy in Japan that it dominates realms of policy formulation that in other countries would be the reserve of the politicians—the second axis of the iron triangle. Tokyo is the seat of government in a non-federal democracy that has been in the hands of the pro-capitalist centre-right Liberal Democratic Party (*Jimintou*) (and its offshoots) ever since 1955. Japanese politicians may initiate less than their peers elsewhere, but they still wield very considerable power in what is a highly clientelist system of power relations.

The third axis of power is arguably the most powerful axis of all: Japanese corporate business. The links between bureaucrats and politicians and Japanese business are extremely (some would say dangerously) close; specifically, Tokyo is the headquarters location for Japan-based manufacturing companies, for a very large part of the domestic and foreign-owned financial sectors and for the Japanese media and cultural industries. Examples of the proximity of these corporations to the bureaucrats and the politicians include the enormous benefit gained by the Japanese construction industry from the use of public funds (such as Post Office savings) for infrastructure investments both in the capital region and in the provinces, and the (unhealthily) close relationship between the government and the media as reflected in the reporter's club (*kisha kurabu*) system. This concentration of wealth and power in the Tokyo Metropolitan Region has changed almost imperceptibly slowly in the last 60 years, and then in the direction of greater concentration rather than less.

What this means for migration is that if you are an able, well-educated and ambitious young adult, you risk losing a great deal of career advancement if you choose to stay in a provincial city or region. Migrate to Tokyo, however, and you insert yourself in an environment that is teeming with all kinds of job opportunities, is criss-crossed with influential social networks and is the location where social, cultural, celebrity and entrepreneurial success, though by no means assured, is far more likely than in other parts of the country (Ishiguro *et al.*, 2012; Matsubara, 2014; Shimizu, 2014). Relative to provincial Japan (and especially so for women), Tokyo is the place for upward socio-occupational mobility—a mechanism for professional and managerial middle (and upper) class formation (Nakagawa, 2000; Noro, 1988; White, 1982). Tokyo is also the place you migrate from when you want to cash in your assets, or when your passion for social and economic advancement is spent. Not surprisingly, flows to and from the Tokyo Metropolitan Region dominate the Japanese migration system (Esaki, 2002; Sakai, 1993; Wakabayashi, 1987), and they do this year-in year-out seemingly unaffected (or only marginally affected) by business cycles or urban and regional restructuring.

It is relevant in this regard to look at the preliminary results from the 2015 Census. They show that only eight prefectures out of 47 experienced population growth in the 2010–2015 period. Four of these were in the Tokyo Metropolitan Region (Tokyo, Saitama, Kanagawa and Chiba) and of these Tokyo's population growth was the highest at 2.7% (ranking 2nd out of 47 prefectures—after Okinawa).

## Internal migration and socio-demographic change in Japan

While few would question the centrality of economic forces and motives in internal migration, there is increasing recognition of the importance of other forces shaping the volume and direction of migration flows. These forces are not, however, always easy to identify. There are two reasons for this: the first is the sheer diversity of motives and circumstances that provoke migration—almost anything that matters to a person could, in a particular situation, trigger the decision to move. Indeed, so complex is the empirical detail of migration flows, especially at the intra-city-region scale, that researchers are tempted to speak of 'churn' or turnover migration (migration that is seemingly random in timing and in length and direction). The second reason is that it is not always as useful to ask people why they migrate as one would hope. Not only are people often partially unaware of the reasons for their actions, but there is also a strong tendency to answer such a question in a manner that reflects well on the respondent, rather than in a way that would attract criticism (for example, for holding racist views) or would bring shame upon him/her or his/her family. Avoiding the expression of confrontational or nonconformist views, and even more so, avoiding saying anything that could damage a family's reputation, matter enormously in Japan. This is important because, to take a non-hypothetical example, the decision to move house to prevent your school-aged child attending a school that had many *burakumin* children in it, would never be discussed with academic strangers, or even for that matter with work colleagues or neighbours (see Tsumaki, 2011–2012). Discussion of the *buraku* issue (the position of the descendants of social outcasts, whose situation in relation to mainstream Japan is not very different from that of the *roma* to wider society in Europe) is essentially forbidden in polite society—it is a taboo subject (Amos, 2011; De Vos and Wagatsuma, 1966).

With these thoughts and qualifications in mind, we can inspect what the Seventh National Survey on Migration conducted in 2011 tells us about contemporary internal migration in Japan (Hayashi *et al.*, 2013):

• The migration rate was lower than in the Sixth Survey in 2006—those who had an address that was different from that five years ago made up 24.7% of the population compared to 28.1% previously, confirming the results from Census and household registration data.
• The shift of peak mobility towards older adult age groups continued and is now clearly located in the 30–34 age group.
• The decline in migration was concentrated at the short distance end (i.e. at the intra-urban, suburban and peri-urban scales); in fact, inter-prefectural migration increased slightly as a proportion of all migration.
• 25.7% of those surveyed lived in the Tokyo Metropolitan Region in 2011, but this proportion was higher for those born abroad (35.4%) and those born in Tohoku (30.4%) and much lower (5.9%) for those born in the Kansai region (Osaka, Hyogo and Kyoto).
• Among the reasons for migration, those relating to housing (35.0%) and family reasons (30.8%) (Shimizu, 1984) dominated the list (but mainly refer to the short-distance component of address-changing), work-related reasons accounted for 14.1% (20.1% for males) and education 5.4% (6.0% for females). Work reasons equalled or exceeded housing reasons for men in their 40s and 50s, education did the same for young people aged 15–24 and the dominant reason for the migration of the very elderly was to move in with or live closer to their children.
• Of all inter-prefectural migrants, a significant (and slightly rising) pro-portion (13.3%) were migrants returning to their prefecture of birth, and an increasing share of these were young elderly (aged 55–74).
• There was a sizeable increase in the proportion of people who were certain that they would not migrate in the next five years (up to 63.7% in 2011, up from 44.7% in 2006). This result implies that it is likely that low mobility has continued well beyond 2011.

Drawing upon these results, but also very aware of their shortcomings, we can now explore what *can* be said about the migration-enhancing and migration-inhibiting socio-demographic changes that have occurred in Japan over the last two generations. This discussion of social and cultural relationships and processes, values and identities, along with the place preferences that they imply, will be organised within a life course framework (Fielding and Ishikawa, 2003; Kawaguchi, 2002; Murayama and Hashimoto, 1994; Otomo *et al.*, 1991).

Children are often falsely regarded as having little or no agency in migration behaviour. In Japan, family migration decisions sometimes reflect the wishes of children. More frequently, however, parents make decisions about where to live on the basis of judgments made by them of what is in the best interests of their children. A map of the migration flows of children shows net gains in suburban areas and losses in remote and peripheral prefectures and in major cities—this pattern is almost identical to the pattern of those aged 35–39 when marriage rates are high (Kawabe and Liaw, 1994). The numbers of children migrating within Japan have declined with the fall in fertility; the total fertility rate bottomed out at 1.26 in 2005 and is now at about 1.42—i.e. still well below the replacement level of 2.1.

People aged 15–19 were shown in Figure 8.2 to have been very highly mobile in the past but much less so today. Today, about half of those leaving high school go on to higher education, many of them to a university or college located in another prefecture. In the past, this student migration was predominantly male, but today female students slightly outnumber their male peers. This transition to university is crucial for successful entry into middle class occupations; the reputation of the university being especially important in the selection of new entrants into big business and public sector jobs. The big change has been the migration behaviours of those going straight into employment after high school graduation. In the past, very many of these young people (especially young men other than first sons), often assisted by family members who had migrated before them, would migrate to distant major cities seeking employment. Today, it is very different. The key to their successful entry into local non-credentialised employment is their local social capital in the form of members of a network of family and friends who can supply relevant information and vouch for their good character, competence and commitment (Kobayashi *et al.*, 2015).

It is also the case, however, that since the early post-war period there has been a revolution in the attitudes of late adolescents and young adults towards contemporary Japanese society. At its most extreme, alienation from this society results in an equally extreme form of immobility—a refusal (mostly affecting young men) to go out, socialise, engage with others or plan for the future—this is the well-publicised *hikikomori* (acute social withdrawal) problem. It is matched by a growing number of older men who are 'solitary non-employed persons' (Genda, 2013). Forms of social withdrawal like this, considered by some to be linked to strong familism, changes in sexual behaviour and a weak 'couple culture', have significant demographic effects (Sato and Iwasawa, 2013). Many young people, for example, share a very negative attitude towards adult society and its burdensome responsibilities. In a survey conducted in the early 1990s (when times were much better than they are now) Kinsella (1995, pp. 241–242) asked her post-teenager young adult respondents 'When you think of adulthood what images come to mind?' She writes 'the great majority of respondents described adulthood as a bleak period of life: 'the harshness of having to make do every day and make a living, the harshness of supporting a family', 'controlled society, responsibility and effort', 'dreams disappear as the necessity to conform comes nearer' and 'strictness day in day out, you can't stop working—life is lonely'. She continues 'the negativity of the answers given is startling. Their assessments of Japanese society were very dark, and their impressions of adult life in that society equally depressing.' And then she adds (p. 244) 'For their part young women—even more than young men—desire to remain free, unmarried and young'.

These strong feelings help us to understand the delays in, and avoidance of, leaving home, entering permanent employment, buying a house, forming partnerships, getting married and starting a family in Japan—these delays and avoidances are either likely to imply, or inevitably imply, lower levels of migration. For those aged 15–19 who do migrate, mostly students, the patterns of their migration are very clear (Figure 8.4a). There are major net gains in university-city prefectures, especially those containing the seven elite national universities: Tokyo, Kyoto, Hokkaido, Tohoku (Miyagi), Nagoya (Aichi), Osaka and Kyushu (Fukuoka) (Liaw and Otomo, 1991).

People in their 20s face a much bleaker world than their fathers and grandfathers (Gottfried, 2014). The lucky and highly successful ones, as before, are recruited by large companies and public sector organisations and they enter what are still effectively life-time employment jobs. At the other extreme, not wanting to engage in the 'rat-race', a small number drop out completely (Miyauchi, 1998). For many in between, reduced job opportunities for graduates and non-graduates results in them becoming 'freeters' (this is short for 'free' i.e. not employed, and 'arbeiters' which is a term that was previously used just for the temporary and part-time jobs that students tended to do). Facing insecurity of employment and the almost impossible hurdle of finding a deposit on a house (25% of the price) results in fewer young adults being able to become home-owners, hence fewer young adults able to become independent of their parents and fewer able to envisage marriage and family formation (Fukuda, 2010). Inter-generational co-residence is very common in Japan (until recently, about two-thirds of the elderly lived with their offspring. See Budak *et al.*, 1996); the tradition used to be, however, that sons other than the eldest son left home early and daughters a little later when they married—say in their early mid-20s (Inagaki, 2003). Instead, 13 million young adults have now become 'parasite singles', living in the parental home into their late 20s, often into their 30s and sometimes into their 40s (Tran, 2006; for the elderly as a social safety-net, see Ogawa, 2008). This extension of dependence on parental financial, housing and emotional support is not, of course, conducive to migration.

The map patterns of net migration for those in their 20s (only that for 20–24 is reproduced here as Figure 8.4b) are very distinctive. The metropolitan regions gain by migration and the rural areas lose. But the differences for those in their early 20s from those in their late 20s are also very significant. In Figure 8.4b it can be seen that there is some continuity of pattern from that in Figure 8.4a; i.e. most major cities with large state universities are still gaining by internal migration and the rural and peripheral regions are still losing (though the rates of loss are even higher than for the 15–19 year olds for the main depopulation prefectures). What is new is the intensity of gain in the three metropolitan regions, most of which is economic migration but some marriage migration is also involved (see Nakagawa, 2001; Suzuki, 1990; Tani, 1997), and the lower gains of the regional capitals, notably Sendai and Fukuoka. For those aged 25–29, the losses in rural areas are smaller and several of the university-city prefectures are now experiencing net migration loss (notably Miyagi, Kyoto and Fukuoka) as students return to their home prefecture (Esaki *et al.*, 2000) or move on to the capital city-region after their studies. Some suburban prefectures (Saitama, Chiba) are showing signs of net gain. So, while work-related migration patterns dominate, it is clear that both student and housing/family migration effects are present in these distributions.

For people in their 30s and 40s the major change since the 1960s–1980s has been the decline in suburbanisation (for a detailed study of the suburbanisation of the three main metropolitan areas in the period 1965–1975, see Ishimizu and Ishihara, 1980). These are relatively migratory age groups, but whereas in the recent past their migrations were to suburban and peri-urban locations resulting in a significant population redistribution, now levels of net migration gain and loss are low: as Figure 8.4c shows, 'churn' has largely replaced centrifugal flow. This is to some extent due to re-urbanisation (especially in Tokyo, see Shimizu, 2004), but it also, of course, lies in the socio-demographic changes outlined above. The combination of changes in values and behaviours around sex, partnership, marriage commitment and family formation have combined with reduced opportunities in the labour and housing markets to produce a preference for city-centre (student-style) lifestyles with their local job and housing mobilities, while holding back the career-building, house upgrading mobilities that would be likely to result in one-way suburban and inter-prefectural migration.

This is not, however, the whole story. For the married majority, one of the key changes to family life has been the increase in the likelihood that the wife will also work (Nishioka *et al.*, 2012). There remain many barriers (social rather than legal) to the treatment of women as equals in the labour market, so it is still unusual for a wife to earn the same or more than her husband. But both middle-class and working-class households are now typically dual earner or, if not as often, dual career households. What this means is that whereas in the high growth period it was only one member of the household who had to change job when moving to a new location, now it is two. This dual-income feature assists entry into house or apartment ownership but puts a brake on migration while also privileging the largest cities, but especially the Tokyo metropolitan region, because such places have a wealth of employment opportunities for both men and women (Izuhara and Forrest, 2011).

Alongside the increase in female employment, there have been changes in information and communication technologies that have made it possible to substitute commuting for migration and working from home for both commuting and migration. Undoubtedly, such substitutions have been occurring, but it needs to be remembered that long distance commuting is not new—it has been a feature of Japanese city-regions throughout the post-war period, and that face-to-face meetings, and the all-important socializing that follows, remain central to business relations and managerial practice in Japan. Furthermore, high-technology start-ups and other branches of employment using distance working (teleworking) tend to be highly concentrated in just one region, the Tokyo Metropolitan area (Nakazawa, 2002).

Figure 8.4c shows the geographical patterning of net migration for those aged 35–39. Its main features are continued low-level depopulation in rural and peripheral prefectures (especially in the Tohoku region of northern Honshu), significant net gains in suburban and peri-urban prefectures (Ibaraki, Tochigi, Gunma, Saitama, Chiba for Tokyo, Shizuoka and Mie for Nagoya and Shiga and Hyogo for Osaka), big net losses for major cities where these cities are the core parts of extended city-regions (especially Tokyo, Osaka and Kyoto) and net gains for some non-metropolitan prefectures in central and western Japan. The map patterns for people in their 40s, many of them settled in their current locations before the end of the bubble economy around 1990, shows much smaller losses in rural prefectures and small gains in suburban prefectures (e.g. Shiga Prefecture located north of Osaka and east of Kyoto).

People in their 50s, 60s and 70s in 2010—born between 1930 and 1960—comprise the lucky generation in many respects. Mostly born after (or too young to remember) the horrors of war, defeat and occupation, they became adults when jobs were secure, plentiful and increasingly well paid, when land and houses were cheap, and when sexual freedom, luxury consumption and a new 'me-ism' peeled away some of the surface layers of the obligation-laden Japanese family and workplace tradition. Apart from their great sadness at having few if any grandchildren (something that matters a lot in a family and ancestor-oriented culture), most of them have enjoyed a life during which their material living standards and social wellbeing have improved. One of the key contributions to this wellbeing has been home ownership; around 60% of households in Japan are living in owner-occupied accommodation, with the rates for older adults, of course, being significantly higher than for younger ones. The migration rates for home-owners are, however, very low. The reason for this is that, unlike the situation in many other countries, housing in Japan is hardly ever used as a tradable

(a)

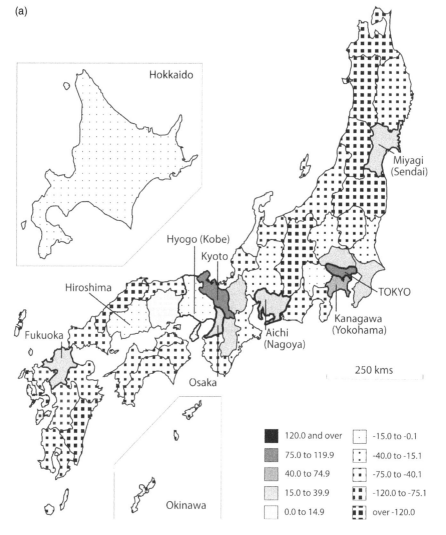

*Figure 8.4* Japan: age-specific net internal migration rate, by prefecture, 2005–2010: (a) 15–19; (b) 20–24; (c) 35–39; (d) 60–64.

Source: Author's calculation from 2010 Census; rate per thousand people in the specified age group in 2010.

(b)

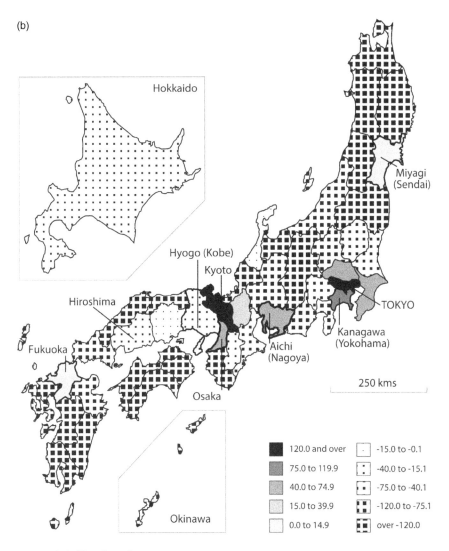

Hokkaido

Miyagi
(Sendai)

Hyogo (Kobe)

Kyoto

Hiroshima

Fukuoka

Aichi
(Nagoya)

TOKYO

Kanagawa
(Yokohama)

Osaka

250 kms

Okinawa

| | 120.0 and over | | -15.0 to -0.1 |
| | 75.0 to 119.9 | | -40.0 to -15.1 |
| | 40.0 to 74.9 | | -75.0 to -40.1 |
| | 15.0 to 39.9 | | -120.0 to -75.1 |
| | 0.0 to 14.9 | | over -120.0 |

*Figure 8.4* (Continued)

(c)

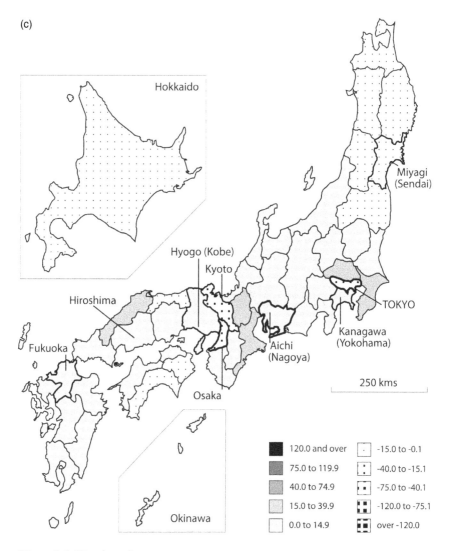

Figure 8.4 (Continued)

(d)

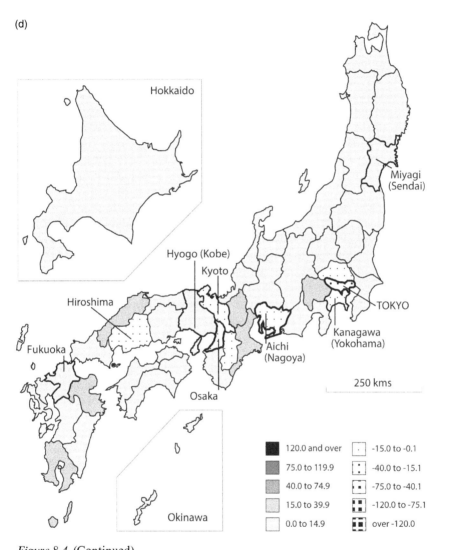

Hokkaido

Hyogo (Kobe)

Kyoto

Hiroshima

Miyagi
(Sendai)

Fukuoka

TOKYO

Aichi
(Nagoya)

Kanagawa
(Yokohama)

250 kms

Osaka

Okinawa

| | |
|---|---|
| 120.0 and over | -15.0 to -0.1 |
| 75.0 to 119.9 | -40.0 to -15.1 |
| 40.0 to 74.9 | -75.0 to -40.1 |
| 15.0 to 39.9 | -120.0 to -75.1 |
| 0.0 to 14.9 | over -120.0 |

*Figure 8.4* (Continued)

asset. Families rarely trade-up in the housing market; they tend not to use housing as a speculative venture (for example, by borrowing to invest in buy-to-let), they do not use their homes for the purpose of equity release and there is no question of using housing equity for care costs in old age. In countries that do use housing as a tradeable asset, the migration rates of home owners are high. In Japan they are extremely low, and the age group that this affects the most is those aged in their 50s, 60s and 70s i.e. before, for most people, the frailties of old age kick in.

By now the migration histories of people born between 1930 and 1960 have largely concluded, so mobility rates are generally very low. But the higher divorce rate at around retirement age means that there is a slight rise in family related reasons for migration at this age, and, much more significant in terms of population redistribution, there is a newly emerging net migration gain to rural and peripheral areas. Much of it, it turns out, is return migration to the home prefectures that had been left behind 40–50 years previously, especially so when both spouses come from the same prefecture (Kishi, 2014; Ishikawa, 2011; Yamamoto, 2013; this retirement migration was foreseen in Morikawa, 1992). The map pattern in Figure 8.4d shows this very clearly. Overall, the rates of net gain and loss are very small, but the larger cities (notably Tokyo and Osaka), and sometimes also their suburban prefectures, lose by internal migration while rural and peripheral regions (sometimes even the most peripheral regions) gain (Inoue and Watanabe, 2014; Tanigawa, 2004).

The migration of those aged over 80 is at a slightly higher rate than that for the young old, and has increased over the years (Tahara and Iwadare, 1999). It is dominated by the need, arising from frailty and loss of independence, to be close to their offspring upon whom there is a very strong social obligation to care (Hirai, 1999, 2007; Sakai, 1989; Tahara, 2002; Yamashita, 1989). Not surprisingly, the map distribution (not reproduced here) shows net gains in suburban prefectures (and in Okinawa) and net losses in remote rural areas and in city-centre prefectures—notably Tokyo.

This section of the chapter has shown that there are strong reasons to suppose that, in addition to the political economic reasons for lower interprefectural migration in Japan, the falling migration rates are also the product of major shifts in Japanese culture and society.

## Conclusions

Standing back from the detail, Japan is unusual in its low rate of internal migration when compared with other high-income industrial capitalist societies (Hayashi, 2014). It is not unusual, however, in its recent history of internal migration rates (both inter- and intra-prefectural); these peaked around 1970, then declined rapidly, then declined slowly right up to 2014. This chapter has attempted to tell the story of this decline: part of that story lies in the changing political economy of urban and regional development in a country that experienced 'miracle' rates of economic growth until about 1990, followed by 'lost decades' of essentially no economic growth since. The other part of the story lies in the social and demographic development of Japan—in its ageing, of course (Ishikawa, 2007), but also in the major shifts in attitudes and beliefs, behaviours and practices that turned a modern mobile society of the 1960s and 1970s into the arguably post-modern, lower mobility society that it is today.

## References

Abe, H. 1997. Regional disparities in employment opportunities and interregional migration in Japan. *Studies in Regional Science*, 28(1), 45–60 [In Japanese].

Amos, T.D. 2011. *Embodying Difference: The Making of Burakumin in Modern Japan*. Honolulu: University of Hawaii Press.

Arai, Y. and Ohki, S. 1999. Recent tendency of personnel transfers in Japan: New tabulation of the data of the Third Migration Survey. *Komaba Studies in Human Geography*, 13, 111–136 [In Japanese].

Arai, Y., Kawaguchi, T. and Inoue, T. (eds) 2002. *Japanese Internal Migration: Lifecourse and Regional Character*. Tokyo: Kokon-Shoin [In Japanese].

Budak, M.-A.E., Liaw, K.-L. and Kawabe, H. 1996. Co-residence of household heads with parents in Japan: A multivariate explanation. *International Journal of Population Geography*, 2(2), 133–152.

De Vos, G. and Wagatsuma, H. 1966. *Japan's Invisible Race*. Berkeley, CA: University of California Press.

Dunford, M. and Fielding, A. 1997. Greater London, the South-east Region and the wider Britain: metropolitan polarisation, uneven development and inter-regional migration. In Blotevogel, H.H. and Fielding, A.J. (eds) *People, Jobs and Mobility in the New Europe*. Chichester: John Wiley, 247–276.

Elis, V. 2011. Rural depopulation and economic shrinkage in Japan: What can affected municipalities do about it? In Coulmas, F. and Lutzeler, R. (eds) *Imploding Populations in Japan and Germany*. Leiden: Brill, 443–460.

Esaki, Y. 2006. *The Future of Tokyo City-Region's Population*. Tokyo: Senshu University Press [In Japanese].

Esaki, Y., Arai, Y. and Kawaguchi T. 2000. Return migration in Japan: A comparative analysis of migrants returned to Nagano and Miyazaki Prefectures. *Human Geography*, 52(2), 80–93 [In Japanese].

Fielding, A.J. and Ishikawa, Y. 2003. Migration and the life course in contemporary Japan. *Geographical Review of Japan*, 2, 246–257.

Fielding, T. 2012. *Migration in Britain: Paradoxes of the Present, Prospects for the Future*. Cheltenham: Edward Elgar.

Fukuda, S. 2010. Leaving the parental home in post-war Japan: Social, economic, and demographic determinants. *Max-Plank Institute for Demographic Research WP* 2010–007.

Fukurai, H. 1991. Japanese migration in contemporary Japan: Economic segmentation and interprovincial migration. *Social Biology*, 38(1–2), 28–50.

Gauthier, H.L., Tanaka, K. and Smith, W.R. 1992. A time series analysis of regional income inequalities and migration in Japan 1955–85. *Geographical Analysis*, 24(4), 283–298.

Gedik, A. 1995. *Changes in the Migratory Patterns in Japan 1955–90*. Nihon University: Institute of Developing Economies, VRF 247.

Genda, Y. 2013. The solitary non-employed persons (SNEPs): A new concept of non-employment. *Japan Labour Review*, 10(4), 6–15.

Gottfried, H. 2014. Precarious work in Japan: Old forms, new risks. *Journal of Contemporary Asia*, 44(3), 464–478.

Hama, H. *et al.* 1995. Tokyo metropolitanisation and inter-regional migration of labour. In Mizuno, A. and Ono, A. (eds) *Constrained Supply of Labour and the Japanese Economy*. Tokyo: Taimeido, 151–181 [In Japanese].

Hayashi, R. 2014. International comparison of migration—a construction of model-mobility using Japanese indicators. *Journal of Population Problems*, 70(1), 1–20 [In Japanese].

Hayashi, R., Chitose, Y., Kojima, K., Shimizu, M., Koike, S., Kishi, M. and Nakagawa, M. 2013. *Overview of the Results of the Seventh National Survey of Migration*. Tokyo: National Institute of Population and Social Security Research.

Higuchi, Y. 2008. Circumstances behind growing regional disparities in employment. *Japan Labour Review*, 5(1), 5–35.

Hirai, M. 1999. Characteristics of in-migration of the elderly in a suburb of Tokyo: A case study of Tokorozawa City, Saitama Prefecture. *Geographical Review of Japan*, 72A(5), 289–309.

Hirai, M. 2006. Population and regional structure in Japan. In Miyakawa, Y. and Yamashita, J. (eds) *Regional Structure and Regional Planning*. Tokyo: Minerva, 2–17. [In Japanese]

Hirai, M. 2007. The regional characteristics of the inter-prefectural migration of the elderly. In Ishikawa, Y. (ed) *Population Decline and Regional Imbalance: Geographical Perspectives*. Kyoto: Kyoto University Press, 129–147. [In Japanese]

Imai, Y. 1989. *A Study of Determinants of Migration between Japanese Prefectures*. Brown University PhD Thesis.

Inaba, H. and Mita, F. 1995. Trend analysis for interprefectural migration in Japan 1954–1993. *Journal of Population Problems*, 51(2), 1–19 [In Japanese].

Inagaki, R. 2003. Migration behaviour of the suburban second generation: A case study of Kozoji, New Town. *Geographical Review of Japan*, 76(8), 575–598.

Inoki, T. and Suruga, T. 1981. Migration, age and education: A cross-sectional analysis of geographical labour mobility in Japan. *Journal of Regional Science*, 21(4), 507–517.

Inoue, T. 2002. A demographic approach to the Japanese migration turnround. In Arai, Y., Kawaguchi, T. and Inoue, T. (eds) *Japanese Internal Migration: Lifecourse and Regional Character*. Tokyo: Kokon-Shoin, 53–70 [In Japanese].

Inoue, T. and Liaw, K-L. 2004. Life-course perspective on some distinctive features of migration from non-metropolitan prefectures in Japan. *Geographical Review of Japan*, 77(12), 765–782.

Inoue, T. and Watanabe, M. 2014. *The Ageing of the Capital City Region*. Tokyo: Hara Shobou [In Japanese].

Ishiguro, I., Lee, Y-J., Sugiura, H. and Yamaguchi, K. 2012. *The Brain Drain: Why Japanese Youth Move to Tokyo*. Tokyo: Minerva Shobou.

Ishiguro, M. 1976. An analysis of out-migrants from the centre of the metropolis: A case study of Naka-ku, Nagoya City. *Human Geography*, 28(3), 27–54.

Ishikawa, Y. 1978. Internal migration in postwar Japan. *Geographical Review of Japan*, 51(6), 433–450 [In Japanese].

Ishikawa, Y. 1992. The 1970s migration turnaround in Japan revisited: A shift-share approach. *Papers in Regional Science*, 71(2), 153–173.

Ishikawa, Y. 1994. *An introductory study of transfer migration.* Nihon University Population Research Institute, RP64.

Ishikawa Y. (ed) 2001. *Studies in the Migration Turnarounds.* Kyoto: Kyoto University Press [In Japanese].

Ishikawa, Y. (ed) 2007. *Population Decline and Regional Imbalance: Geographical Perspectives.* Kyoto: Kyoto University Press [In Japanese].

Ishikawa, Y. 2011. Recent in-migration to peripheral regions of Japan in the context of incipient national population decline. In Coulmas, F. and Lutzeler, R. (eds) *Imploding Populations in Japan and Germany.* Leiden: Brill, 421–442.

Ishikawa, Y and Fielding, A.J. 1998. Explaining the recent migration trends of the Tokyo metropolitan area. *Environment and Planning A*, 30, 1797–1814.

Ishimizu, T. and Ishihara, H. 1980. The distribution and movement of the population in Japan's three metropolitan areas. In Association of Japanese Geographers (ed) *Geography of Japan.* Tokyo: Teikoku Shoin, 347–378.

Isoda, N. 1993. The inter-regional migration since the 1970s: A case study of the Chugoku district. *Human Geography*, 45(1), 24–43 [In Japanese].

Isoda, N. 1995. Japanese internal migration pattern and industrial restructuring in the latter half of the 1980s. *Economic Geography Yearbook*, 41(2), 1–17 [In Japanese].

Ito, K. 1998. An examination of regional economic differentials as explanatory factors of equilibrium phenomena of internal migration after the bubble period in Japan. *Studies in Regional Science*, 29(3), 71–87.

Itoh, T. 1984. Recent trends in internal migration in Japan and 'potential life time out-migrants'. *Research on Population Problems*, 172, 24–38.

Ito, T., Naito, H. and Yamaguchi, F. 1979. *The Regional Structure of Population Movement.* Tokyo: Daimeidou [In Japanese].

Iwao, S., Saito, H. and Fukutomi, M. (eds) 1991. *Tanshinfunin.* Tokyo: Yuhikaku [In Japanese].

Izuhara, M. and Forrest, R. 2011. Housing histories and intergenerational dynamics in Tokyo. *Social Science Japan Journal* 22. accessed 9 June 2011.

Kajita, S. 1998. Migration and occupation careers of young male employees in a remote mountainous area: A case study of Wara village. *Geographical Review of Japan*, 71A(8), 573–587 [In Japanese].

Kakiuchi, G.H. and Hasegawa, M. 1979. Recent trends in rural to urban migration in Japan: The problem of depopulation. *Tohoku University SR(G)*, 29, 47–61.

Kamo, H. 1999. Characteristics of the local labour market and return migration in a peripheral region of Japan: A case study of the Aira area in Kagoshima Prefecture. *Human Geography*, 51(2), 24–47 [In Japanese].

Kawabe, H. 1980. Internal migration and the population distribution in Japan. In Association of Japanese Geographers (ed) *Geography of Japan.* Tokyo: Teikoku-Shoin, 379–389.

Kawabe, H. 1985. Some characteristics of internal migration observed from cohort-by-cohort analysis. *Journal of Population Problems*, 175, 1–15 [In Japanese].

Kawabe, H. and Liaw, K.-L. 1994. Selective effects of marriage migration on the population redistribution in a hierarchical region system in Japan. *Geographical Review of Japan*, 67(B)(1), 1–14.

Kawada, T. 1993. The entrance behaviour of university candidates and occupation behaviour of university graduates from Saku region, Nagano Prefecture. *Geographical Review of Japan*, 66A(1), 26–41.

Kawaguchi, T. 2002. Residential migration in the major metropolitan regions). In Arai, Y., Kawaguchi, T. and Inoue, T. (eds) *Japanese Internal Migration: Lifecourse and Regional Character*. Tokyo: Kokon-Shoin, 91–112 [In Japanese].

Kinsella, S. 1995. Cuties in Japan. In Skov, L. and Moeran, B. (eds) *Women, Media and Consumption in Japan*. Richmond: Curzon Press, 220–254.

Kishi, M. 2014. In-*migration to the Tokyo metropolitan area and return to birth prefecture among those born in non-metropolitan areas*. Journal of Population Problems, 70(4), 441–460 [In Japanese].

Kobayashi, J., Kagawa, M. and Sato, Y. 2015. How to get a longer job? Roles of human and social capital in the Japanese labour market. *International Journal of Japanese Sociology*, 24, 20–29.

Koganezawa, T. 1987. Recent changing patterns of 'dekasegi' seasonal migration in the South Yokote Basin, northern Japan. *Tokyo Metropolitan University, Geography Reports*, 22, 85–97.

Kondo, K. and Okubo, T. 2015. Interregional labour migration and real wage disparities: Evidence from Japan. *Papers in Regional Science*, 94(1), 67–88.

Liaw, K.-L. 1992. Interprefectural migration and its effects on prefectural populations in Japan: An analysis based on the 1980 census. *Canadian Geographer*, 36(4), 320–335.

Liaw, K.-L. 2003. Distinctive features in the sex ratio of Japan's interprefectural migrants: An explanation based on the family system and spatial economy in Japan. *International Journal of Population Geography*, 9, 199–214.

Liaw, K.-L. and Otomo, A. 1991. Interprefectural migration patterns of young adults in Japan: An explanation using a nested logit model. *Journal of Population Studies*, 14, 1–20.

Lutzeler, R. 2008. Population increase and 'new-build gentrification' in central Tokyo. *Erdkunde*, 62(4), 287–299.

Matsubara, H. 2014. Industrial structural changes in the Tokyo Metropolitan Area. *Journal of Geography*, 123(2), 285–297 [In Japanese].

Matsuda, M. 1986. The development of seasonal labour migration and the labour market of migrants employed in construction works in Matsudai, Niigata Prefecture. *Geographical Review of Japan*, 59A(5), 243–260.

Matsuhashi, K., Mizuno, M., Kashima, H. and Oda, H. 2013. A review of geographical studies on manufacturing industries in Japan. *Geographical Review of Japan*, 86B(1), 82–91.

Mera, K. 1988. The emergence of migration cycles? *International Regional Science Review*, 11(3), 269–275.

Miyauchi, H. 1998. An analysis of characterisitcs of newcomers from other prefectures in Zamami village, Okinawa Prefecture. *Geographical Sciences*, 53(4), 23–36 [In Japanese].

Miyazaki, Y. 1998. Chain migration of the masters of the public bath houses to Osaka Prefecture: A case study of migrants from South Kaga, Ishikawa Prefecture. *Human Geography*, 50(4), 80–96 [In Japanese].

Mizutani, T. 1989. Processes of decrease: Migration and recovery in urban population after disasters. *Geographical Review of Japan*, 62A(3), 208–224.

Morikawa, H. 1992. The 1985–90 net migration of age groups within Hyogo Prefecture. *Human Geography*, 44(4), 1–19 [In Japanese].

Murayama, Y. 1990. Space-time analysis of internal migration in postwar Japan. *Human Geography Research*, 14, 169–188 [In Japanese].

Murayama, Y. and Hashimoto, Y. 1994. The spatial structure of age and sex-specific migrations in the Nagoya metropolitan area. In Takahashi, N. and Taniuchi, T. (eds) *Japan's Three Major Metropolitan Regions*. Tokyo: Kokon-Shoin, 141–165 [In Japanese].

Nakagawa, S. 1990. Changing segregation patterns by age group in the Tokyo Metropolitan Area. *Geographical Review of Japan*, 63B(1), 34–47.

Nakagawa, S. 1996. Changing distribution of university graduates in Japan from a cohort-by-cohort perspective. *Journal of Population Problems*, 52, 41–59 [In Japanese].

Nakagawa, S. 2000. *Internal migration in today's Japan. Geographia Polonica*, 73(1), 127–140.

Nakagawa, S. 2001. Unbalanced spatial distribution of gender and 'migration for marriage' in Japan. *Journal of Population Problems*, 57(1), 25–40 [In Japanese].

Nakagawa, S. 2006. *Migration by Social Class in Japan*. Paper presented at the Liverpool meeting of the Population Geography International Conference, 20 June.

Nakazawa, T. 2001. The R&D labour market for newly graduated engineering students: the process of concentration in the Tokyo metropolitan area). *Economic Geography Yearbook*, 47(1), 19–34 [In Japanese].

Nakazawa, T. 2002. The life course of R&D staff. In Arai, Y., Kawaguchi, T. and Inoue, T. (eds) *Japanese Internal Migration: Lifecourse and Regional Character*. Tokyo: Kokon-Shoin, 149–168 [In Japanese].

Nishihara, J. and Saito, H. 2002. Coal mine closures in the late 1980s and the reaction of redundant mining workers: The case of Takashima coal mine, Nagasaki Prefecture. *Human Geography*, 54(2), 1–22 [In Japanese].

Nishioka, H. *et al.* 2012. The family changes in contemporary Japan: Overview of the results of the Fourth National Survey on Family in Japan (2008). *Japanese Journal of Population*, 10(1), 1–31.

Noro, Y. 1988. Interregional migration and social class mobility. In *1985 Social Class and Social Mobility: A National Survey*, Vol 1, 219–249. Tokyo: Tokyo University Press [In Japanese].

Oe, M. 1995. Cohort analysis of population distribution change in Japan: Processes of population concentration to the Tokyo region and its future. *Journal of Population Problems*, 51(3), 1–19 [In Japanese].

Ogawa, N. 1986. *Internal migration in Japanese postwar development*. Nihon University Population Research Institute RP 33.

Ogawa, N. 2008. The Japanese elderly as a safety net. *Asia-Pacific Population Journal*, 23(1), 105–113.

Okahashi, H. 1996. Development of mountain village studies in postwar Japan: Depopulation, peripheralisation and village renaissance. *Geographical Review of Japan*, 69B(1), 60–69.

Oshiro, K. 1976. Post war seasonal labour migration from the rural areas of Japan. *Tohoku University SE(G)*, 26, 7–36.

Otomo, A. 1990. Japan. In Nam, C. *et al.* (eds) *International Handbook on Internal Migration*. New York: Greenwood Press, 257–274.

Otomo, A. 1996. Trends in mobility and flows of spatial moves of population in postwar Japan. *Demographic Research*, 19, 5–17.

Otomo, A., Liaw, K.-L. and Abe, T. 1991. Departure and destination choice processes in Japanese interprefectural migration: A characterisation of overall and age-specific patterns. *Geographical Review of Japan*, 64B(1), 1–23.

Ozeki, Y. 1997. *The Distribution Pattern of Urban Migration in the Kanto Region*. Internal report. Tsukuba: Tsukuba University, Institute of Geosciences, SRA18, 1–36.

Saito, T. and Higashi, K. 1978. Structure and its change of inter-prefectural migration in Japan. *Geographical Review of Japan*, 51(12), 864–875.

Sakai, H. 1989. The elderly migration: Characteristics and reasons. *Journal of Population Problems*, 45(3), 1–13 [In Japanese].

Sakai, H. 1991. Change in sex ratio among migrants in Japan. *Journal of Population Problems*, 46(4), 1–13 [In Japanese].

Sakai, T. 1993. An analysis of determinants of internal migration in Japan. *Memoirs of Nara University*, 21, 167–173 [In Japanese].

Sakuno, H., 1994. Regional differences in the depopulation process in mountainous settlements of Hiroshima Prefecture. *Human Geography*, 46(1), 22–42 [In Japanese].

Sato, R. and Iwasawa, M. 2013. Single, sexless and infertile: Sexuality aspects of very low fertility in Japan. Paper presented at the British Society for Population Studies Annual Conference, 9–11 September.

Shimizu, H. 1984. Introduction to the study of 'family reasons' in geographical mobility. *Journal of Population Problems*, 169, 17–30 [In Japanese].

Shimizu, M. 2004. An analysis of recent migration trends in the Tokyo City core three wards. *Japanese Journal of Population*, 2(1), 1–16.

Shimizu, M. 2014. Residences by life stage and population structures of metropolitan residents. *Journal of Population Problems*, 70(1), 44–64 [In Japanese].

Suzuki, T. 1990. Inter-regional marriage in Japan. *Journal of Population Problems*, 46(2), 17–32 [In Japanese].

Tabuchi, T. 1986. Inter-regional income differential and inter-regional migration. *Studies in Regional Science*, 17, 215–226 [in Japanese]

Tachi, M. 1971. The interregional movement of population as revealed in the 1970 census. *Area Development in Japan*, 4, 3–24.

Tahara, Y. 2002. The migration of the elderly. In Arai, Y., Kawaguchi, T. and Inoue, T. (eds) *Japanese Internal Migration: Lifecourse and Regional Character*. Tokyo: Kokon-Shoin, 169–190 [In Japanese].

Tahara, Y. and Iwadare, M. 1999. Where the elderly move to: A review and study of elderly migration flows in Japan. *Komaba Studies in Human Geography*, 13, 1–53 [In Japanese].

Tanaka, K. and Nakano, N. 1998. Change of migration behaviour and expansion of commuting areas from Kono village, Fukui Prefecture. *Fukui University: Memoirs of the Research and Education Centre for Regional Environment*, 5, 119–134.

Tanaka, Y. 1991. *Research on Leaving Alone for a New Post*. Tokyo: Chuo Keizaisha.

Tani, K. 1997. An analysis of residential careers of metropolitan suburbanites: A case study of Kozoji new town in the Nagoya metropolitan suburbs). *Geographical Review of Japan*, 70A(5), 263–286.

Tanigawa, N. 2004. Life history of migrants, life story of a community: A case study of the Osumi-shoto islands). *Human Geography*, 56(4), 63–79 [In Japanese].

Tran, M. 2006. Unable or unwilling to leave the nest? An analysis and evaluation of Japanese parasite single theories. *Electronic Journal of Contemporary Japanese Studies*, Discussion Paper 5.

Traphagan, J.W. 2000. The liminal family: Return (internal) migration and intergenerational conflict in Japan. *Journal of Anthropological Research*, 56(3), 365–385.

Tsumaki, S. 2011–12. Local manifestations of poverty and social exclusion: Destabilisation of life among the urban buraku. *Sociological Review*, 62(4), 489–503 [In Japanese].

Tsutsumi, K. 1987. Analysis of out-migration from a mountain village: A case study of Kamitsue-mura in Oita Prefecture. *Human Geography*, 39(3), 1–23.

Tsuya, N.O. and Kuroda, T. 1989. Japan: The slowing of urbanisation and metropolitan concentration. In Champion, A.G. (ed) *Counterurbanisation*. London: Edward Arnold, 207–229.

Uchino, S. 1982. Behaviour of moving and non-moving: with special reference to migration survey in Sendai and Kumamoto. *Journal of Population Problems*, 164, 1–18 [In Japanese].

Uchino, S. 1990. Trends and characteristics of internal migration in postwar Japan. *Journal of Population Problems*, 46(1), 16–34 [In Japanese].

U.S. Bureau of the Census (John F. Long). 1990. Comparing migration measures having different intervals, in Perspectives on migration analysis. *Current Population Reports: Series P-23 Special Studies*. Washington DC: Government Printing Office.

Wakabayashi, K. 1987. Population migration and regional policy: The Fourth Comprehensive National Development Plan. *Journal of Population Problems*, 182, 18–35 [In Japanese].

Wang, D. 1991. The internal migration in modern Japan. *Asian Geographer*, 10(1), 89–101.

Watanabe, M. 1989. Internal migration and regional economic differentials in postwar Japan. *Demographic Research*, 12, 11–24 [In Japanese].

White, J.W. 1982. *Migration in Metropolitan Japan: Social Change and Political Behaviour*. Berkeley, CA: University of California Press.

Wiltshire, R. 1980. Research on reverse migration in Japan III: return migration of university graduates. *Tohoku University SR(G)*, 30, 65–75.

Wiltshire, R. 1995. *Relocating the Japanese Worker*. Folkestone: Japan Library.

Yahata, S. 1997. Regional mobility of Japan's workers: making the U-, J- and I-turn. *Japan Labour Bulletin*, 1 April.

Yamagami, T. 2003. Spatio-temporal structure of population growth in the metropolitan areas of Japan. *Geographical Review of Japan*, 76(4), 187–210 [In Japanese].

Yamaguchi, Y. 2002. The 'era of localities' and stay-behind youth. In Arai, Y., Kawaguchi, T. and Inoue, T. (eds) *Japanese Internal Migration: Lifecourse and Regional Character*. Tokyo: Kokon-Shoin, 35–52 [In Japanese].

Yamaguchi, Y., Arai, Y. and Esaki, Y. 2000. Migration trends of young jobseekers in and around peripheral regions in Japan. *Economic Geography Yearbook*, 46(1), 43–54 [In Japanese].

Yamamoto, T. 2013. *Return (U-turn) Migration and the Sociology of Rural and Remote Areas*. Tokyo: Gakubunsha [In Japanese]

Yamashita, J. 1989. The migration characteristics of elderly people in mountainous areas. *Regional Investigation Report* 11, 107–111 [In Japanese].

Yano, K. 1989. Age-disaggregated intra-metropolitan migration in Tokyo Metropolitan area. *Geographical Review of Japan*, 62A(1), 269–288 [In Japanese].

Yano, K. 2007. Spatial age-specific migration patterns of the economic downturn years. In Ishikawa, Y. (ed) *Population Decline and Regional Imbalance: Geographical Perspectives*. Kyoto: Kyoto University Press, 91–128 [In Japanese].

# 9 Sweden

## Internal migration in a high-migration Nordic country

*Ian Shuttleworth, John Östh and Thomas Niedomysl*

This chapter takes a Scandinavian perspective, dealing with Sweden which, along with its Nordic neighbours, is a relatively high-mobility country in present-day terms. According to the IMAGE-based results presently earlier in this volume (see Figure 4.1 in Chapter 4), Sweden comes close to the top of the European countries for one-year address-changing, alongside Finland, Denmark and Iceland, all with rates of 14% or higher, with Norway not far behind. This puts Sweden on a par with the rates shown there for the USA and Australia and significantly ahead of those for the UK, Germany, Japan and especially Italy. In terms of recent trends, according to Chapter 4, Sweden's rate of address-changing has been fairly stable in the present century, in contrast to the falls in rates shown there for the USA, Australia, Japan and Italy but similar to the pattern for the UK and Germany.

Sweden also makes for an interesting case study because it is representative of the 'Nordic' social democratic welfare regime and has only recently imported some of the free-market features of neo-liberalism such as the privatisation of activities and institutions formerly run by the state. On this basis, it might be expected to have experienced a different migration regime and trajectory from representatives of the 'Anglo-American' liberal school of Capitalism identified by Esping-Andersen (1990), notably the USA and the UK. Sweden is also distinctive, compared to the USA and UK at least, in its economic and social history, with its industrialisation coming in the late nineteenth century and being led by the exploitation of natural resources and by large industrial combines. Also, a particular shape of historical social class and political conflict fashioned the contours of Swedish society through the twentieth century, leading to the political and economic settlement that emphasised social democracy and welfarism in the post-war period.

This chapter has several objectives. Firstly, it aims to situate the most recent migration trends in their longer-term context, taking advantage of several sources of data, one of which takes the picture back to the dawn of the twentieth century. Secondly, it provides more detail about Sweden's economy and society as they have evolved since World War II and reviews the literature on the links between these developments and the longer-term trends in the country's internal migration patterns. Thirdly, it presents the results of an original analysis that focuses on how migration rates in Sweden have altered since the early 1990s and disaggregates these changes into that arising from changes in the migration behaviour of individual population subgroups and that explained by the changes in the socio-demographic composition of the population. Before developing these three topics, however, the chapter begins with a description and assessment of the data available for studying Swedish internal migration.

## Swedish migration data

Those with a research interest in population change and migration in Sweden are very fortunate in the range and depth of data that are available. Sweden was one of the first European states to collect comprehensive data on its population, with parish registration starting in the late seventeenth century followed by the first census in 1749. This was a consequence of the strength of the 'enlightenment state', the power of the monarchy and the pivotal role played by the established (Lutheran) church in Swedish society. Historical population data on births, deaths, marriages and confirmations are available from Parish Books (*Kyrkböcker*), which have been digitised to make them available to contemporary researchers. Some of these data can be used for migration research; for example, it is possible to trace lifetime migration by comparing place of birth with place of residence for named individuals (DDSS, 2016). Of course, there are severe limitations with these historical data. It is impossible in most cases to know when migration events took place or to know how many migration events happened given the restriction imposed by only having place of birth and current residence. Nevertheless, these data can provide valuable insights into population migration during the onset of Swedish industrialisation, with a good evidence base to show the increasing shares of people living in major cities and evolving industrial regions that had been born elsewhere. Indeed, it is probable that, taking a very long-term perspective, some of the highest rates of Swedish internal migration happened in the mid to late nineteenth century, mirroring the experience of Germany in the same phase of industrialisation (Langewiesche and Lenger, 1987; see also Chapter 10 in this volume).

Sweden's migration statistics became more regularised in the later nineteenth century, following the establishment of Statistics Sweden in 1858, and expanded further during the twentieth century, with migration statistics collected by population and housing censuses conducted every five or 10 years (Statistics Sweden, 2008). Additionally, migration information was obtained annually from registration undertaken within parishes and (after 1967) at the county level (Statistics Sweden, personal communication). This made it possible to measure migration not just in terms of a transition between place of birth and current residence but also as a series of events experienced by an individual and recorded at a much finer temporal resolution. There are data series from Statistics Sweden that provide information on inter-parish migration events from 1900 onwards, well over a century of data. From the 1960s, this included information on the distance of address-changing as identified from boundary crossing, such as moving between parishes or between counties. At the same time, it is important to note that this is not a perfect measure of distance as these spatial units vary in size in different parts of the country. Also, the published data do not allow for the fact that the geography of these administrative units has changed over the years, so they must be treated with some caution.

The last population and housing census in Sweden was undertaken in 1990, after which, in common with other Nordic countries, Sweden moved to a register-based digital system for the collection of population data. This system allows longitudinal individual-based statistics with fine spatial resolution to be made available for research purposes, with customised extracts sent to different Swedish universities. Unfortunately, due to the high level of detail in the digital registers, for reasons of confidentiality it is not possible to link them to the statistics from the earlier censuses and achieve the full potential benefits of long-duration longitudinal data, meaning that the more detailed and disaggregated analyses that this system permits cannot go back before 1990. As an example, the PLACE dataset held at Uppsala University contains statistics on all residents in Sweden for the period 1990–2010, with the usual address of each resident on 31st December each year being stored as coordinates aggregated to grids of 100 by 100 m. This is a greater spatial resolution than for most of the other university extracts, allowing the description and analysis of geographically finely grained migration events, measured as all address-changes of greater than 100 m (Niedomysl and Fransson, 2014).

This PLACE dataset is used for the analyses reported later in the chapter. Its benefits are twofold. Firstly, it is possible to measure migration using linear distances and not just as flows between administrative or statistical output geographical units. This frees the analysis from a dependency on the variable size of the units used to capture flows and the fact that these units can change over time, as just mentioned (see also Niedomysl et al., 2017, and the discussion in Chapter 3). The georeferencing also provides the possibility of estimating distances on the basis of road or railway networks, but early studies (e.g. Olsson, 1965) concluded that more complex approaches did not add much to migration analysis. The second benefit is that it allows individual-level analysis in which the characteristics of people can be taken into account in ways that are far more difficult using aggregated data. On the downside, the 1990–2010 time span mentioned above is because 2010 is the latest year currently available in the PLACE database.

In sum, there is a richness, temporal depth and individual-level availability in Swedish population and migration data that exceed what is available in most other countries. However, the Swedish data are not immune to the problems that also occur in other national contexts, as outlined in Chapter 3 and mentioned in the other case study chapters where appropriate. In particular, there are some caveats to do with the changing geographies of statistical reporting units and variations between datasets in definitions and temporal coverage that mean that judgement must be exercised when measuring and interpreting trends in Sweden's internal migration rates through time.

## Trends in Swedish internal migration since 1900

The longest series of data, dating back to the year 1900, concerns moves between parishes as measured by annual migratory events. These are shown in Figure 9.1 as a percentage rate based on the total population in each year. For more than a century this rate has oscillated between 7% and 10%, with peaks in the 1930s and immediately after the Second World War and with a lesser peak around 1970. There is, of course, no easy way to relate inter-parish moves to specific economic, social and political events nor to know the individual motivations that led to someone to move (or to stay put) at specific times. However, these historic peaks conform to the broad economic and political periods that shaped the world economy during the twentieth century, with recession in the 1930s and then the post-war readjustments perhaps explaining the higher rates in these years. Likewise, the economic turbulence of the later 1960s and 1970s with boom and bust could plausibly be related to the peaks and troughs then.

What the between-parish migration data series in Figure 9.1 also shows is a fairly steady increase in migration rate in the recent past from lower

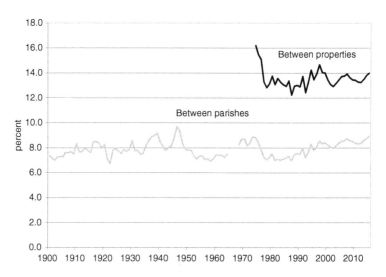

*Figure 9.1* Sweden: rates of migration between properties and between parishes, 1900–2015.

Source: Authors' calculations from Statistics Sweden data.

levels in the 1980s, such that it has now returned to the peak reached in the 1970s. Since parishes became fewer (and larger) after 2000 with the disestablishment of the Swedish Church, this cannot be the reason for this uplift because this change would have reduced the number of moves between the parishes. Rather, it is highly likely that the series underestimates the increase that would have been seen if a consistent parish geography has been used. Also, Figure 9.1 shows the annual trend in the rate of people moving between properties from when that series commenced in 1974, which by definition will have been unaffected by the parish changes. This reveals fairly similar annual fluctuations since the sharp decline in rate in the early 1970s, though with a somewhat more subdued performance since 2000 than for the between-parish rate. Nevertheless, the key point remains: overall, there are no signs of internal migration falling over the long-term since 1900, nor even—unlike in the US case discussed in Chapter 5—in more recent decades. Indeed, there are indications that migration rates have actually increased since the 1980s. This observation is consistent with those made by Lundholm (2007) on Swedish migration trends since the 1970s.

Is this pattern true for all distances of within-Sweden address-changing? Data availability restricts the answer to this question just to the 1960s onwards for most types of boundary crossing and to 1974 onwards for the most localised moves between properties, as just seen in Figure 9.1, with the results presented in Figure 9.2. It is the latter—people moving between properties within the same parish—that displays the most distinctive pattern,

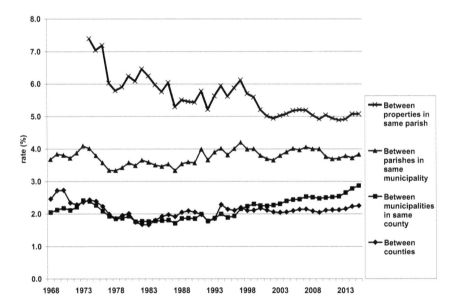

*Figure 9.2* Sweden: migration rates, 1968–2015, by type of move.
Source: Authors' calculations from Statistics Sweden data.

with an overall downward trend in rate through to the end of the 1990s after which the rate stabilised at around the level of 5% of the Swedish population undertaking this type of move each year. The latter tends to confirm that the change in parish geography in 2000 had no appreciable effect on this series, in that while this change would be expected to increase the amount of moving within the same parish, there is no apparent discontinuity in the series at this time. Anyway the trend since 2000 closely follows the rate for moves between parishes within municipalities, which would be expected to be negatively affected by the change.

All the other three types of migration shown in Figure 9.2 conform to the picture of stable or rising rates since the 1980s, albeit to varying extents. The rate for moves between parishes within the same municipality rose fairly steadily between the mid-1980s and the mid-1990s, dipped a little through to 2002, rose again for a few years, dipped again after 2009 and has shown only marginal recovery since then. Meanwhile, the rate for moves involving changing municipalities but staying in the same county display a steady increase through the whole period including since 2000, with an especially large uplift in the last three years of the series. Finally, the annual rate of migration between counties, which include all the longest-distance moves, appears to have stayed around 2% of the Swedish population since the general fall of migration rates in the 1970s (Figure 9.2).

An alternative way of looking at these trends is through calculating the percentage change in rates by type of migration over time, using five-year averages to smooth out the annual fluctuations. Taking 1981–1985 as the reference period, the rate of moving between properties within the same parish had declined by 18% by the latest five-year period, 2011–2015, dropping from 6.1 to 5.0% of the Swedish population. This contrasts with a 3% increase in the total migration rate (i.e. for moves over any distance) between these two periods. The latter was possible because of the rise in the rates for the three longer-distance types of migration, with the rate for moves taking place between parishes within the same municipality up by 6%, that for inter-county moves up by 24% and that for moves between municipalities within the same county up by fully 50%. Cumulatively starting with the inter-county rate rise of 24%, the rise in all inter-municipality moves between 1981–1985 and 2011–2015 was 37% and that for all inter-parish moves was 21%.

It is therefore abundantly clear that there has been no systematic downward trend in Swedish internal migration rate since the fall that occurred in the latter half of the 1970s. What has become less common over this period is local residential mobility in the form of moves taking place within the same parish, but the effect of this has been more than offset by increases in the rates of the three longer-distance types of migration, most notably those intermediate-distance moves which involve a change of municipality but do not cross a county boundary. The between-county migration rate has tended to flatline since recovering somewhat in the late 1980s, while the rate for moves between parishes within the same municipality has been rather more variable following its recovery in the early 1990s, with dips around 2000 and since 2009. Overall, there is little evidence of migration rate decrease in Sweden but rather the opposite appears to be taking place, albeit varying according to the spatial scale of move and the period used as the benchmark.

## Explaining the long-term trend in Sweden's internal migration

The existing literature on Sweden can be drawn on to help explain all but the latest decade of trends in Swedish internal migration. In particular, Holmlund (1984), Edin and Holmlund (1994) and Nilsson (1995) have examined the fall in internal migration rate that occurred in the 1970s and 1980s from high levels in the 1960s, while Lundh (2006), Westerlund (2006) and Eliasson *et al.* (2007) have also covered its subsequent stabilisation and increase during the 1990s. One important reason for the migration boom during the late 1960s was the execution of the *"miljonprogrammet"*, the label originally given to a social democratic campaign for a large-scale housing project that eventually resulted in the building of around a million new homes, especially in the suburbs on the outskirts of the major urban areas.

The 1960s was also a period of strong economic growth for Sweden as for much of the rest of Europe at this time, which in particular spurred strong rural-to-urban migration (SOU, 2007a). This was a period of rapid shrinkage in rural jobs, especially in forestry and farming, with employment there contracting by around 50% in 20 years, a trend that has continued to the present day (SCB-AKU, 2016). By contrast, in urban areas there was substantial growth in labour market opportunities, including both blue-collar jobs in manufacturing and white-collar jobs in the growing public administration sector. This urban growth was reinforced by international labour immigration during the 1950s and 1960s, especially from Finland and the Balkans (Hedberg, 2004), as these immigrants moved almost exclusively towards the major industrial regions in the central and southern parts of the country such as Västerås, Köping and Eskilstuna.

During and after the 1970s, the geography of economic growth changed again and with it the patterns of migration. This change involved a decrease in industrial jobs often located in the smaller urban areas and an increase in service-sector employment and knowledge-oriented jobs most concentrated in the larger cities, especially in the three metropolitan areas of Stockholm, Gothenburg and Malmo (Eriksson and Hane-Weijman, 2017; Östh *et al.*, 2015a). As previously, this reorientation was reinforced by a change in the nature and motives of international immigration, which switched from being predominately labour-related to become much more focussed on refugees who were equally, if not more, prone to settle in the major urban concentrations with their greater employment and housing opportunities (Edin *et al.*, 2003; Åslund *et al.*, 2010).

The demographics of migrating groups have also changed over time. According to Lundholm (2007), a driver in the change towards a reduction in migration rates during the 1980s and 1990s was the transformation of Sweden from a predominately male and single-breadwinner labour market into a society with a dual-breadwinner labour market and an increasing demand for skilled labour which reduced the main age span for longer-distance labour migration as well as reducing the scope of the areas to which migration could take place. As shown by Lundholm (2007), during the last three decades of the twentieth century Swedish migration became more concentrated on ages 18–29, while rates of migration for those in their 30s and 40s, as well as the under-16s, shrank markedly. This trend can also be linked directly to the growth of migration for higher education over this period, which additionally explains a relatively greater increase in the number of females involved in these flows since more women than men proceed to education at university level (Eliasson, 2006). Hansen and Niedomysl (2009) also show that migration for higher studies and migration to first job after completing higher studies makes up the great part of all migratory moves among the higher educated.

Others (see, for instance, Lundh, 2006) have placed emphasis on decreasing regional wage differentials and general labour market issues as factors in the decline in Swedish migration in the 1970s. The chief development was the alleviation of post-war labour shortages with time. As mentioned above, labour shortages in urban industrial areas in the post-war period played an important role in motivating individuals to migrate from rural to urban areas in the 1960s. However, for later decades, other drivers for migration differentials would seem to have become more important (SOU, 2007a) because of changing sectoral and geographical distributions of economic activity. The most important reason for this change is that the rural population in Sweden, having been severely reduced by the earlier decades of migration, has become unable to contribute to any larger contemporary rural-to-urban migration flows. In fact, less than 15% of the Swedish population is now (in 2015) listed as residing outside an urban area (World Bank, 2016) and large shares of the rural population live within commutable distances from economically resilient and major urban areas (Östh and Lindgren, 2012; Östh *et al.*, 2015b). The growth in commuting from rural areas in relative close proximity to major employment centres in Sweden is advocated by public agencies, primarily to increase skill-matching on the labour market and to make the labour market more resilient (Östh and Lindgren, 2012). However, the partition of Sweden into greater-sized labour market regions has been criticised as wishful thinking rather than a substantial change of the labour market, since much of the commuting increase is due to complex patterns of local commuting rather than an increase in inter-regional commuting (Amcoff, 2009).

This section has reviewed the evidence for changes in Swedish internal migration in the post-war period. The overall picture is one of buoyant economic growth, labour shortages and state housing policy underpinning rising levels of migration in the 1960s. At this time, migration often took the form of moves from rural areas to industrial districts. The 1970s and 1980s saw this process slow down. This was partly due to the shrinkage of the reservoir of rural dwellers who were potential migrants but also to the slowing of the economy then and to the restructuring of society and the labour market in other ways such as the growth of two-earner households, all of which acted to lower migration rates. Since then, by contrast, migration rates have been rising for all but the most local moves. To get a better understanding of what lies behind this more recent uplift, the next section looks in more depth at the 1990s and 2000s using the Swedish Population Register.

## Perspectives from the PLACE database

As mentioned above, PLACE is a digital register-based system that allows for longitudinal, individual-level analysis and contains statistics for all residents in Sweden spanning 1990–2010, with details of the whereabouts of each resident on the last day of each calendar year with a spatial accuracy of 100 m. For present purposes, migration is conceptualised as (a) all address-changes (in practice, all moves of more than 100 m) and (b) moves of 100 km or more, as identified as taking place between these annual observations. It needs to be stressed that this is a different approach from the aggregate data from Statistics Sweden used in the chapter thus far, in that the latter record all migration events, not just the one-year transitions of the PLACE dataset. Another difference is that, in what follows, attention is focused on those aged 17 and over rather than all ages. We also take advantage of the considerable range of information available in the PLACE dataset on the characteristics of residents. We begin by examining the variation in migration rates by type of person and over time since 1990, as well as looking at Sweden's socio-demographic structure, and then use Blinder-Oaxaca modelling (Jann, 2008; Cooke, 2011) to separate out the effect of changing population composition on overall change in migration rate from that of change in each type of person's migration behaviour.

*Descriptive analysis*

The first step towards a better understanding of the changes in Swedish internal migration since 1990 is to undertake a descriptive analysis that disaggregates migration by age. Figures 9.3a and 9.3b show the annual trend between 1990 and 2010 in the rates of all moves (i.e. of 100 m and over) and of 100 km and over respectively. Two observations can be made from these charts. Firstly, there are marked variations in internal migration rates by age and these are very much as expected from the migration literature; namely that younger age groups (those aged up to 44) were more mobile than the mean for the population aged 17 and over, whereas those aged 45 and older were less mobile. Generally, the rates fall with age for both types of migration, apart from the all-distance rate for those aged 85 and over who tend to move locally on the death of a partner or on becoming unable to continue living by themselves. Secondly, the year-on-year changes over this 20-year period broadly parallel the trends shown above by Statistics Sweden's aggregate series—despite the definitional differences—with a general uplift in rates since the early 1990s, peaking in the mid-2000s and then falling back somewhat to 2010. This picture applies to all five age groups for all distances of migration, though in relative terms more for the three aged 45 and over (Figure 9.3a), while for moves of 100 km and over there is less of a peaking in the mid-2000s and less of a subsequent decline by 2010 (Figure 9.3b).

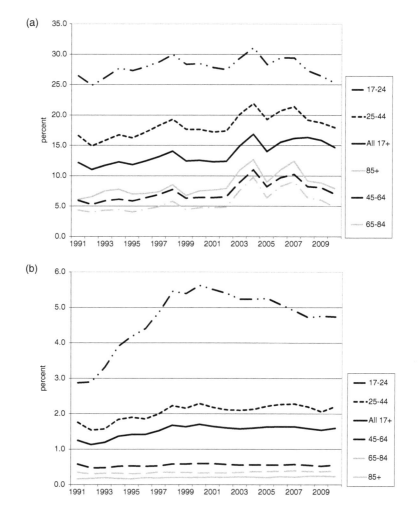

*Figure 9.3* Sweden: migration rate, 1991–2010: (a) all recorded moves; (b) moves of 100 km and over.

Source: Authors' calculations from PLACE database, Uppsala University.

As a second step, we take advantage of the wide range of information on personal characteristics in addition to age available in the PLACE data in order to see how migration rates vary and change over time for different types of person, again looking separately at all distances of move and those of at least 100 km. For this purpose, we have chosen to compare two five-year periods 1991–1995 and 2001–2005, where a move is identified by comparing the two addresses on the first and last days of each of these periods. This allows us to examine the relative differences between types of people in both periods, then see whether these relativities have altered between the two and finally to discover the extent to which the change in rate for each type diverges from the overall population trend.

The choice of these two periods was determined by a number of factors. One is the temporal scope of the PLACE dataset which limits us to the 1990–2010 window. Within this span, analyses of the relationship between job vacancies and unemployment indicates that the start and end years of the two periods are similarly situated in terms of the economic cycle, meaning that the results and interpretations are less likely to be affected by time-specific deviations in supply of and demand for labour (SOU, 2007b). In fact, the chosen starting years (1991 and 2001) were both times of worsening labour market conditions but unemployment figures were not yet as alarming as they would be in the years to follow, while the chosen ending years (1995 and 2005) both depict times when the labour market was recuperating from recession. If we had used a later five-year time period for analysis we would have been comparing the early five-year period, struggling with the effects of the Swedish bank crisis, to a period experiencing steady economic growth and increasing shares of employment (see Eriksson and Hane-Weijman, 2017, Figure 1, for over-time employment trends).

We focus on age, education, economic activity, housing tenure and gender. Whilst not inclusive of all the determinants of migration, these variables have been shown to be important as drivers of migration in the literature (Coulter and Scott, 2015; Lundholm, 2007; Niedomysl, 2011) and incorporate some of the most important predictors of whether someone stays put or moves. Table 9.1 shows the proportion of each category of person that was living at a different address at the end of the five-year period from that at the start for each of the two periods 1991–1995 and 2001–2005. The table also illustrates changes in socio-demographic structure by comparing the composition of the population aged 17 and over between 1991 and 2001 on the basis of these categories.

Looking first at compositional change between 1991 and 2001 (in the first three data columns of Table 9.1), it is apparent that the Swedish population has been changing in ways that are very similar to other more advanced countries. The population is ageing, as the proportion of those aged 17–24 has fallen substantially whereas those aged 75 and over and especially those aged 55–64 grew between 1991 and 2001. Other notable changes include growth in the proportions of single and divorced people, growth in the share

Table 9.1 Sweden: population composition and migration rates, 1991–1995 and 2001–2005, for all aged 17 and over

| Population characteristic | Composition at start of period | | | Address-change rate for period | | | 100 km+ migration rate for period | | |
|---|---|---|---|---|---|---|---|---|---|
| | 1991 | 2001 | %pt change | 1991–1995 | 2001–2005 | %pt change | 1991–1995 | 2001–2005 | %pt change |
| Total | 100.0 | 100.0 | 0.0 | 31.7 | 38.0 | 6.3 | 3.8 | 5.1 | 1.3 |
| *Age* | | | | | | | | | |
| 15–24 | 15.7 | 13.5 | -2.2 | 67.8 | 71.8 | 4.0 | 10.8 | 15.7 | 4.9 |
| 25–34 | 18.1 | 17.4 | -0.7 | 47.5 | 56.3 | 8.8 | 5.4 | 8.4 | 3.0 |
| 35–44 | 18.4 | 18.0 | -0.4 | 25.4 | 35.1 | 9.7 | 2.3 | 3.2 | 0.9 |
| 45–54 | 16.9 | 17.6 | 0.7 | 18.6 | 26.7 | 8.1 | 1.6 | 2.3 | 0.7 |
| 55–64 | 12.2 | 15.4 | 3.2 | 14.7 | 21.7 | 7.0 | 1.6 | 2.3 | 0.7 |
| 65–74 | 11.5 | 10.0 | -1.5 | 13.1 | 21.1 | 8.0 | 1.1 | 1.6 | 0.5 |
| 75+ | 7.2 | 8.1 | 0.9 | 18.8 | 25.2 | 6.4 | 0.8 | 1.0 | 0.2 |
| *Gender* | | | | | | | | | |
| Male | 48.8 | 49.3 | 0.5 | 32.3 | 38.2 | 5.9 | 3.8 | 5.1 | 1.3 |
| Female | 51.2 | 50.7 | -0.5 | 31.2 | 37.8 | 6.6 | 3.7 | 5.2 | 1.5 |
| *Economic status* | | | | | | | | | |
| Employed | 64.9 | 60.1 | -4.8 | 32.2 | 37.9 | 5.7 | 3.2 | 4.6 | 1.4 |
| Unemployed | 9.9 | 11.7 | 1.8 | 48.1 | 53.1 | 5.0 | 10.0 | 12.2 | 2.2 |
| Inactive | 25.2 | 28.2 | 3.0 | 24.3 | 31.9 | 7.6 | 2.8 | 3.4 | 0.6 |
| *Marital status* | | | | | | | | | |
| Single | 36.1 | 39.1 | 3.0 | 51.6 | 53.7 | 2.1 | 7.0 | 9.2 | 2.2 |
| Divorced | 8.3 | 11.1 | 2.8 | 35.0 | 38.4 | 3.4 | 3.2 | 3.9 | 0.7 |

| | | | | | | | | |
|---|---|---|---|---|---|---|---|---|
| Married | 49.0 | 43.9 | −5.1 | 18.2 | 25.4 | 7.2 | 1.8 | 2.3 | 0.5 |
| Widowed | 6.7 | 5.9 | −0.8 | 19.2 | 26.8 | 7.6 | 1.1 | 1.4 | 0.3 |

*Education*

| | | | | | | | | |
|---|---|---|---|---|---|---|---|---|
| Compulsory | 41.6 | 38.2 | −3.4 | 28.0 | 34.5 | 6.5 | 3.0 | 3.4 | 0.4 |
| Intermediate | 36.4 | 33.9 | −2.5 | 35.9 | 39.1 | 3.2 | 5.0 | 4.0 | −1.0 |
| Post-upper secondary | 21.9 | 27.9 | 6.0 | 31.2 | 39.9 | 8.7 | 2.9 | 7.8 | 4.9 |

*Housing tenure*

| | | | | | | | | |
|---|---|---|---|---|---|---|---|---|
| Public | 16.2 | 13.9 | −2.3 | 48.1 | 50.4 | 2.3 | 4.6 | 7.2 | 2.6 |
| Bostadsrätt | 13.1 | 14.5 | 1.4 | 30.9 | 41.7 | 10.8 | 3.3 | 4.6 | 1.3 |
| Owner | 49.6 | 50.7 | 1.2 | 22.2 | 27.5 | 5.3 | 3.2 | 3.8 | 0.6 |
| Renting | 21.2 | 20.9 | −0.3 | 41.2 | 51.9 | 10.7 | 4.3 | 7.5 | 3.2 |

Source: calculated from PLACE database, Uppsala University.

Notes: $N$ = 6,515,353 for 1991–1995 and 6,735,105 for 2001–2005. Bostadsrätt is a type of housing cooperative (see text).

of people with post-upper secondary education (i.e. university education or equivalent) and decrease in the proportion living in public-sector housing. Also, the proportion of economically inactive (i.e. outside the labour force) increased strongly, no doubt as a result of the growing older population, while the proportion unemployed rose somewhat.

Turning to the change in the five-year migration rates between the two periods, the overall picture for the two types of moves shown in Table 9.1 is for rising rates, as would be expected from the previous commentary. The rate of all address-changing rose from 31.7% for 1991–1995 to 38.0% for 2001–2005, an increase of 6.3% points, while for moves of 100 km or more the rate rose from 3.8 to 5.1%, up by 1.3 points. Disaggregating by characteristic, there are the expected differentials. For example, a general decrease in migration with increasing age is apparent, but when comparing migration rates for the two periods it is striking that all age groups are more migratory in the latter period regardless of distance moved. With reference to economic activity, the migratory behaviour of those who were unemployed at the start of the period is notable with around 12% moving over 100 km between 2001 and 2005. Once again, however, across all economic activity types there is a general increase in migration through time for both distance bands. Single people were the most migratory group by marital status whereas the widowed were the least mobile, but it is interesting once again that an increase in spatial mobility is apparent across all groups. That divorced individuals are more migratory compared to married individuals yet less migratory than singles suggests that double breadwinner households are less able to migrate and that divorced individuals often have parenting responsibilities, especially making longer migrations difficult.

Increased education is associated with increased migration as might have been expected and the gradient is more pronounced for longer-distance address-changes. Tenants in public housing and rented accommodation are more migratory than those who are owner occupiers or in the *bostadsrätt* category (which refers to housing cooperatives where members formally own the right to their apartment, which can be bought and sold on the open market) and, if anything, this relative gap increases through time. The descriptive analysis shows that increases in migration, whether measured by all address-changes or housing moves of 100 km or more, are common across our social and demographic categories since the early 1990s.

Thus the PLACE dataset provides little evidence of the sort of migration slowdown in Sweden that has been observed across several different geographical scales in the USA over the past two decades. On the contrary, there are some grounds for arguing that the Swedish population has become more geographically mobile over the last two decades, confirming the description of recent trends in the aggregate data series reported earlier. Having that said, a recent study suggests that migration towards the larger urban areas is starting to slow down as a result of a relatively low rate of new housing construction which in turn makes it increasingly difficult for people to afford housing in attractive regions (Boverket, 2016).

*Decomposition analysis*

These changes in total migration rates between 1991–1995 and 2001–2005 can, in the broadest terms, be deemed to result from the combination of two factors; changes in the socio-demographic composition of the population and changes in the propensity of each of the population groupings to migrate. In other words, some groups may become more or less migratory, and more or less migratory population groups can increase or decrease in their relative size, so there could be quite complex combinations of effects that drive aggregate rates in populations. So far we have approached this problem indirectly through our examination of Table 9.1, but it is possible to calculate the separate roles of the rate and composition effects using the Blinder-Oaxaca method.

Blinder-Oaxaca analysis is a regression decomposition procedure that separates the change in a rate between two time periods into three parts (for the mathematical details, see Jann, 2008; Cooke, 2011). The first part, termed the composition effect, is an estimate of the effect of the changing composition of the population on the total change in rate between the two time periods. For example, in this context the procedure could identify an age-specific composition effect which would likely be negative—reflecting the fact that, as the age composition of the population shifts toward older, less migratory groups, this will have a negative effect on the overall migration rate. The second part, termed the rate effect, is an estimate of the effect of the changing behaviour of specific population sub-groups on the total change in the rate between the two time periods. For example, in this context the procedure could identify a positive rate effect among young adults as the expansion of higher education between the two periods has caused an increase in migration among this age group, thereby contributing to an increase in the overall migration rate between the two periods. The third part is an interaction effect. Defined as a residual, it occurs because the rate and composition effects may act in concert. In reality, these effects are typically quite small (see Cooke, 2011) and, indeed, in other versions of the Blinder-Oaxaca regression decomposition procedure this effect is allocated to the rate- and composition-effects (see Jann, 2008). Following Cooke (2011), a linear specification was used to decompose the difference in overall rate between the two five-year periods into these effects, together with the interaction effect of the two operating in tandem.

Table 9.2 summarises the results of this analysis for our two measures of migration. Looking first at all distances of address-changing, the 6.3% point increase in the five-year migration rate between 1991–1995 and 2001–2005, noted above in Table 9.1, was entirely due to the rate effect of 7.1% points. This means that if the 1991–1995 population had the same migration rates/ coefficients as in 2001–2005 the aggregate migration rate would be 7.1% points higher than it actually was. However, the composition effect counterbalanced this by 0.6% points meaning that if the 1991–1995 population had the same composition as that in 2001–2005—with a shift on balance to less migratory groups—that its migration rate would be 0.6% points lower than it was in reality. The interaction effects contributed just another 0.2 points. The equivalent results for moves of 100 km or more (Table 9.2, final column) reveals that the increase in rate of 1.4% points between the two periods breaks down into a 1.1% point increase due to higher rate effects in 2001–2005, 0.3% point increase for composition effects as the 2001–2005 population's composition fell into groups that are more mobile over this distance, and a 0.1% point increase for the interaction between the two.

These summary findings suggest that, for both types of migration, most of the increase in Swedish migration can be attributed to the various types of people moving more frequently on average. By contrast, any shifts in population composition between higher- and lower-mobility types would seem to balance out to give only much small effects and also ones that differ in direction between longer- and shorter-distance migration. These aspects are explored further in the fully disaggregated results below.

Tables 9.3 and 9.4 unpack the summary results just described and show the separate contribution of the explanatory variables to the 6.3 and 1.4% point uplifts in the rates of, respectively, all address-changing and moves of 100 km and over. All coefficients (except the interaction term for gender) are statistically significant at the 5% level at least. As regards all address-changing, the first data column of Table 9.3 dealing with the composition effects indicates that between the two periods there was an increase in those age groups that are less migratory, which has acted to reduce migration by 1.2 points overall, seemingly the result of the ageing of the population. The shifts in housing tenure have had a similar, though smaller, impact. The combined effect of these two aspects of population structure, however, has been largely counterbalanced by the other four structural changes, most notably in education levels where the increasing share of people with higher qualifications has acted to push migration rates up and in marital status with the increase in share of single and divorced people at the expense of the married group. The other main positive effect of compositional change is for economic activity with increases in the relative size of the groups that are more mobile.

*Table 9.2* Sweden: migration rate change, 1991–1995 to 2001–2005, summary of Blinder-Oaxaca decompositions (%)

|  | *All address-changes* | *Moves of 100 km and over* |
| --- | --- | --- |
| 1991–1995 | 31.71 | 3.76 |
| 2001–2005 | 37.98 | 5.14 |
| Difference, of which: | 6.27 | 1.38 |
| Composition effects | −0.59 | 0.26 |
| Rate effects | 7.06 | 1.06 |
| Interaction effects | −0.20 | 0.05 |

Source and notes: As for Table 9.1.

*Table 9.3* Sweden: Blinder-Oaxaca decomposition of change in rate of all address-changing, 1991–1995 to 2001–2005 (%)

| Characteristic | Composition effect | Rate effect | Interaction effect | Total |
|---|---|---|---|---|
| *Age* | | | | |
| 15–24 (ref) | | | | |
| 25–34 | 0.15 | 0.51 | −0.02 | 0.64 |
| 35–44 | 0.15 | 0.51 | −0.01 | 0.64 |
| 45–54 | −0.31 | 0.18 | 0.01 | −0.12 |
| 55–64 | −1.53 | 0.13 | 0.04 | −1.36 |
| 65–74 | 0.75 | 0.45 | −0.06 | 1.15 |
| 75+ | −0.43 | 0.16 | 0.02 | −0.24 |
| Sum | −1.20 | 1.94 | −0.03 | 0.71 |
| *Gender* | | | | |
| Male (ref) | | | | |
| Female | 0.00 | 0.06 | 0.00 | 0.06 |
| Sum | 0.00 | 0.06 | 0.00 | 0.06 |
| *Economic activity* | | | | |
| Employed (ref) | | | | |
| Unemployed | 0.07 | −0.08 | −0.02 | −0.03 |
| Inactive | 0.09 | −0.54 | −0.06 | −0.51 |
| Sum | 0.16 | −0.62 | −0.08 | −0.55 |
| *Housing tenure* | | | | |
| Public (ref) | | | | |
| Bostadsrätt | −0.19 | 1.12 | 0.12 | 1.05 |
| Owner | −0.24 | 1.99 | 0.04 | 1.79 |
| Renting | 0.01 | 1.37 | −0.02 | 1.37 |
| Sum | −0.41 | 4.48 | 0.14 | 4.22 |
| *Marital status* | | | | |
| Single (ref) | | | | |
| Divorced | 0.18 | −0.02 | −0.01 | 0.15 |
| Married | 0.31 | 1.61 | −0.17 | 1.76 |
| Widowed | 0.02 | 0.24 | −0.03 | 0.23 |
| Sum | 0.51 | 1.83 | −0.20 | 2.14 |
| *Education* | | | | |
| Compulsory (ref) | | | | |
| Intermediate vocational | −0.03 | 0.46 | 0.02 | 0.45 |
| Intermediate theoretical | 0.07 | −0.53 | −0.02 | −0.48 |
| Post-upper | 0.32 | −0.07 | −0.03 | 0.22 |
| Sum | 0.36 | −0.14 | −0.03 | 0.19 |
| Constant | | −0.50 | | −0.50 |
| Total sum | −0.59 | 7.06 | −0.20 | 6.27 |

Source and notes: As for Table 9.1. All coefficients except the interaction term for gender are statistically significant at the 5% level.

The second column of Table 9.3 deals with rate effects and shows why it is not so much population composition change but the increasing propensity to change address that has pushed migration rates up. It is the independent roles of housing tenure, age and marital status that have contributed most towards this overall outcome. All age groups have become more migratory relative to the base category and this has more than compensated for the ageing of the population. By contrast, the rate effects associated with economic activity and educational qualifications (relative to their base categories) have worked in the opposite direction to the others, but merely served to dampen those positive effects. The constant term and the interaction effects are very small compared with the other elements of the results. In sum, the portion of the differences in total migration rates that can be accounted for by population compositional changes is small in total: most of the difference between the two time periods is down to altered behaviour.

Table 9.4 presents the results of the equivalent analysis for moves of 100 km or over, unpacking the 1.4% point rise in this migration rate between the two periods. In terms of population composition, the ageing of the population again has the overall effect of pushing migration down, but it is interesting to note the positive compositional effects of higher levels of education in the population. The rate effects relative to their respective base categories differ from those shown for all distances of address-changing in Table 9.3, in that more are negative. However, there are other differences between the two tables that reflect the different correlates of migration over different distances. Unlike in Table 9.3, the compositional effect in Table 9.4 is positive, perhaps reflecting the importance of the growth in education as a positive effect on longer-distance moves in particular. The constant term has a very small negative effect for all address-changes but a relatively large positive effect for moves of 100 km or more. One interpretation of this (Cotton, 1988) is that, while a small unmodelled effect is acting to reduce migration as measured by all address-changes, there is a comparatively large effect unaccounted for in the analysis that is tending to increase migration rates over longer distances.

## Discussion

The chapter shows that Sweden differs from the experience of the USA over the long term and from the UK in the short term. The overview of internal migration from 1900 shows that, at some spatial scales, rates have fluctuated with no definite trend. There are some signs of migration decreases during the 1970s and 1980s, and it is from these relatively lower levels that the later analysis, drawing on the PLACE database, detects an increase in migration from the early 1990s in Sweden. This contrasts with recent American migration trends where there were decreases through the 1990s and the early 2000s. There are country-specific reasons for this which might include government centralisation of jobs and services and the expansion of university education over the past decade (Östh et al., 2015a). However, as with all temporal analyses, there are questions concerning the

*Table 9.4* Sweden: Blinder-Oaxaca decomposition of change in rate of moves of 100 km and over, 1991–1995 to 2001–2005 (%)

| Characteristic | Composition effect | Rate effect | Interaction effect | Total |
|---|---|---|---|---|
| *Age* | | | | |
| 15–24 (ref) | | | | |
| 25–34 | 0.03 | −0.56 | 0.02 | −0.51 |
| 35–44 | 0.03 | −0.84 | 0.02 | −0.79 |
| 45–54 | −0.05 | −0.80 | −0.03 | −0.89 |
| 55–64 | −0.24 | −0.50 | −0.13 | −0.87 |
| 65–74 | 0.14 | −0.32 | 0.04 | −0.14 |
| 75+ | −0.09 | −0.13 | −0.02 | −0.24 |
| Sum | −0.18 | −3.16 | −0.10 | −3.44 |
| *Gender* | | | | |
| Male (ref) | | | | |
| Female | 0.00 | 0.01 | 0.00 | 0.01 |
| Sum | 0.00 | 0.01 | 0.00 | 0.01 |
| *Economic activity* | | | | |
| Employed (ref) | | | | |
| Unemployed | 0.09 | −0.07 | −0.01 | 0.01 |
| Inactive | 0.09 | −0.40 | −0.05 | −0.36 |
| Sum | 0.18 | −0.47 | −0.06 | −0.35 |
| *Housing tenure* | | | | |
| Public (ref) | | | | |
| Bostadsrätt | −0.02 | −0.06 | −0.01 | −0.08 |
| Owner | −0.01 | −0.37 | −0.01 | −0.39 |
| Renting | 0.00 | 0.16 | 0.00 | 0.16 |
| Sum | −0.03 | −0.27 | −0.02 | −0.32 |
| *Marital status* | | | | |
| Single (ref) | | | | |
| Divorced | 0.01 | −0.02 | −0.01 | −0.02 |
| Married | 0.04 | −0.21 | 0.02 | −0.15 |
| Widowed | 0.01 | −0.02 | 0.00 | −0.01 |
| Sum | 0.06 | −0.25 | 0.02 | −0.18 |
| *Education* | | | | |
| Compulsory (ref) | | | | |
| Intermediate vocational | −0.04 | 0.27 | 0.01 | 0.24 |
| Intermediate theoretical | 0.05 | 0.26 | 0.01 | 0.33 |
| Post-upper | 0.22 | 0.44 | 0.19 | 0.86 |
| Sum | 0.23 | 0.98 | 0.22 | 1.43 |
| Constant | | 4.22 | | 4.22 |

Source and notes: As for Table 9.1. All coefficients are statistically significant at the 5% level except the rate and interaction terms for gender.

start and end point of the analysis. If there were register data running back to the 1960s, we might interpret the growth with more certainty during the study period, perhaps not as an increase but as a return to previous levels after a decline.

Despite this, the Swedish experience would still differ from other countries such as the United States which have seen a long-term migration decline from the late 1960s. In the Blinder-Oaxaca analyses, the net effect of structural population change varies according to the distance of move, with population ageing having a large negative effect for all address-changes. This is not quite outweighed by increased education and shifts in marital status whereas for moves of 100 km or more there is a comparatively large positive effect arising from better education which seems to outweigh other factors that might push migration rates at this spatial scale down. More importantly, rate effects (an increasing propensity to change address across many demographic groups) appear to be the main driver of increasing population mobility in Sweden.

One interpretation of the Swedish migration patterns calls upon the particular economic, social and political history of Sweden. Migration is high because individuals do not need to invest time and effort in the local community to 'earn' the right to welfare and also because welfare is perceived as equally distributed regardless of choice of destination due to the well-developed welfare state. Accordingly, people have been, and remain, relatively footloose. An alternative interpretation of increased migration, especially for young people, relates to the expansion of higher education opportunities which are unevenly distributed, with the largest three regions having too few university places compared to the demand there. This gives rise to additional moves as young people move to university towns to obtain education and then, upon graduation, move elsewhere to find suitable jobs (Amcoff and Niedomysl, 2013).

## Conclusion

Regardless of analytical approach for the analysis of Swedish migration rates in recent decades, the results show that migration has not fallen over time. In fact, the opposite seems to be happening where individuals in corresponding age cohorts, gender, economic statuses, marital statuses, levels of education and housing tenures have a greater propensity to move in 2001–2005 compared to 1991–1995. But why do we see this trend in migration? There is of course no simple answer to this question but there are general tendencies in the Swedish case that can be used to shed light on some of the migration patterns and offer wider lessons.

Sweden changed economically from what can be perceived as a predominately industrial country to a globalised post-industrial economy, though the change did not happen suddenly, but gradually. The effects were profound on the labour force: blue-collar jobs were widespread and relatively abundant in 1991, and approximately 82% of the workers had no post-upper secondary education. In later years, the compulsory educated workers (note that these numbers are constrained to the working population) became fewer, especially among the younger age cohorts (Östh *et al.*, 2011). In 2008 as many as 45% of the all workers in ages 25–30 years had a higher education (Östh *et al.*, 2011). To stay competitive in the labour market, both workers and employers needed to become more skill-oriented and this gradually led to the shutting down of low-skilled plants and workshops (most were moved abroad) and growth of jobs that in those days (2001) were part of what was termed *the new economy*. This change was just as much a change of the economic geography where smaller industrial based municipalities started to lose jobs while larger urban areas (often administrative centres) started to gain and/or create jobs.

The widely observed migration peak for young adults in 2001 serves to exemplify how migration is becoming important for getting a first job and accessing higher education. A substantial amount of the young migrants moved to university cities and either stayed in the vicinity or moved for jobs after finishing exams. However, the relatively high degree of mobility may also partly be attributable to factors embedded in the Nordic welfare regime. In comparison to many other countries, social and economic inequality is limited, with access to social services such as education and health care being relatively equally distributed geographically. This makes it easier to resettle regardless of origin and destination. That said, recently social and economic inequality has been increasing in Sweden and rental housing is becoming less common (often due to transformation of flats to housing cooperatives) which may ultimately slow down migration rates. Revisiting Sweden and trends in migration in the future will tell if Sweden remains a highly migratory country or whether more neoliberalism acts to reduce spatial mobility—in contradiction to its rhetoric.

## References

Amcoff, J. and Niedomysl, T. 2013. Back to the city: Internal return migration to metropolitan regions in Sweden. *Environment and Planning A*, 45(10), 2477–2494.

Amcoff, J. 2009. Rapid regional enlargement in Sweden: A phenomenon missing an explanation. *Geografiska Annaler: Series B, Human Geography*, 91(3), 275–287.

Åslund O., Östh, J. and Zenou, Y. 2010. How important is access to jobs? Old question, improved answer. *Journal of Economic Geography*, 10(3), 389–422.

Boverket. 2016. *Housing, Internal Migration and Economic Growth in Sweden*. Rapport 2016:13. Karlskrona: Boverket.

Cooke, T.J. 2011. It is not just the economy: Declining migration and the rise of secular rootedness. *Population, Space and Place*, 17(3), 193–203.

Cotton, J. 1988. *On the Decomposition of Wage Differentials*. Boston: Economics Faculty Publication Series, University of Massachusetts. http://citeseerx.ist.psu.edu/viewdoc/download?doi=10.1.1.917.9126&rep=rep1&type=pdf/

Coulter, R. and Scott, J. 2015. What motivates residential mobility? Re-examining self-reported reasons for desiring and making residential moves. *Population, Space and Place*, 21(4), 354–371.

DDSS. 2016. *Demographic Database for Southern Sweden*. www.ddss.nu/ [In Swedish].

Edin, P-A., Fredriksson, P. and Åslund, O. 2003. Ethnic enclaves and the economic success of immigrants: evidence from a natural experiment. *The Quarterly Journal of Economics*, 118(1), 329–357.

Edin, P-A. and Holmlund, B. 1994. Unemployment and the functionality of the labour market. *Bilaga 8 till Långtidsutredningen 1994*. Stockholm: Fritzes [In Swedish].

Eliasson, K. 2006. *College Choice and Earnings among University Graduates in Sweden*. PhD thesis, Institute for National Economics, Umeå University.

Eliasson, K., Westerlund, O. and Åström, J. 2007. Migration and Commuting in Sweden. *Bilaga 3 till Långtidsutredningen 2008*. Stockholm: SOU, 35 [In Swedish].

Eriksson, R. and Hane-Weijman, E. 2017. How do regional economies respond to crises? The geography of job creation and destruction in Sweden (1990–2010). *European Urban and Regional Studies*, 24(1), 87–103.

Esping-Andersen, G. 1990. *The Three Worlds of Welfare Capitalism*. Princeton, NJ: Princeton University Press.

Hedberg, C. 2004. The Finland-Swedish wheel of migration identity, networks and integration 1976–2000. *Geografiska regionstudier*, ISSN 0431-2023; 61. Uppsala: Kulturgeografiska institutionen.

Holmlund, B. 1984. *Labour Mobility: Studies of Labour Turnover and Migration in the Swedish Labour Market*. Stockholm: Industriens utredningsinstitut (IUI).

Jann, B. 2008. The Blinder-Oaxaca decomposition for linear regression models. *The Stata Journal*, 8(4), 453–479.

Lundh, C. 2006. Workforce mobility and the organisation of the labour market 1850–2005. In Rauhut, D. and Falkenhall, B. (eds) *Arbetsrätt, rörlighet och tillväxt*. Östersund: Institutet för tillväxtpolitiska studier, 17–62 [In Swedish].

Lundholm, E. 2007. Are movers still the same? Characteristics of interregional migrants in Sweden, 1970–2001. *Tijdschrift voor Economische en Sociale Geografie*, 98(3), 336–348.

Hansen, H. and Niedomysl, T. 2009. Migration of the creative class: Evidence from Sweden. *Journal of Economic Geography*, 9(2), 191–206.

Langewiesche, D. and Lenger, F. 1987. Internal migration: Persistence and mobility. In Ritter, G.A., Poels, W. and Nicholls, A.J. (eds) *Population, Labour and Migration in 19th- and 20th-Century Germany*. Leamington Spa: Berg Publishers Limited, 87–100.

Niedomysl, T. 2011. How migration motives change over migration distance: Evidence on variations across socioeconomic and demographic groups. *Regional Studies*, 45(6), 843–855.

Niedomysl, T. and Fransson, U. 2014. On distance and the spatial dimension in the definition of internal migration. *Annals of the Association of American Geographers*, 104(2), 357–372.

Niedomysl, T., Ernstson, U. and Fransson, U. 2017. The accuracy of migration distance measures. *Population, Space and Place*, 23(1), e1971. doi:10.1002/psp.1971

Nilsson, C. 1995. *Interregional Relocation in Sweden: Consequences of Labour Market Changes, Labour Politics and Household Costs*. EFA-rapport 33. Stockholm: Arbetsmarknadsdepartementet [In Swedish].

Olsson, G. 1965. Distance and human interaction. A migration study. *Geografiska Annaler. Series B, Human Geography*, 47(1), 3–43.

Östh, J., Niedomysl, T., Amcoff, J., Ander, L. and Hedberg, S. 2011. *A Labour Market Under Change*. Rapport 2011:12B. Stcokholm: Arbetsmiljöverket.

www.av.se/arbetsmiljoarbete-och-inspektioner/kunskapssammanstallningar/
arbetsmarknad-i-forandring----en-analys-av-regionala-branschforandringar-rap-
201112-kunskapssammanstallning/ [In Swedish].

Östh, J. and Lindgren, U. 2012. Do changes in GDP influence commuting distances?
A study of Swedish commuting patterns between 1990 and 2006. *Tijdschrift voor
Economische en Sociale Geografie*, 103(4), 443–456.

Östh, J., Niedomysl, T. and Amcoff, J. 2015a. Depopulation and overheating, a result of public management. http://equipop.kultgeog.uu.se/Press.pdf [In Swedish].

Östh, J., Reggiani, A. and Galiazzo, G. 2015b. Spatial economic resilience and
accessibility: A joint perspective. *Computers, Environment and Urban Systems*,
49(1), 148–159.

SCB-AKU. 2016. *Labour Force Studies.* www.statistikdatabasen.scb.se [In Swedish].

SOU. 2007a. *Migration and Commuting in Sweden.* Attachment 3 to Långtidsutredningen 2008. Governmental report 2007:35. Stockholm: Fritzes. www.regeringen.
se/contentassets/4b92473a96d544c68b9dfb4e86cbb013/sou-200735-flyttning-och-
pendling-i-sverige [In Swedish].

SOU. 2007b. *Labour Market Training for Unemployed Youth and for Labour Train-
ing in Employment Sectors Lacking Professionals.* Governmental report 2007:18.
Stockholm: Fritzes. www.riksdagen.se/sv/Dokument-Lagar/Dokument/SOU/
Arbetsmarknadsutbildning-for-bristyrken-och-insatser-for-arbetslosa-ungdomar_
GVB318/?html=true [In Swedish].

Statistics Sweden. 2008. *Summing Up the Sums—the First 150 Years of Statistics
Sweden.* Stockholm: Statistics Sweden [In Swedish].

Westerlund, O. 2006. Workforce mobility and the organisation of the labour market 1850–2005. In Rauhut, D. and Falkenhall, B. (eds) *Arbetsrätt, rörlighet och
tillväxt.* Östersund: Institutet för tillväxtpolitiska studier, 93–111 [In Swedish].

World Bank. 2016. *Urban Population.* http://data.worldbank.org/indicator/SP.URB.
TOTL

# 10 Germany
## Internal migration within a changing nation

*Nikola Sander*

Germany is currently characterised by a medium intensity of internal migration according to the IMAGE-based rates in Chapter 4 (see Figure 4.1), with its estimate of around 9% of people changing address in a year putting it midway between the high-mobility Nordic countries like Sweden (Chapter 9) and the low-mobility ones of southern Europe like Italy (Chapter 11). Meanwhile, Germany was one of the countries to have recorded a decline in migration rates since the year 2000 according to Table 4.7, but the decreases shown there were very small and anyway some decline might have been expected following the decade of upheaval arising from the reunification of the country in 1990. However, while the world is becoming increasingly globalised, national context is of equally high importance for understanding the mechanisms driving human mobility. In the case of Germany, a variety of features, including its political framework, settlement pattern, economic structure, education system and housing market, need to be taken into account when comparing its internal migration patterns and trends with those of other countries.

The high volume of migration from the eastern to western states of Germany following reunification is a prime example of how disparities in economic development, labour market structure and perceived job prospects affect migration and vice versa (Kemper, 1997; Wendt, 1993). Then, as the disparities were reduced in the years following reunification, the westward flow also declined, with the net loss from East to West becoming negligible by the mid-2000s (Sander, 2014). Looking further back in time, the distinctive sequence of events affecting this nation over much of its modern existence present a major challenge to any attempt at detecting a long-term secular trend in the intensity of its address-changing behaviour, hence the considerable amount of attention given to historical background and national context in what follows.

Against the background of a review of historical patterns of Germany's internal migration, this chapter aims to discover whether there has been a general reduction in migration rate since reunification paralleling the decline in East-West migration and to identify the relative contribution of these other types of migration to the overall trend observed. It begins with a brief account of the principal sources of data available for studying Germany's internal migration patterns and trends. The following section sets out the distinctive features of the national context that can be expected to help explain differences in internal migration between Germany and other Western societies. After this, the chapter provides an historical account of internal migration in Germany from the pre-industrial era through to the present day. The remainder of the chapter largely focuses on the post-reunification period so as to take advantage of an integrated dataset spanning the whole country for 1995–2010. This is used to analyse the changing volume and patterns of inter-regional migration flows across Germany for this 15-year period. The concluding section summarises the main findings of this analysis and places them in wider context.

## Data on internal migration in Germany

The availability of data on internal migration and population structure in Germany is rather limited compared to most other countries. The lack of spatially detailed population data can be traced to the country's history and the misuse of population statistics that aided the Nazi persecution of Jews. Moreover, the 16 federal states of Germany have the prerogative to decide on the ways of collecting and storing population statistics, resulting in fragmented statistics and difficulties in collating data for studies of population at the national level.

A particular deficiency is the dearth of data on the total number of people changing their usual residence. The ranking of Germany on the overall intensity of address-changing, mentioned at the outset, was possible only because the IMAGE team estimated this from between-area flow data using the Courgeau method (see Chapter 4). Having said that, one source does provide some information on all changes of address, namely the Socio-Economic Panel (SOEP), which is a nationally representative longitudinal survey of the adult population in Germany. The SOEP's first wave was conducted in 1984, and every year since then respondents have been asked a wide range of questions on personal and household characteristics, including whether they have moved home since the last interview. The SOEP's small sample size, however, means that its spatial detail is limited to 96 planning regions (*Raumordnungsregionen*).

The principal source of data on internal migration in Germany is the national population register. This captures changes of permanent address that cross the boundary of a municipality, which is the smallest administrative entity in Germany. Currently, 11,313 municipalities (*Gemeinden*), of which 2,060 are in cities (*Städte*), are nested within 402 counties (*Landkreise & kreisfreie Städte*) and 16 states (*Länder*). The migration data derived from this source is of high quality in terms of coverage and accuracy.

Nevertheless, there are certain limitations in this register-derived data that need to be noted. Besides not covering within-municipality movements, the register allows only limited insight into the characteristics of migrants since it cannot be linked to other national datasets. Secondly, the register does not cover those students who choose not to register their primary residence at the place of study. While most large university towns have implemented special regulations aimed at encouraging students to do this, the proportion that do so is unclear.

In addition, time-series analysis is complicated by the numerous boundary changes that have occurred, most notably in the years following reunification. As mentioned in Chapter 4, comparisons of migration patterns and intensities across counties and over time require the development of temporally consistent geographies. Yet these are almost impossible to generate at the municipality level for the whole of present-day Germany, while even data at the county level cannot be reliably compared between before and after 1995. It is for this latter reason that the study of internal migration trends and patterns presented later in this chapter uses a customised dataset derived from the population register.

## National context

Germany's distinctiveness (besides in its political and economic history just recounted, but also partly linked to it) revolves primarily around its fairly balanced settlement structure, its relative lack of economic disparities and its rental-heavy housing market. For a long part of its history, Germany consisted of many independent states, which eventually united under Prussian leadership in 1871. Today, the country is a federal republic with several almost equally strong regions that provide high living standards as well as good education and employment opportunities.

The settlement structure resembles Zipf's rank-size distribution remarkably closely, with many medium-size towns, a smaller number of larger cities, and the capital Berlin as the most populous city. In contrast to London or Paris, however, Berlin is not hugely dominant, with Hamburg, Munich, Frankfurt, Stuttgart and Cologne-Dusseldorf being almost as important as the capital as centres of economic growth, service provision and innovation. Hence, a large number of destinations can be regarded as being of similar attractiveness for migrants, potentially lowering the prevalence of longer-distance migration.

Secondly, regional economic disparities in Germany are less pronounced than in many other Western societies such as the US, the UK and France and have been shrinking recently. According to the OECD (2011), regional GDP per capita has converged somewhat since reunification as income levels in the country's most affluent southern regions decreased from five times larger than that of the poorest eastern regions to just three times larger. This convergence of the past two decades was at least in part due to national and European Union funding for infrastructure development and urban renewal.

The *Mittelstand* is a key factor in reducing economic disparities between regions and between rural and urban areas. This term refers to the large number of small- and medium-size businesses that are mostly located in medium-size towns outside the major urban areas. Companies like Miele and Bosch, for example, are a key driver of Germany's economic strength and have remained at the cutting edge of global manufacturing.

The economy contracted more than those of most other European countries during the world recession of 2008–2009, but the increase in unemployment was much less pronounced, with the unemployment rate peaking at 7.9% in late 2009. The *Mittelstand*, coupled with the dual vocational education system which combines classes at a vocational school and on-the-job training at a company, provides young adults with a well-established alternative to university higher education. Many vocational schools and participating companies are located outside the largest cities and tend to be more evenly scattered across the country than universities. Hence, education-related movements among younger adults can be expected to be less focussed on larger cities than in, for instance, English-speaking countries.

Finally, the German housing market is characterised by rising prices, a low ownership rate, decreased social housing production, high transaction costs and a disparity between rural vacancies and urban supply shortages (see, for instance, Clark and Drever, 2000). Average house prices were rather stable in real terms between 1970 and 2010, but since then they have been increasing by more than 5% annually. Moreover, differences in property prices are widening between the larger cities and more remote rural areas. The rise in property prices in urban areas has been paralleled by rent increases, prompting the federal government to impose rent-rise caps on inner-city properties in 2015.

Germany has the greatest proportion of home-renters in Europe, and its homeownership rate ranks among the lowest in the developed world. Germany's rental-heavy real-estate market has its origins in the housing crisis inherited from World War II and the subsequent housing policies that incentivised renting. The rental market has been more robustly regulated than in most other Western countries, the regulations are quite favourable to renters, and longer-term renting tends to be as much in demand as home ownership. Compared to the USA or the UK, German house prices and rental markets are significantly less volatile. Nevertheless, the small share of owner occupation has tended to provide relatively few incentives for residential mobility, and the recent soaring of inner-city rental prices will likely suppress housing-related moves even further.

## Historical perspective on Germany's internal migration

Before going into detail about post-unification migration trends, by way of context this section sets out the longer-term record on Germany's migration trends. In pre-industrial days, labour mobility was a normal part of agriculture in the context of low population growth and localised production-consumption patterns (Moch, 2011). The German citizenship registers

(*Bürgerbücher*) from these times are indicative of an annual migration rate of between 2% and 8%, with migrants from the countryside making up about half of the population in cities like Frankfurt am Main and Würzburg (Hochstadt, 1983).

During the second half of the eighteenth century, population growth accelerated due to declining mortality. At the same time, rural manufacturing expanded. In North Rhine-Westphalia, textile and linen production increased and employed the most workers, especially women. The growth of rural industry provided non-agricultural employment outside the cities, reducing the rural-urban economic disparities and slowing out-migration from the countryside. This early industrialisation was characterised by the growth of manufacturing in mid-sized towns and by population increase in the peripheries of cities, with high mortality dampening population growth in the larger cities (Moch, 2011).

During the nineteenth century, the era of industrialisation unfolded in the context of crises in rural economies and accelerating population growth. Crop failures and the onset of capitalist agriculture forced people to leave the countryside in search of employment in the cities. This period saw an unprecedented rate of urbanisation coupled with migration rates that reached their historical maximum before World War I (Moch, 2011). The industrial cities, especially Duisburg, Essen and Dortmund in the area that is commonly referred to as the *Ruhrgebiet*, were then attracting labour migrants from the de-industrialising countryside, producing a shift in migration patterns towards rural-to-urban movement. World War I marked the end of a 60-year period of urbanisation, industrialisation and strong migration to the cities of western Germany (Hubert, 1998).

As the twentieth century unfolded, migration intensities declined in response not just to World War I but also to the Great Depression, the collapse of the Weimar Republic, the Nazi regime and then World War II (Hochstadt, 1999; Moch, 2011). But migration accelerated again after this. During the immediate post-war years, this was fuelled by the forced migration of people expelled from their homes through territorial exchange and the displacement of those who lost their homes during the war. By the 1950s, West Germany's rapid economic recovery (*Wirtschaftswunder*) caused a growing demand for labour that began to outstrip the capacity of the native labour force, leading to the initiation of a guest worker programme designed to meet the labour demand in agriculture and heavy industry and shifting political attention from internal to international migration.

The general patterns of internal migration in West Germany over the next four decades were dominated by urbanisation tendencies in the 1950s and 1960s, followed by sub- and counter-urbanisation patterns in the 1970s and early 1980s (Kontuly, 1991). However, it is difficult to quantify these trends and follow them through confidently to the end of the century due to the problems of developing a longer time series at a fine spatial scale described in the previous section, notably the separation and reunification of East and West Germany and the extensive changes to administrative geographies since reunification. The lack of adequate time series on between-area migration flows has meant that the existing literature has focused largely on the impact of net migration on the settlement structure, with much less attention being given to trends in migration intensity (see, for example, Kontuly and Vogelsang, 1989).

One exception is the work of Bucher and Heins (2001), which covers the period 1950–1998 and attempts to circumvent these issues by restricting analysis to the state (*Länder*) level and by dealing with just West Germany till 1990. On this basis, in the 1950s and 1960s, migration between the states was relatively high compared to earlier decades, running at an annual average of around 1.8%. The large volume of movement at this time was mostly due to people returning from the countryside to the cities because of the rapid recovery of the large industrial centres, most notably the Rhine-Ruhr region. By the 1970s, Munich had emerged as a centre of strong economic growth, causing the dominant migration pattern to shift from a westward to a southward direction. Although the latter became even more dominant as economic growth in the south continued throughout the 1980s, the overall level of inter-state migration declined over most of this period in response to slowing economic growth and tightening labour markets. The rate of inter-state migration in West Germany dropped to 1.3% in 1975, stayed at this level for about five years, and then declined further to barely 1% in the mid-1980s before moving back up to 1.3% in 1990 (Bucher and Heins, 2001).

Research by Milbert and Sturm (2016) allows a more up-to-date picture of trends in net and gross migration rates at the county level. The estimates of annual gross migration rates for the whole of Germany for 1974–2013 are displayed in Figure 10.1. They reveal that until 1990 the trend (for just West Germany) followed closely that for inter-state migration just described, with an initial fall in rate to 3.8% in the mid-1970s, a further fall after 1982 to 2.7% and then a degree of recovery between 1987 and 1990. After reunification, when their data incorporates the former German Democratic Republic, there is a new low starting point for the time series that reflects the traditionally lower mobility there due to tight governmental controls of housing and labour markets and the concentration of housing in the cities. During the early years after reunification, no doubt responding to the economic and political transformation of eastern Germany, the all-Germany rate rose substantially, up from 2.7% in 1991 to 3.2% in 1998, before subsiding slightly in the early 2000s and finally returning to around 3.2% by 2013.

Finally, one study has attempted to estimate the overall address-changing rate in post-reunification Germany, albeit just for 1997–2006. Drawing on longitudinal survey data from the SOEP (see above), Kemper (2008) concluded that the all-moves rate experienced a steady decline from 12% to 10% over this period. This reduction in rate contrasts with the relatively stable intensities of inter-county migration then, suggesting that the decline is restricted to local residential mobility. The latter can at least in part be traced to a tightening housing market and an insufficient supply of affordable housing in the larger cities of western Germany.

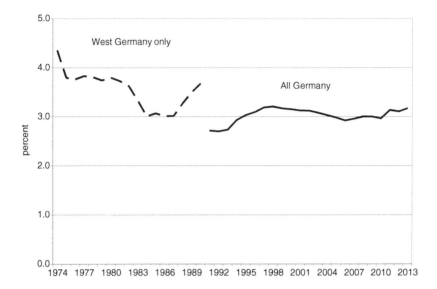

*Figure 10.1* Germany: inter-county migration rate, 1974–2013.
Source: Author's estimates based on data from Milbert and Sturm (2016).

## An analysis of migration trends and patterns for 1995–2010

This section focuses on recent patterns and trends of between-area migration for the whole of reunified Germany. It uses the German Internal Migration (GIM) database which contains inter-county migration flow matrices for each year from 1995 to 2010, allowing the analysis to depart from the common practice of analysing internal migration in Germany at the broader scale of just the 16 federal states. The choice of start year is due to the separation of Germany before 1990 and the changes in administrative geography in the years after reunification making comparisons between before and after 1995 extremely difficult in any meaningful and robust way. An update to 2015, originally planned for this study, had to be abandoned because of the federal statistical office introducing a substantial fee for data provision.

Even during this 16-year period, however, there were multiple changes in the county geography of Germany for which the flow matrices are given. The GIM database overcomes this methodological challenge using the "Update to contemporary zones" approach discussed in Blake *et al.* (2000). Sander (2014) provides a more detailed summary of the GIM database and the method for harmonising the county geography into 397 regions with temporally consistent boundaries, as shown in Figure 10.2. This map also classifies the regions into a three-way typology of rural, urban and intermediate ('hinterland'), which was originally developed by the Federal Institute for Research on Building, Urban Affairs and Spatial Development (BBSR, 2009).

This approach is not without limitations. In the first place, in using counties as the unit of analysis, the analysis cannot circumvent the issue of heterogeneity in the size and form of the administrative areas. It is not possible to eliminate the resulting distortions in the region-level analyses of migration

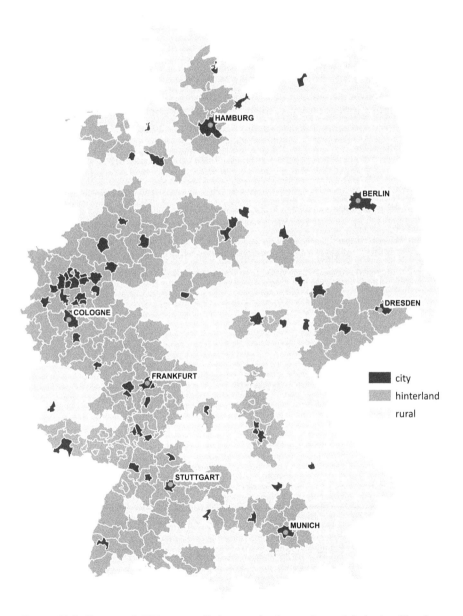

*Figure 10.2*  Germany's 397 temporally harmonised counties and their classification into city, hinterland and rural.

Source: Redrawn by the author from a more detailed classification.

flows; generally referred to as the modifiable areal unit problem (MAUP, see Openshaw, 1983). Secondly, the classification into rural, hinterland and urban cannot fully capture the underlying settlement patterns. For instance, the counties surrounding larger urban areas are often too large in size to effectively capture the hinterland. Hence, it is unavoidable that several counties bordering larger urban centres (e.g. Berlin) are classified as 'rural'. Nevertheless, this allows some light to be shed on differences in the intensity of migration by distance moved and by type of movement for 1995–2010. The flow data are also disaggregated into six age groups.

The absolute number of inter-county movements totalled 2.54 million in 1995 and was barely unchanged in 2010, at 2.55 million. Figure 10.3 depicts the annual trend transformed into crude rates by reference to the mid-year population, showing that the level at the end of the period was the same as at the start, at 3.1%. Over the 16-year period, there was just a barely perceptible rise to 3.2% at the end of the 1990s and a shallow trough around 3.0% in 2004–2006. It is, however, a considerably different picture for the six separate age groups, especially for the two younger adult ones. The rate for 25–29 year olds rose fairly steadily over the period from 7.4 to 9.8%, an increase by almost a third. That for the 18–24s rose almost as much, up from 7.0 to 9.0% and with a fairly steadily uplift apart from a period of stability in the mid-2000s. At the same time, the rate for the under-18s declined from 2.9 to 2.4%, a relative decline by one-fifth. The rate for those aged 65 and over

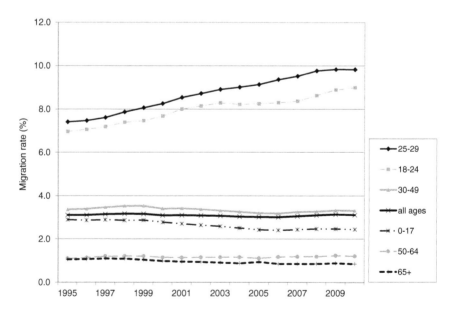

*Figure 10.3* Germany: inter-county migration rate, 1995–2010, by age group.

Source: Author's calculations based on harmonised inter-county migration data described in Sander (2014).

also dropped by around one-fifth, down from 1.1 to 0.9%. Meanwhile, there was virtually no change in the rates for those aged 30–49 and 50–64. The stable trend in the overall intensity thus disguises two distinct age-specific trends, which are likely to also impact on the spatial pattern of flows and hence population redistribution.

The GIM database also permits disaggregation of these inter-county address-changes by distance of move, calculated as the Euclidian distance between the geographical centroids of the counties. Table 10.1 presents the volume of flows recorded in the first and last years of the period, distinguishing those of less than (roughly) 50 km from the longer ones, again for all ages together and for the six age groups separately. This shows that the number of longer-distance inter-county moves increased by 46% among 18–24 year olds. For all other age groups, changes in the intensity by distance moved were negligible between 1995 and 2010. These findings suggest that the increase in overall migration intensity among 18–29 year olds discussed earlier is predominantly driven by an increase in longer-distance moves, while the decline in overall intensity among children and adults aged 30 years and over is not distance-specific.

The rise in longer-distance moves among young adults might be linked to the expansion of higher education, in which case larger university towns (e.g. Berlin, Bonn, Cologne, Dresden, Heidelberg, Frankfurt, Muenster, Munich, Tübingen) should have become more popular destinations for migrants from rural areas. This idea is tested by grouping the 397 counties on the basis of the three-way typology shown in Figure 10.2, with Table 10.2 depicting the volume of gross movements between the three types of counties in 1995 and 2010 for all ages combined and separately for 18–24 year olds and for 30–49 year olds.

Table 10.2 reveals that in 1995 the dominant flows across all ages were within hinterlands, between cities and their hinterlands, and within rural areas. In 2010, the flows within the city category and flows from rural areas to cities had increased from 1995, whereas flows down the urban hierarchy

*Table 10.1* Germany: number of inter-county moves, 1995 and 2010, by distance moved, thousands

| Age group | 1995 | | 2010 | | % change | |
|---|---|---|---|---|---|---|
| | *<50 km* | *≥50 km* | *<50 km* | *≥50 km* | *<50 km* | *≥50 km* |
| All ages | 1,156 | 1,381 | 1,085 | 1,457 | −6.2 | 5.5 |
| 0–17 | 215 | 244 | 152 | 173 | −29.3 | −28.9 |
| 18–24 | 196 | 257 | 222 | 383 | 13.8 | 49.2 |
| 25–29 | 227 | 269 | 189 | 297 | −16.7 | 10.7 |
| 30–49 | 386 | 438 | 358 | 421 | −7.2 | −3.7 |
| 50–64 | 76 | 95 | 95 | 103 | 25.3 | 8.1 |
| 65+ | 57 | 79 | 66 | 77 | 16.9 | −1.9 |

Source: Author's calculations.

*Table 10.2* Germany: flows between 397 counties classified by type, 1995 and 2010, for selected age groups, thousands

| 1995 | | | | 2010 | | | |
|---|---|---|---|---|---|---|---|
| Origin type | Destination type | | | Origin type | Destination type | | |
| *All ages* | | | | | | | |
| | City | Hinterland | Rural | | City | Hinterland | Rural |
| City | 202.3 | 455.3 | 197.3 | City | 277.4 | 405.0 | 171.6 |
| Hinterland | 365.2 | 523.3 | 198.5 | Hinterland | 428.7 | 481.0 | 161.4 |
| Rural | 147.9 | 179.3 | 268.0 | Rural | 211.9 | 164.3 | 240.9 |
| *Aged 18–24* | | | | | | | |
| | City | Hinterland | Rural | | City | Hinterland | Rural |
| City | 36.8 | 65.5 | 26.9 | City | 65.0 | 72.8 | 34.4 |
| Hinterland | 79.0 | 87.9 | 32.1 | Hinterland | 126.6 | 101.3 | 32.5 |
| Rural | 38.2 | 35.9 | 49.9 | Rural | 77.0 | 40.1 | 55.8 |
| *Aged 30–49* | | | | | | | |
| | City | Hinterland | Rural | | City | Hinterland | Rural |
| City | 73.1 | 161.7 | 66.9 | City | 94.1 | 141.6 | 54.3 |
| Hinterland | 116.9 | 170.4 | 61.2 | Hinterland | 120.9 | 152.8 | 48.9 |
| Rural | 41.5 | 52.9 | 78.9 | Rural | 49.9 | 47.5 | 69.5 |

Source: Author's calculations.

from cities to rural areas had decreased in comparison to the patterns in 1995. As regards the two selected age groups, the panel for the 18–24 year olds shows a marked increase in moves to the cities from both hinterlands and rural areas, whereas moves out of the cities declined for 30–49 year olds. The overall pattern appears to be one of increasing movement up the urban hierarchy (i.e. from rural to hinterland to cities), although significant counter-movements dampen the effectiveness of population redistribution.

The net impact of these flow changes on the three types of counties is traced on an annual basis from 1995 through to 2010 in Figure 10.4. The increase in movement up the urban hierarchy across all ages seen by comparing the two years in the top panel of Table 10.2 is reflected in a pronounced shift in net flow over the 15-year period. Cities shifted from net losses to net gains, while rural areas and hinterlands shifted from net gains in 1995 to net losses in 2010 (Figure 10.4a).

As with the trends over time in overall intensities, the changes in net migration by county type vary considerably by age group (Figure 10.4b–f). Among children and adults aged 30 and over, net migration converged towards zero as more and more families chose to stay in the cities rather than undertaking a counter-urban move. At the other end of the spectrum, net migration among 18–24 year olds rose substantially for cities and declined for rural counties and hinterlands, suggesting that the appeal of the urban labour market and the desire to attend higher education institutions have become important drivers of migration to the cities. The changes in trends

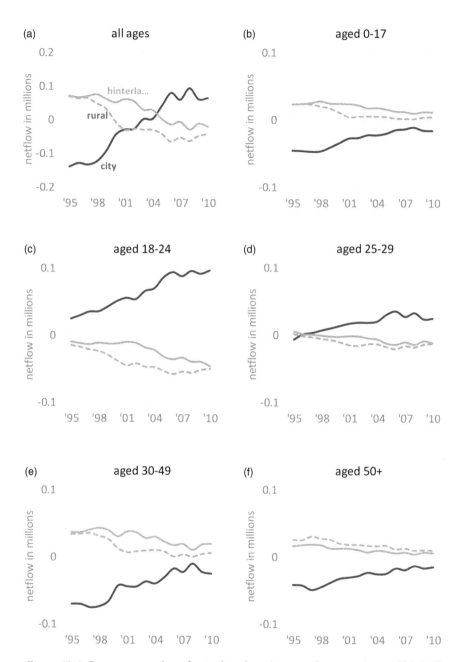

*Figure 10.4* Germany: number of net-migrations, by type of move and age, 1995–2010.
Source: Author's calculations.

of net migration among 25–29 year olds were similar to those of 18–24 year olds, although the increase in net gains in cities was less pronounced and levelled off in the mid-2000s. It appears that urban centres in general have become increasingly attractive across several life course stages, ranging from entry into education and entry into the labour market to marriage and family formation.

The question that now arises is whether the trend is spatially homogeneous: do all cities have the same degree of attractiveness to migrants? This can be answered by looking more closely at the rates of net migration for the counties in which the 40 largest German cities are located. Their net migration rates are shown in Figure 10.5, revealing that most of the largest cities have indeed experienced a growing attractiveness. The highlighted case of Muenster, one of the country's largest university towns, underwent a steady transition from net loss in the late 1990s to net gains in the late 2000s. This transition has been most pronounced among East German towns like Halle and Magdeburg, which recorded the strongest net losses in the late 1990s. However, some cities like Karlsruhe, also highlighted in Figure 10.5, have not performed so strongly, with their recoveries peaking in 2007.

An examination of the age-specific net migration rates for these cities (not shown here) reveals that, while net migration among students (i.e. the 18–24 year olds) is still positive, it is the negative net migration for those entering the labour market (i.e. the 25–29 year olds) and also for 30–49 year olds that cause the overall net migration losses. One may speculate that students who graduate in these cities move to other urban centres with more attractive job markets and better urban amenities. The 37% increase in inter-city

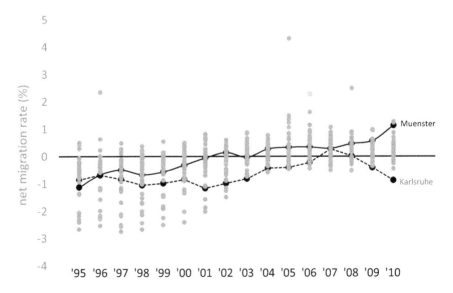

*Figure 10.5* Net migration rates for Germany's 40 largest cities, all ages, 1995–2010.
Source: Author's calculations.

*Table 10.3* Correlation between net migration rate for 18–24 year olds and selected indicators, 2010, for Germany's 40 largest cities

| Variable | R |
| --- | --- |
| studio rent (€/m²) | 0.33* |
| students in % of population | 0.48* |
| unemployment rate | −0.35* |
| GDP per capita | 0.14 |

Source: Author's calculations. Asterisks indicate significance at the 5% level.

movements over the period 1995–2010 shown in Figure 10.3 supports this notion.

Explanations for these differences in migration attractiveness can be sought through association with four variables that can be expected to influence the intensity of migration to and from larger cities: GDP per capita, unemployment, rent per square meter for a studio apartment, and the size of the student population as an indicator of higher education. For the analysis presented here, the association of net migration rates among 18–24 year olds with these indicators was explored. The Pearson correlation coefficients given in Table 10.3 show a moderate positive association of net migration with rent prices and the size of the student population. Hence, the larger the student population and higher the rent (the former can be assumed to be at least in part causing the latter), the higher the net gain among 18–24 year olds. The level of unemployment is negatively associated with net migration, whereas GDP per capita is no significant predictor of migration.

While the correlations in Table 10.3 shed some light on the drivers of net migrations gains in Germany's larger cities, the four selected indicators clearly fall short of fully explaining the differences and similarities in migration intensities across space and time, even though three of the four indicators are significant predictors of net migration to cities. Moreover, the relative importance of each indicator can be expected to vary across places. In Berlin, for example, net migration among 18–24 year olds increased steadily from 1.6% in 1995 to 4.1% in 2010, despite a relatively small student population (4% of its county's population) and relatively high unemployment (14.7%) in 2010.

## Conclusions

Despite the important consequences of changes in the intensity of migration flows on socio-economic, political and demographic conditions, few attempts have been made to identify the trends in migration intensity in Germany below the state level. The analysis presented in the previous section has drawn on the annual inter-county migration flow matrices for the period 1995–2010 in the GIM database (Sander, 2014) and a regional classification of the BBSR to determine whether Germany has followed the US trend of declining migration rates. It has been found that, unlike in the

USA, the overall intensity of internal migration was highly stable over this period and has been proceeding at a significantly lower level than in many other Western societies, including the USA, Australia, the UK and Sweden (Bell *et al.*, 2015).

The stability in the intensity of inter-county migration in Germany may be traced to the settlement pattern, the housing market conditions and government policy regimes that aim at the even growth of regions. Swings in the housing market, which in part triggered the decline in migration intensities in the USA, are much less pronounced in Germany. The real estate market is quite highly regulated, so that average house prices have been relatively stable in real terms between 1970 and 2010. Moreover, high transaction costs and discounted longer-term rental contracts tend to inhibit frequent housing-related moves.

The stability of overall migration, however, masks substantial variations in the age-specific trends. The analysis of movement intensities by region type, distance moved and life course stage (using age as a proxy) shows that since 1995 two broad trends have combined to keep the overall intensity of migration at a stable level: an increase in longer-distance migration to the larger urban centres among 18–24 year olds, and the widespread decline in movements over shorter distances down the urban hierarchy among 30–49 year olds. The two processes are interrelated, as young adults are attracted to move to the cities and families are increasingly inclined to stay in the cities for similar reasons related to education, employment and services such as childcare.

The capital Berlin has seen a particularly strong increase in inflows among young adults from a diverse array of origin regions, coupled with a decline in outflows among families to the surrounding hinterland. However, the growing attractiveness of cities as a place to live and raise a family is not universal. In cities like Munich and Hamburg, families continue to leave the city for destinations in the hinterland as the city's high property prices cause even dual-worker couples to prefer to commute to work. But in recent years the housing market has also tightened in many other larger cities, potentially leading in the future to a slowing of the re-urbanisation trend and the displacement of city-dwellers with lower incomes to more affordable places outside the cities. These ebbs and flows of migration to the larger cities highlight the fact that there is no longer a universal increase or decrease of intensity, nor a unidirectional shift of population up or down the urban hierarchy.

# References

BBSR. 2009. *Siedlungsstrukturelle Kreistypen*. Bonn: Bundesinstitut für Bau-, Stadt- und Raumforschung. www.bbsr.bund.de/cln_032/nn_340582/BBSR/DE/Raumbeobachtung/Raumabgrenzung/SiedlungsstrukturelleGebietstypen/Kreistypen/Downloadangebote

Bell, M., Charles-Edwards, E., Ueffing, P., Stillwell, J., Kupiszewski, M. and Kupiszewska, D. 2015. Internal migration and development: comparing migration intensities around the world. *Population and Development Review*, 41(1), 33–58.

Blake, M., Bell, M. and Rees, P. 2000. Creating a temporally consistent spatial framework for the analysis of interregional migration in Australia. *International Journal of Population Geography*, 6(2), 155–174.

Bucher, H. and Heins, F. 2001. Internal migration between the German states. In Nationalatlas Bundesrepublik Deutschland, Band 4: Bevölkerung, published by Institut für Länderkunde Leipzig. Heidelberg: Spektrum, 108–111 [In German].

Clark, W.A. and Drever, A.I. 2000. Residential mobility in a constrained housing market: Implications for ethnic populations in Germany. *Environment and Planning A*, 32(5), 833–846.

Hochstadt, S. 1999. *Mobility and Modernity: Migration in Germany, 1830–1989.* Ann Arbor: The University of Michigan Press.

Hochstadt, S. 1983. Migration in pre-industrial Germany. *Central European History*, 16, 195–224.

Hubert, M. 1998. *Germany in Transition: The History of the German Population since 1815*. Stuttgart: Franz Steiner Verlag [In German].

Moch, L.P. 2011. Internal migration before and during the Industrial Revolution: The case of France and Germany. In European History Online (EGO), published by the Institute of European History (IEG), Mainz. www.ieg-ego.eu/mochl-2011-en

Kemper, F.-J. 2008. Residential mobility in East and West Germany: mobility rates, mobility reasons, reurbanisation. *Zeitschrift für Bevölkerungswissenschaft*, 33(3–4), 293–314.

Kemper, F.-J. 1997. Regional change and population geography disparities in Germany. Internal migration and interregional decentralisation of population in the old West German states. In Akademie für Raumforschung und Landesplanung (ed) *Räumliche Disparitäten und Bevölkerungswanderungen in Europa. Regionale Antworten auf Herausforderungen der europäischen Raumentwicklung.* Forschungs- und Sitzungsberichte Akademie für Raumforschung und Landesplanung 202. Hannover: ARL, 91–101 [In German].

Kontuly, T. 1991. The deconcentration theoretical perspective as an explanation for recent changes in the West German migration system. *Geoforum*, 22, 299–317.

Kontuly, T. and Vogelsang, R. 1989. Federal Republic of Germany: The intensification of the migration turnaround. In Champion, A.G. (ed) *Counterurbanisation. The Changing Pace and Nature of Population Deconcentration.* London: Edward Arnold, 141–161.

Milbert, A. and Sturm, G. 2016. Internal migration in Germany between 1975 and 2013. *Informationen zur Raumentwicklung*, 2, 121–144 [In German].

OECD. 2011. *Regional Outlook: Building Resilient Regions from Stronger Economies. IV, Country Notes: Germany.* Paris: Organisation for Economic Co-operation and Development, 248–249.

Openshaw, S. 1983. *The Modifiable Areal Unit Problem*. Concepts and Techniques in Modern Geography 38. Norwich: Geo Books.

Sander, N. 2014. Internal Migration in Germany, 1995–2010: New Insights into East-West Migration and Re-urbanisation. *Comparative Population Studies*, 39(2), 217–246.

Wendt, H. 1993. Migration to and within Germany under special consideration of East-West migration. *Zeitschrift für Bevölkerungswissenschaft*, 19(4), 517–540 [In German].

# 11 Italy

## Internal migration in a low-mobility country

*Corrado Bonifazi, Frank Heins and Enrico Tucci*

This final chapter in the national case-studies section of the book takes the case of Italy. This is traditionally considered a 'rooted' society, but with some historical exceptions like the periods of intense international emigration overseas between 1890 and 1914, the guest worker migration to other European countries between 1950 and the early 1970s (Bonifazi, 2013a; Pugliese, 2002) and the concurrent internal movements from the South to the Centre-North of Italy (Bonifazi, 2013a; Bubbico, 2012). As noted in Chapter 4, internal migration in Italy continues to be a relatively rare event compared to most other countries, but, as we shall see, the frequency of people changing address has not fallen much since the early 1980s and indeed has risen somewhat in recent years.

Impressively, as shown below, one of Italy's migration-data time series spans the best part of a century, with data on inter-municipality moves starting in 1929. The length of this series provides an excellent basis for trying to separate out the role of short-term period effects from any longer-term trend in migration intensities. This chapter therefore aims to describe the long-term trends of migration within Italy and identify the main factors that explain the intensity of internal migration and its changes, but with a particular focus on the developments taking place in the last 25 years.

Our chapter starts by setting out the distinctive features of the national context, including the tradition of low migration rates and the evolution of migration patterns over the decades since records began. There follows a brief description and assessment of the data sources available for monitoring internal migration for Italy. The third section examines in more detail the levels and trends in internal migration over the last quarter of a century. The fourth section assesses the various potential factors responsible for the relatively low migration propensities as well as for the most recent trends observed. The chapter concludes with a discussion of the role of substitutes for internal migration, such as international migration, temporary movements and other forms of mobility.

### National context

The evidence for Italy currently being a low-mobility country is presented in Table 11.1, which we have compiled from the latest round of population censuses (and survey data for the USA) for a selection of economically more developed countries that include counts of all changes of address between dwellings and allow the separation of national and foreign populations. Of the seven countries shown, Italy displays a far lower internal migration propensity than the others, with just 5.5% changing address compared to the 11–15% of the others. To underline the 'national' dimension behind this, the table also demonstrates that it is the Italians that are driving this, with foreigners having a rate of address-changing which, at 10.6%, is almost twice that of the former.

Moreover, this low-mobility behaviour is not a recent phenomenon reached after years of continuous decline in rate and then accentuated by the financial crisis of 2008, as in the USA. Instead, in terms of the rates of inter-municipality migration shown for 1929–2013 in Figure 11.1, for much of the last 30 years the level has been fluctuating between 2.0 and 2.3%, much the same level as recorded during World War II. On the other hand, this range of rates is below that experienced in the 1930s and during 1954–1974 when it was mainly between 2.5% and 3.0% and peaked at over 4.0% in 1962. At the same time, it is also clear from Figure 11.1 that since 1990 the general trend has been upwards rather than downwards, though the sudden blip in 2012 should be ignored because it was a temporary feature resulting from administrative changes (see the next section).

What lies behind these fluctuations in inter-municipality migration and the move to the considerably greater degree of stability after the early 1970s? During the 1930s, internal migration increased in Italy, as for many European countries (Millward and Baten, 2010), as a result of the decrease of international emigration arising from the almost total closure of US

*Table 11.1* Proportion of people living in the same dwelling as 1 year ago, for selected countries, by citizenship, percent

| Country (date of census/ survey) | Total population | National population | Foreign population |
|---|---|---|---|
| Italy (09/10/2011) | 94.5 | 94.8 | 89.4 |
| France (01/01/2010) | 89.1 | 89.1 | 88.4 |
| United Kingdom (27/03/2011) | 89.0 | 89.8 | 78.6 |
| Portugal (21/03/2011) | 88.8 | 89.4 | 73.4 |
| Sweden (31/12/2011) | 88.6 | 89.1 | 81.2 |
| Denmark (01/01/2011) | 87.9 | 88.3 | 81.5 |
| United States (2013) | 85.5 | 85.6 | 84.0 |

Source: For Italy, Istat Data warehouse of the population and housing census 2011 [http://dati-censimentopopolazione.istat.it/]; for the other European countries that report all changes of residence, Eurostat Population and Housing Census Database [http://ec.europa.eu/eurostat/web/population-and-housing-census/census-data/database]; for USA: American Community Survey 2013 [www.census.gov/acs]. See C. Bonifazi and F. Heins, 'Poco, ma si muove', in Neodemos, July 3rd 2015 [www.neodemos.info/poco-ma-si-muove].

Note: The data exclude those born, and the persons who emigrated and immigrated, during the previous year.

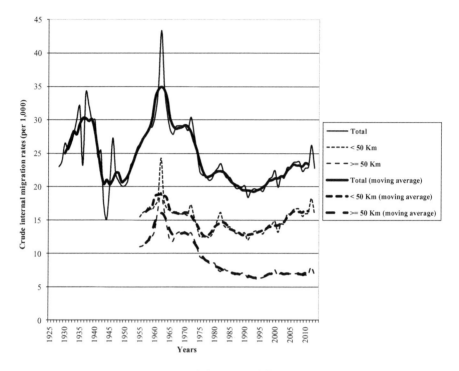

*Figure 11.1* Italy: crude rates of inter-municipality migration, by distance, 1929–2013.

Source: Authors' calculations from Istat data: 'Iscrizioni e cancellazioni all'anagrafe per trasferimento di residenza'; population estimates.

Note: The distance is estimated at the provincial level: before 1995 for 95 and thereafter for 107 provinces. The moving average is calculated over 5 years.

borders. In the case of Italy, this coincided with the creation and development of national industrial districts and occurred despite the anti-migration law introduced by fascists to control and reduce rural-urban migration. In fact, fascist policy against migration culminated in Law 1092 of July 6th 1939 (the law to counter moves to urban centres), which limited the ability to transfer someone's residence to municipalities with more than 25,000 or to industrial municipalities if this was not clearly motivated by public obligation (Treves, 1976; Gallo, 2012).

Then, for more than 20 years following World War II and its immediate aftermath, Italy attained extraordinary and unexpected levels of economic growth, especially in the northern part of the country. For the first time in Italian history, internal migration became a viable alternative to emigration and played, as in other countries during this period, a crucial role. Millions of Italians moved from inland to coastal areas, from the hilly and mountainous zones to the plains, from rural to urban areas, from small to large cities, from the North East to the North West and from the *Mezzogiorno* (the southern areas of Abruzzo, Apulia, Basilicata, Calabria, Campania, Molise, Sardinia and Sicily) to the Centre and North. During this period, long-distance moves were of great importance, with the Fordist economy leading to considerable flows into the triangle of the North West formed by Turin, Milan and Genoa. Meanwhile, the economy of Rome was driven by the expansion of the central administration. During the 1950s rural-urban flows, over short as well as long distances, were also important.

The two oil-price hikes of the 1970s brought this growth phase to an end and led to a significant and enduring drop of longer-distance internal migration, as shown by the rate for moves of 50 km and over in Figure 11.1. The background to this development is that during the 1980s clusters of small family conducted firms developed in parts of central and north-eastern Italy comprising Tuscany, Umbria, Marche, Emilia-Romagna, Veneto, Friuli Venezia Giulia and Trentino and South Tyrol—the so-called 'Third Italy' coming after the industrial and Fordist first Italy and the less developed South (Dunford and Greco, 2005). This part of Italy started to attract more migrants, but never at the sorts of levels observed during the period of mass production (Bonaguidi and Terra Abrami, 1996; Bonifazi, 2013a; Bonifazi and Heins, 2000; Coorti and Sanfilippo, 2009; Gallo, 2012; Ginsborg, 1989; Pugliese, 2011; Rees *et al.*, 1998).

By contrast, shorter-distance inter-municipality moves reached a minimum in 1977 and, following a short-lived uplift, began rising steadily from the mid-1990s onwards. As a result, their share of all inter-municipal changes of residence has increased from its 1950s level of under 60% to close to 70% in recent years. The new upward trend was driven initially by short-distance flows, and especially by the within-Italy moves of foreigners. More recently, the latter has grown further in importance, with their contribution rising to 18.3% of all internal flows by 2013, which is well in excess of their 6.8% share of total population at the latest census. The contribution of high-school and university graduates has also increased (Impicciatore and Tuorto, 2011; Piras and Melis, 2007; Viesti, 2005). Besides these demographic factors, attention has also been paid to the sociological (Pugliese, 2002, 2011) and economic (Basile and Causi, 2007; Bonifazi, 2013b; Mocetti and Porrello, 2010) underpinnings of this resurgence in migration, including the role of labour and housing markets (Angelini *et al.*, 2013; Caldera Sánchez and Andrews, 2011).

This new rise of internal migration has been the focus not just of academic study but also of public interest, especially the movements from the South to the Centre and the North of Italy. In public debates, the increase in internal migration out of the South, like emigration abroad, was mostly seen as being due to a failure of the Italian economy and a negative response to the socio-economic disparities in Italy. We provide a fuller discussion of both the theoretical significance and policy implications of this development later in the chapter, after presenting a fuller account of the migration trends and—before that—the data sources used to monitor them.

## Data available for monitoring internal migration

There are two main sources of internal migration data for Italy: population registers and the housing and population censuses. This section describes the nature of the data that they provide and assesses their strengths and weaknesses.

The continuous recording of Italian migratory flows is based on the population registers (*anagrafi*) that are kept at municipal level, where the administration collects the forms that people who change their municipality of residence have to fill out. The information based on this source pertains to the main features of the changes of residence as well as to the characteristics of the person involved. This provides the data on annual inter-municipality flows plotted for 1929–2013 in Figure 11.1.

The process of producing internal migration statistics based on the register-based records is carried out by the Italian National Institute of Statistics (Istat). It has been extensively revised over the decades and the quality of the data has been considerably improved in terms of timeliness, coverage and consistency with other statistical and administrative data sources. These improvements affect the quality of the reporting of migration events, but the volume of migration flows is felt to be quite well reported even in the past. One point to note, however, is that the data refer to the *de jure* legally resident population, and not to the *de facto* one. This aspect is not always fully consistent with the international concepts and definitions as set out by the United Nations (1998).

Another point to note is that the number of changes of residence displays short-term fluctuations that can be linked in some cases to the decennial population censuses and the operations of aligning the census and the population registers. For instance, the temporary jump in the number of inter-municipal moves in 2012, evident in Figure 11.1, results from the introduction of new regulations regarding their registration. Starting in May 2012, the time to conclude the procedure was considerably shortened, leading to a significant, one-off increase in their number in the rest of that year. Instead of registering the change of residence at the date when the entire administrative procedure is completed, the date now refers to the moment of declaration of the change of residence. The new procedure, termed 'changes of residence in real time' (Circular 9/2012 of the Ministry of the Interior), allows Istat to receive the information almost in real time, with a marked reduction of the time lapse between the notification of the move and its transmission for statistical purposes. In addition, the total elimination of paper forms now allows a faster and more efficient communication between municipalities and improves the quality of the data transmitted from the *anagrafi* to Istat. The inclusion of the five-year running averages in Figure 11.1 helps to portray the general trends eliminating the short-term fluctuations.

The other main source of migration data is the population and housing census carried out by Istat every 10 years, most recently in October 2011.

This source is important for statistics on internal migration and it provides the starting point for demographic accounting. It gives direct information on the previous place of residence one year and five years prior to the census, allowing estimations of overall population movements in the preceding year or in the five years prior to the enumeration. One further valuable feature is that the census-based data includes moves between dwellings and so provides data on shorter-distance address-changing that does not involve a switch of municipality, unlike the population register and other sources.

On the downside, census data tend to underestimate the number of changes of residence since only the residence one and/or five years earlier is recorded, with the result that multiple and return migrations taking place in the reference period are indiscernible. On the other hand, as mentioned in Chapter 3's discussion about the difference between this type of transition data and the event data provided by registers, the difference between the two for a one-year reference period is likely to be small and will certainly not undermine the finding of Table 11.1 that the address-changing rate shown for Italy is much lower than for the other six countries there.

## Trends in overall rate of internal migration

This section goes into more detail about Italy's internal migration trends drawing on the two sources just described. First, it examines the trends in inter-municipality migration rates using the population registers as in Figure 11.1, but concentrating on the period since 1981 and breaking down the rates by gender and by whether Italian or foreign, the latter just for the latest 10 years as the denominator became available only after 2003 (see below). Then it presents evidence from the population censuses of 2001 and 2011, which, as mentioned above, include within-municipality address-changing.

### *Annual trends in inter-municipality migration since 1980*

Figure 11.2 focuses in on the post-1980 section of Figure 11.1 but, instead of crude migration rate, it uses the alternative measure of the gross migraproduction rate (GMR) (Rogers, 1975; Bell *et al.*, 2002), which standardises for change in the age structure of the population in the same way as a life table does for mortality and survivorship and as the Gross Reproduction Rate does for fertility. Here, GMR is the sum of the age-group-specific migration rates for the age range 0–79 years and can be interpreted as the number of changes of residence over that life span that can be expected on the basis of the propensities of the year under observation assuming survival to age 80 (even if in 2013 this was the case only for 57% of men and for 75% of women). This is done separately for men and women. As in Figure 11.1, a distinction is made between longer- and shorter-distance migration flows based on the 50 km criterion.

*Figure 11.2*  Italy: gross migraproduction rate (GMR) for the 0–79 age span, by gender and distance, 1980–2013.

Source: Authors' calculations from Istat data: 'Iscrizioni e cancellazioni all'anagrafe per trasferimento di residenza'; inter-census population estimates by gender and age.

Note: GMR is the sum of the age-group-specific migration rates for the age range 0–79 years
and can be interpreted as the number of moves per 1,000 people over that life span expected
on the basis of the years under observation and assuming survival to age 80. Distance is estimated at the provincial level: before 1995 for 95 and thereafter for 107 provinces.

The main picture shown by Figure 11.2 is, in fact, not much different
from that in Figure 11.1. The total GMR indicates 1.5–1.6 inter-municipality
moves per person between the ages or 0 and 79 in the early 1980s, a slight
fall to 1.4–1.5 in the early 1990s and then a rise to 1.7–1.8 moves by the 2010s
(discounting the peak in 2012, see previous section). The economic crisis
that started in 2008 led to a contraction of internal mobility in 2009, but this
drop was only temporary despite the persistence of the precarious economic
situation in Italy.

Figure 11.2 also shows that the trend differs by distance of inter-
municipality move. While the longer-distance rate stayed at around 0.5–0.6
moves over this quarter century, the under-50 km rate reached around 1.3
moves per person by 2010, a third higher than that recorded up to the early
1990s. As a result, the short-distance moves increased their share of all

inter-municipality movement from 63% in 1981 to about 70% by 2013. By contrast, there is very little difference between the genders, though the rates for men are generally a little higher, with the main exception of shorter-distance moves in the 1980s.

With the arrival of an increasing number of foreigners since the beginning of the millennium, especially in northern Italy, the role of foreign citizens in the internal migration process—long-distance and especially short-distance—is increasing (Bonifazi *et al.*, 2012). A consistent time series of migration rates by citizenship can be estimated only after 2003, because in that year the official number of regularly present foreign residents increased by more than 40% from 1.80 to 2.57 million. This exceptional increase was mainly due to the huge regularisation process that took place in 2002 and 2003. The GMRs for foreign citizens are compared with those for Italians in Table 11.2, disaggregated by gender and by distance of inter-municipality move as in Figure 11.2.

Table 11.2 shows that the rates for foreigners have changed considerably during this decade. In 2004–2006 foreign citizens had a propensity to move 3.4 times higher than Italian citizens, whereas by 2011–2013 this value had fallen to 2.3. Foreigners continue to display significantly higher internal

*Table 11.2* Italy: gross migraproduction rate (GMR), 2004–2006 and 2011–2013, by gender, citizenship and distance, per 1,000 people

| Category of migration flow | Average for 2004–2006 | | | Average for 2011–2013 | | |
|---|---|---|---|---|---|---|
| | Italians | Foreigners | Ratio Foreigners/ Italians | Italians | Foreigners | Ratio Foreigners/ Italians |
| *All migration flows* | | | | | | |
| Total | 1,627 | 5,517 | 3.4 | 1,770 | 4,124 | 2.3 |
| Men | 1,653 | 5,889 | 3.6 | 1,795 | 3,866 | 2.2 |
| Women | 1,603 | 5,166 | 3.2 | 1,749 | 4,276 | 2.4 |
| *Flows of less than 50 km* | | | | | | |
| Total | 1,107 | 3,842 | 3.5 | 1,212 | 2,881 | 2.4 |
| Men | 1,115 | 4,089 | 3.7 | 1,218 | 2,692 | 2.2 |
| Women | 1,102 | 3,608 | 3.3 | 1,208 | 2,994 | 2.5 |
| *Flows of 50 km and more* | | | | | | |
| Total | 520 | 1,675 | 3.2 | 559 | 1,243 | 2.2 |
| Men | 539 | 1,800 | 3.3 | 577 | 1,174 | 2.0 |
| Women | 501 | 1,558 | 3.1 | 541 | 1,281 | 2.4 |

Source: Authors' calculations from Istat data: 'Iscrizioni e cancellazioni all'anagrafe per trasferimento di residenza'; inter-census total and foreign population estimates by gender and age; total and foreign population by gender and age.

Note: GMR is the sum of the age-group-specific migration rates for the age range 0–79 years and can be interpreted as the number of moves per 1,000 people over that life span expected on the basis of the years under observation and assuming survival to age 80. Distance is estimated at the provincial level for 107 provinces.

mobility than Italians, primarily due to their precarious position in the labour and housing market, but nevertheless a tendency towards convergence is clearly observable. The main factor in the drop in the overall rate for foreigners can be seen to be short-distance moves, dropping from around four to well below three moves in the 0–79 life span.

The changes over time and differences by citizenship are also reflected in calculations of the mean distance of move. Overall, the average distance migrated declined from about 160 km in the early 1980s to about 120 km in 2013, with generally lower values for women than for men and a considerable difference in the mean distance migrated between foreigners (91 km) and Italians (125 km) (see also Bonifazi *et al.*, 2014). A considerable share of long-distance migration flows involves the two major Italian macro areas of the Centre-North and the *Mezzogiorno*. In 2013 40.9% of these long-distance flows fell into this category, 15.4% from North to South and 25.5% from South to North.

As regards the rates by gender shown in Table 11.2, it is found that the differential between men and women for foreign citizens changed radically during this period. Whereas between 2004 and 2008 foreign men displayed the higher values, since 2009 the rate has been higher for foreign women. This might be caused by a considerable number of foreign women, especially from Ukraine, working as *badante*—a person taking care of an elderly individual and sharing their accommodation.

The data on age in the population register system also allows the calculation of the mean age of movers and how this has altered over time, enabling linkage to the life course. In the case of Italian citizens in the 0–79 age span, short-distance migrants have a lower mean age than long-distance ones: for the former, 30.6 years for women and 32.5 for men, compared to 32.1 and 33.6 years respectively for long-distance moves. Regarding the trend for Italians, the mean age for long-distance flows has declined slightly since the 1980s for men but the trend is less regular for women, while for their short-distance flows the mean age has increased since the 1990s for both genders. For foreign citizens, the picture is rather different: mean age at move exceeds 37 years for women for both short- and long-distance moves, but the mean age for men is lower at 31.5 on average for recent years for short-distance movers and 34.7 for long distance movers. The average distance of migration flows is closely linked to the age structure of the migrants and the age-specific migration intensities. Short-distance or residential moves are typical for families with children and do not involve a change of the daily activity space. They might be motivated not only by the aim to improve the residential situation but also by the loss of a lease, in other words forced on the individuals and families. By contrast, longer-distance moves tend to be driven more by personal reasons or by motivations linked to educational or labour market opportunities, as well as having much less involvement of families with children.

*All distances of address-changing from the 2001 and 2011 censuses*

The migration data from the census extends the picture by including dwelling changes within the same municipality. Between the latest two censuses a slight increase in address-changing was observed. As shown in the top panel of Table 11.3 (first four data columns), 94.3% of people in the 2011 Census (and alive 12 months earlier) had not changed addresses in the pre-census year compared to 94.9% in the 2001 Census. Of the address-changer share, in the year before the 2001 Census 2.7% of the population changed dwelling within the same municipality during the previous year, and 2.4% moved to a different municipality, whereas the respective rates from the 2011 census were 3.1 and 2.6%.

The 2011 rates shown in Table 11.3 also confirm the considerable divergence in rates between Italians and foreigners: 8.0% of foreigners changed dwelling in the same municipality, compared to only 2.8% of Italians, while 4.5% opted for a different municipality as against 2.4% of Italians. These differences are even more apparent when extending the comparison to five years prior to the 2011 Census, shown in the last four data columns in Table 11.3's top panel: whereas 83.6% of the Italian population still lived in the same dwelling as five years before, this was the case for only 56.5% of the foreign resident population.

The three lower panels of Table 11.3 display the equivalent data for the three 10-year age groups with the highest address-changing rates. The 15–24 age group is distinctive in seeing no change between the two censuses in their one-year migration rate, which stayed at 6.8%, whereas increases in rate are seen for the two older age groups shown, as is also the case for the other three age groups up to 65–74 (not shown in the table). This general pattern of increase is the result of rises in the rates of moving both within and between municipalities. The reason why the rate for the 15–24 year olds did not change is that an increase in within-municipality moves was offset by a reduction in between-municipality moves. This could be a clear indication of a late age schedule of internal migration intensity of the Italian population. A similar comparison between censuses cannot be done for five-year address-changing because this information is not available from the 2001 Census.

The age-group panels of Table 11.3 also show that the differential in migration rates between Italians and foreigners evident in the aggregate data from the 2011 Census largely persist across the age groups. Generally (and this includes the three older age groups not shown), the foreigners' one-year moving rates are at least twice as high as those for Italians. The sole exception is provided by the 25–34 year olds, for whom the foreigner rate is only one-third higher than that for Italians, this being due to the between-municipality migration rate for foreigners being lower than for Italians. This exception is, however, not found in the five-year rates, where the foreigners' rate is always the higher, with the relative difference being greatest for the 15–24 year olds for both within- and between-municipality moves.

Table 11.3 Italy: extent and nature of address-changing during the 1 and 5 years before the 2001 and 2011 Censuses, percent

| Age group | One-year migration | | | | Five-year migration | | | |
|---|---|---|---|---|---|---|---|---|
| Year and citizenship | Same dwelling | Changed address | Moved within munici-pality | Changed munici-pality | Same dwelling | Changed address | Moved within munici-pality | Changed munici-pality |
| **All ages** | | | | | | | | |
| 2001 total | 94.9 | 5.1 | 2.7 | 2.4 | | | | |
| 2011 total | 94.3 | 5.7 | 3.1 | 2.6 | 82.2 | 17.8 | 10.5 | 7.3 |
| 2011 Italians | 94.8 | 5.2 | 2.8 | 2.4 | 83.6 | 16.4 | 9.6 | 6.8 |
| 2011 Foreigners | 87.5 | 12.5 | 8.0 | 4.5 | 56.5 | 43.5 | 25.7 | 17.8 |
| **15–24** | | | | | | | | |
| 2001 total | 93.2 | 6.8 | 2.7 | 4.1 | | | | |
| 2011 total | 93.2 | 6.8 | 3.2 | 3.6 | 83.9 | 16.1 | 10.2 | 5.9 |
| 2011 Italians | 93.8 | 6.2 | 2.7 | 3.5 | 85.4 | 14.6 | 9.4 | 5.2 |
| 2011 Foreigners | 86.4 | 13.6 | 8.6 | 5.0 | 59.1 | 40.9 | 24.8 | 16.1 |
| **25–34** | | | | | | | | |
| 2001 total | 89.4 | 10.6 | 5.3 | 5.3 | | | | |
| 2011 total | 87.2 | 12.8 | 6.3 | 6.5 | 62.7 | 37.3 | 20.2 | 17.1 |
| 2011 Italians | 87.7 | 12.3 | 5.8 | 6.5 | 64.3 | 35.7 | 19.3 | 16.4 |
| 2011 Foreigners | 83.6 | 16.4 | 10.2 | 6.2 | 47.5 | 52.5 | 29.3 | 23.2 |
| **35–44** | | | | | | | | |
| 2001 total | 94.2 | 5.8 | 3.3 | 2.5 | | | | |
| 2011 total | 92.6 | 7.4 | 4.2 | 3.2 | 72.4 | 27.6 | 16.0 | 11.6 |
| 2011 Italians | 93.2 | 6.8 | 3.7 | 3.1 | 73.9 | 26.1 | 15.1 | 11.0 |
| 2011 Foreigners | 87.7 | 12.3 | 7.9 | 4.4 | 54.9 | 45.1 | 26.5 | 18.6 |

Source: Authors' calculations from Istat data: general population census 21/10/2001 and 09/10/2011.

Note: All ages refers to 1 year old and over for one-year migration and 5 years and over for 5-year migration. The percentage values refer to the population eligible (age, residency in Italy): persons not born or not resident in Italy 1 year or 5 years before the population census are excluded. No 5-year data for 2001.

Figure 11.3 adds a gender dimension to this picture. The age schedule of foreign women stands out because they have, compared to their male counterparts, lower rates in the peak years of mobility and a second hump with higher rates for the older adult age groups. This unusual age schedule can be attributed to the above-mentioned group of foreign caregivers, mostly from Eastern Europe, living in with the person for whom they are caring. These *badanti* have to change residences (and employers) frequently because of the death of the person for whom they care.

What lies behind the patterns in Figure 11.3? Intra-municipal moves are, especially in the case of the Italian population, more family oriented and hence have a more important component in the first years of life, a somewhat delayed schedule in adult life and a significant component in the last years of life when the elderly are moving in with one of their children. The family orientation might also be the cause of the slight delay of the age schedule of men considering the usual age differences between partners. However, older men seem to move less, probably because they are cared for in their own home by the partner, or more bluntly, they die before having to move. In contrast, the age schedule of inter-municipal moves is narrower with a higher peak for young adults. Similar gender differences are observed.

Compared to other countries (Bernard *et al.*, 2015), the Italian age schedules reported in Figure 11.3 show a generalised delay in reaching the high-mobility peak age. This delay will be explored in detail below since it partially explains the low overall levels of internal migration in Italy.

Finally, it should be noted that the number of inter-municipal moves in population registers and the one that can be deduced from population census are quite different. This difference is, as indicated earlier, to be expected given that the population register is based on the administrative principle of legal (*de jure*) residency and implies an act of registration in the municipal population register, whereas the population census relies on the statistical concept of change of usual dwelling. It also should be noted that the census encountered major difficulties in counting the foreign residents.

## Explaining the observed trends

The basic finding from the empirical evidence examined above is that Italian internal mobility was already very low by the 1980s, such that the slight increase in internal migration propensities since the beginning of the 1990s, which seems to be less marked in the last few years, would be only a kind of 'normalisation' of the Italian situation. Based on a review of the literature, two major strands can be put forward to account for this: one focuses on the importance of regional socio-economic disparities in Italy and the other relates to individual behaviour, most notably the phenomenon of delayed transitions to adulthood typically observed in Italy. Obviously, this latter strand will also lead to an explanation at the macro level, since the determinants of late nest-leaving are anchored in the socio-economic and socio-cultural factors of Italian society. The two strands are now examined

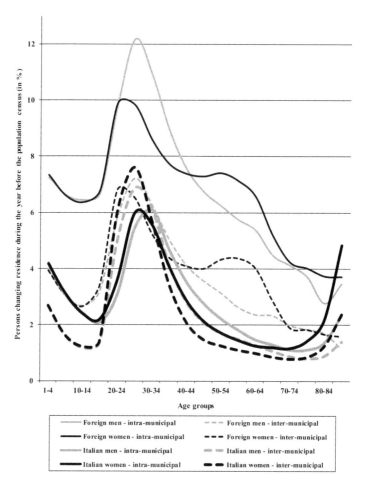

*Figure 11.3* Italy: persons changing dwelling during the year before the 2011 population census, by age, gender, citizenship and type of move, percent.

Source: Authors' calculations from Istat data: general population census 09/10/2011.

Note: Percentage values refer to the eligible population: persons not born or not resident in
Italy 1 year prior to the population census are excluded.

within the general framework of all possible drivers of internal mobility and
the constraints at work set out in Chapter 2.

Traditionally, Italy's internal migration processes were observed through
the lens of the migration flows between the country's two major territorial divisions: the *Mezzogiorno*, characterised by unfavourable economic conditions
like low gross domestic product and high unemployment rates, and the generally well-off Centre-North. Before the mid-1970s, the migration rate from
the former to the latter was generally above 10.6 per thousand. Thereafter it

declined, averaging 5.7 per thousand during 1980–2014, though fluctuating between 4.2 and 7.2 per thousand with a peak around the year 2000, while the reverse flow was quite constant at 3.4 per thousand. The age pattern of this mostly long-distance internal migration was always very selective with the net loss from the *Mezzogiorno* predominantly comprising young adults. Only for a few age groups, especially older adults, were the flows between these two divisions more or less in balance, partly due to their low out-migration from the South and to return migration from the Centre-North. This pattern of movement is an expression of the continuation of regional disparities in internal migration rates in Italy that are caused by persistent differences in economic performance and unemployment (Bonifazi, 2013a; Bubbico, 2012; Mocetti and Porrello, 2010), even if this general characterisation has its exceptions since some areas of the South have been showing some growth potential.

The 2008 global financial crisis will no doubt have intensified the push to change residence. However, the effect of short-term economic cycles on changes of residence in Italy are difficult to confirm, since fluctuations linked to administrative procedures governing the population registers seem to be more important. For example, there was a reduction in changes of residence in 2009, especially in the case of short-distance migration flows. However, a detailed analysis of long-distance migration flows conducted by the authors shows that at this time flow intensity also increased in specific cases, probably linked to return migration—for example between the Bologna area (an archetype of the Third Italy) and Naples. In similar vein, Impicciatore and Strozza (2015, p. 110) have observed:

> With regard to internal mobility, the crisis has had a more modest effect than on international migration. On the whole, only a slight inflection is noted in the short-range mobility that is widely tied to the phases of the individual and family life cycle and to residential choices.
>
> (authors' translation)

The demographic factors that have contributed to the observed changes in internal mobility are perhaps the most obvious, notably relating to age structure and citizenship. In Italy, the share of the 20–34 year olds—the most mobile age group—peaked in the mid-1990s at 23.8% and had declined to 17.0% in 2011 (see Table 11.4) and to 16.6% by 2014, so a decrease in the overall number of changes of residence is expected. While the ageing of the Italian population has had this negative effect on migration rates, the increase in the number of foreigners in the population and a decrease in average family size have both had a direct and positive effect.

Aside from the role of demographic structure, a key factor in Italy's low migration rates is the migration behaviour of young adults. Livi Bacci (2008) argues that in Italy the share of young adults in the total population is relatively low and relationships within the family are traditionally strong.

*Table 11.4* Italy: key indicators of socio-demographic change, 1981–2011

| Characteristic | 1981 | 1991 | 2001 | 2011 |
|---|---|---|---|---|
| Total population (1,000) | 56,556.9 | 56,778.0 | 56,995.7 | 59,433.7 |
| Foreign population (%) | 0.4 | 0.6 | 2.3 | 6.8 |
| Population 20–34 (%) | 21.2 | 23.4 | 21.4 | 17.0 |
| Population 6 years and older with at least upper secondary education (%)* | 14.3 | 22.4 | 33.4 | 41.4 |
| Population 20–34 in private households living in the parental home (%) | n.a. | 46.9 | 53.2 | 53.7 |
| Population 15+ economically active (%) (LFS) | 50.4 | 50.1 | 49.8 | 48.4 |
| Population 15+ occupied in agriculture (%) (LFS) | 13.3 | 8.4 | 4.6 | 3.7 |
| Population 15+ unemployed (%) (LFS) | 8.4 | 10.9 | 9.0 | 8.4 |
| Average family size | 3.0 | 2.8 | 2.6 | 2.4 |
| Owner occupied dwellings (%) | 58.9 | 68.0 | 71.4 | 73.2 |

Source: Istat, Serie Storiche [http://seriestoriche.istat.it/]. Data are based on the General Population Census, except in the cases marked LFS = Labour Force Survey.

Note: *having at least a 'diploma di scuola secondaria superiore'.

The family has to fill many gaps and the decision of independence and leaving the parental home is postponed. Even so, while young adults should have good opportunities because of the favourable ratio between the parent and the child generations, the well-known delay of Italian young adults in leaving the parental home and entering the labour market are obvious signs that in reality these good opportunities do not exist.

There is now a large literature on the phenomenon of late nest leaving in Italy (for example, Aassve *et al.*, 2002; Billari and Tabellini, 2010; Giuliano, 2007; Reher, 1998). Billari and Tabellini (2010, pp. 371–372) put it this way:

One of the key features of Italy's low fertility is its connection with a late transition to adulthood. In order to get a comparable tertiary degree, young Italians tend to study longer than their counterparts in other nations. They enter the labour market later. They live with their parents longer than their peers elsewhere. They form a partnership via marriage or cohabitation later, and now they also tend to have their first child later.

It is obvious that the word 'fertility' can be substituted by the word 'internal mobility' since the same mechanism is leading to lower and later internal mobility.

To explain this delay, cultural factors and the role of the economy and the welfare system are often cited. Prolonged co-residence with parents seems a rational choice when facing the insecurity of the job market, the absence of welfare provisions for young adults and the complexity of the housing market. Probably the factors that most influence the relative late nest leaving of young Italian adults are cultural ones relating to the strong family ties that Reher (1998) postulates for most Mediterranean societies. In these societies,

> ... the definitive departure of young people tends to coincide more or less closely with their marriage and finding a stable job. The years between adolescent maturity (ages 18–20 years) and marriage are spent largely within the parental household. If a person gets a job during this period, he or she normally continues to live at home, a strategy that enables the young adult to save for his or her own marriage.
>
> (Reher, 1998, pp. 204–205)

This system continues in contemporary Italy, allowing young adults a drawn-out passage to adulthood that leads inevitably to a lower number of changes of residence. According to Reher (1998, p. 212), '... the family is seen as defending its members against the difficulties imposed by social and economic realities. A child receives support and protection until he or she leaves home for good, normally for marriage, and even later.' Pursuing a higher education degree could result in leaving the parental home, but in many cases this would be a temporary move that may not be counted officially in the population registers.

The major difficulties which young adults continue to face are the inflexibility of the labour markets. As a result of the 2008 crisis, youth unemployment reached new heights and makes it ever more desirable to remain in the parental home. Whereas the total unemployment rate reached 8.4% in 2011 (see Table 11.4) and 12.7% in 2014, unemployment in 2014 was at 39.3% for 20–24 year olds and 18.6% for 25–34 year olds, with corresponding values for the *Mezzogiorno* of over 50% and 31%. Italy lacks the attractive local labour markets that are looking for workers. A rigid labour market is associated with low long-distance migration rates and only the highly qualified tend to be migratory over longer distances.

Housing markets are similarly characterised by high rigidity and renting is often difficult. Family oriented values and high inflation rates in the past led to a high propensity for owner-occupied housing, this sector accounting for 73% of dwellings at the 2011 Census, rising from 40% in 1951 and 59% in 1981. Today the role of public and social housing is rather limited, because several of these programmes have been ended or cut back, but even during the more flourishing periods of social housing people's mobility was rather low since nobody would be willing to give up a once-acquired entitlement.

The high level of homeownership has a very different meaning according to the place of residence. In metropolitan areas, it blocks the housing market because of the costs involved with a change of residence, whereas in many other areas, especially rural areas and the South, the housing seems affordable but there are no jobs that would allow young adults to start living on their own. So, in effect, the rigidity of the housing markets impedes internal mobility. Obviously homeownership by itself does not impede changing homes; it might even facilitate it, since homeownership is also an expression of financial well-being. However, it certainly slows the process of changing residence, since a person would think twice before selling their home and buying a new one, with all the transaction charges and the costs of refurbishing and possibly rebuilding it.

As regards the role played by the emergence of more complex family structures, this is rather difficult to estimate. Certainly the number of single households is also increasing in Italy, but in most cases these are elderly people with a relative low propensity for internal migration. Internal moves linked to divorces might not be so important in Italy. Also the importance of dual-career and dual-earner households might be limited in the case of internal mobility in Italy, as the characteristics of the Italian labour market make moving residence to pursue a career an exception. Regarding short-distance moves, the increase in the number of two-earner couples might be a prerequisite to move, because a certain financial base is needed to purchase a new home, even when being a home owner.

Internal mobility is often linked to participation in higher education. Several studies have followed young Italian adults in their passage to university and into the labour market. As is the case in many other countries, tertiary education is a prime motive for moving and it is often the first step of leaving the parental home (Impicciatore and Tuorto, 2011). However, its effect is somewhat muted in Italy for two reasons. In the first place, as Table 11.4 shows, the share of population with at least upper secondary qualifications—while it has increased greatly over the last three decades— is still only two-fifths, and even in 2011 only 26% of these had participated in higher education (see https://ec.europa.eu/CensusHub2), both relatively low rates compared to other European countries. Secondly, the creation of new universities, especially in southern Italy, now makes it less necessary to change residence to attend university, allowing more young adults to stay in the parental home during their studies.

Finally, we have already mentioned the role of the ever increasing number of foreign residents in leading to an increase in internal mobility, even despite their declining propensity to change residence. Even if rigid housing and labour markets set the context for the foreign population as for Italians, they might operate differently for the two groups. Italy is characterised by an occupational concentration of foreigners in low-status occupations, with the result that their housing demand is towards lower quality, less well-equipped homes, often located in the peripheries of metropolitan areas. Their precarious housing situation leads to more short-distance address-changing and, as already shown, specific professional groups, e.g. *badanti*, change residence rather frequently.

In sum, whereas increasing 'rootedness' was recently introduced to explain the decline of internal mobility in a traditional high mobility country like the USA (Cooke, 2011), it might be considered a traditional value of Italian society. Even in an era facilitating greater mobility and flexibility, we observe in Italy a cultural desire for spatial 'anchoring' or 'rootedness' and it might explain to a degree the low Italian internal migration in the past. Even if public interest in migration flows from the South to the North of Italy continues to be high, only very few policy initiatives concerning the determinants and consequences of internal migration are put into practice.

## Conclusions

This chapter has followed the internal migration in Italy from the boom of internal migration flows in the 1950s and 1960s to the lower levels of the next two decades and to the slight rise since the mid-1990s. We have presented several explanations for the low intensity of internal migration in Italy compared to other countries, but most can be related to the familial values of Italian society. For decades, young Italian adults have postponed the moment of leaving the parental home. At the same time, the stage of studying and professional formation is prolonged, leading to the decision of cohabiting with a partner or marrying being further postponed. These are all factors that contribute to a further reduction in the intensity of internal mobility.

Looking ahead, it would seem unlikely that we will observe a further decline as in other countries like the USA and the UK (Cooke, 2013; Champion and Shuttleworth, 2016a, 2016b). The period of relative intense internal mobility of the 1950s was probably an exception in Italian recent history and linked to a Fordist economy in the absence of emigration of Italians to other countries. In fact, the economic growth in the Third Italy in the 1980s and 1990s could not repeat the pull effect that was generated by the demand for labour under the Fordist system. The decline of the traditional industrial economy and the associated economic restructuring stopped the 'massive' internal migration flows, as also has the contraction of the public sector since its major expansion in the 1960s and 1970s.

The social and political valuation of internal migration depends on the type of society: a flexible and dynamic society might value interregional mobility positively, whereas a traditionally familial one like Italy tends to take a negative view of interregional migration, as well as international emigration. For young Italians, however, internal and international migration becomes an increasingly obvious necessity: the lack of economic opportunities in Italy prompts them to consider international emigration as an alternative to interregional moves. In recent years, also due to the freedom to cross European borders in search of work and educational opportunities, destinations like London and Berlin became very attractive in comparison to destinations within the country and often it seems easier to move abroad, than having to face Italy's labour and housing market rigidities. At the same time, it has been shown that foreign citizens resident in Italy tend to display a higher propensity to change address.

Finally, it is also the case that in recent decades the low propensities of Italians to change residence have led to increases in more temporary forms of mobility; for example, daily or weekly long-distance commuting (Crisci and Di Tanna, 2016; Pugliese, 2011). Indeed, as stated in SVIMEZ (2015), commuting often hides forms of real out-migration. The phenomenon continues to be important despite the continuing economic crisis that has also affected the Centre-North: in 2014, about 120 thousand workers are estimated to be living in the South but working in the rest of the country or abroad.

## References

Aassve, A., Billari, F.C., Mazzucco, S. and Ongaro, F. 2002. Leaving home: A comparative analysis of ECHP data. *Journal of European Social Policy*, 12(4), 259–275.

Angelini, V., Laferrère, A. and Weber, G. 2013. Home-ownership in Europe: How did it happen? *Advances in Life Course Research*, 18(1), 83–90.

Basile, R. and Causi M. 2007. The determinants of migration flows in the Italian provinces: 1991–2001. *Economia e Lavoro*, XLI(2), 139–159 [In Italian].

Bell, M., Blake, M., Boyle, P., Duke-Williams, O., Rees, P., Stillwell, J. and Hugo, G. 2002. Cross-national comparison of internal migration: Issues and measures. *Journal of the Royal Statistical Society A*, 165(3), 435–464.

Bernard, A., Bell, M. and Charles-Edwards, E. 2015. Life-course transitions and the age profile of internal migration. *Population and Development Review*, 41(2), 213–239.

Billari, F. and Tabellini, G. 2010. Italians are late: does it matter? In Shoven, J.B. (ed) *Demography and the Economy*. Chicago: University of Chicago Press, 371–412.

Bonaguidi, A. and Terra Abrami, V. 1996. The pattern of internal migration: The Italian case. In Rees, P., Stillwell, J., Convey, A. and Kupiszewski, M. (eds) *Population Migration in the European Union*. Chichester: John Wiley, 231–245.

Bonifazi, C. (ed) 1999. *Mezzogiorno and Internal Migrations*. Irp-Cnr Monografia 10. Rome: Irp-Cnr [In Italian].

Bonifazi, C. 2013a. *The Italy of Migrations*. Bologna: Il Mulino [In Italian].

Bonifazi, C. 2013b. Mobile by force: Population movements in the Italy of crisis. *Il Mulino*, LXII(5), 798–805 [In Italian].

Bonifazi, C. and Heins, F. 2000. Long-term trends of internal migration in Italy. *International Journal of Population Geography*, 6(2), 111–131.

Bonifazi, C., Heins, F. and Tucci, E. 2012. The internal migrations of foreign citizens at a time of immigration. *Meridiana: Rivista di storia e di scienze sociali*, 75, 173–190 [In Italian].

Bonifazi, C., Heins, F. and Tucci, E. 2014. Internal migrations in Italy in 2011–12. In Colucci, M. and Gallo, S. (eds) *L'Arte di Spostarsi. Rapporto 2014 Sulle Migrazioni Interne in Italia*. Rome: Donzelli, 3–20 [In Italian].

Bubbico, D. 2012. Internal migration from the Mezzogiorno between search for work and occupational mobility. *Meridiana. Rivista di storia e di scienze sociali*, 75, 149–172 [In Italian].

Caldera Sánchez, A. and Andrews, D. 2011. Residential mobility and public policy in OECD countries. *OECD Journal: Economic Studies*, 2011(1), 1–22. doi:10.1787/19952856

Champion, T. and Shuttleworth, I. 2016a. Is longer-distance migration slowing? An analysis of the annual record for England and Wales since the 1970s. *Population, Space and Place*. doi:10.1002/psp.1024

Champion, T. and Shuttleworth, I. 2016b. Are people changing address less? An analysis of migration within England and Wales, 1971–2011, by distance of move. *Population, Space and Place*. doi:10.1002/psp.1026

Cooke, T.J. 2011. It is not just the economy: Declining migration and the rise of secular rootedness. *Population, Space and Place*, 17(3), 193–203.

Cooke, T.J. 2013. Internal migration in decline. *The Professional Geographer*, 65(4), 664–675.

Coorti, P. and Sanfilippo, M. (eds) 2009. *Migrations, History of Italy, Annals* 24. Grandi Opere. Turin: Einaudi [In Italian].

Crisci, M. and Di Tanna, B. 2016. Flexible mobility for unstable workers: South-north temporary migration in Italy. *Polis*, XXX(1), 145–176.

Dunford, M. and Greco, L. 2005. *After the Three Italies: Wealth, Inequality and Industrial Change*. Oxford: Blackwell.

Gallo, S. 2012. *Without Crossing the Borders: Internal Migrations from Italy's Unification to the Present*. Rome and Bari: Laterza [In Italian].

Ginsborg, P. 1989. *Italian History from the Post-war to the Present*. Turin: Einaudi [In Italian].

Giuliano, P. 2007. Living arrangements in Western Europe: Does cultural origin matter? *Journal of the European Economic Association*, 5(5), 927–952.

Impicciatore, R. and Strozza, S. 2015. International and internal migrations of Italians and foreigners. In De Rose, A. and Strozza, S. (eds) *Rapporto sulla Popo-lazione: L'Italia nella Crisi Economica*. Bologna: Il Mulino, 109–140 [In Italian].

Impicciatore, R. and Tuorto, D. 2011. Internal migration and higher education: Family/individual resources and opportunities for social mobility. In *Sociologia del Lavoro*, 121, 51–78 [In Italian].

Livi Bacci, M. 2008. *Forward Youngsters to the Rescue: How to Get out of the Crisis of Youth in Italy*. Bologna: Il Mulino [In Italian].

Millward, R. and Baten, J. 2010. Population and Living Standards 1914–45. In Broadberry, S. and O'Rourke, K. (eds) *Cambridge Economic History of Modern Europe 1700–2000*. Cambridge: Cambridge University Press, 232–263.

Mocetti, S. and Porrello, C. 2010. The mobility of labour in Italy: New evidence on migration dynamics. *Questioni di Economia e Finanza*. Occasional Paper 61. Rome: Banca d'Italia [In Italian].

Piras, R. and Melis, S. 2007. Developments and trends of internal migration. *Econo-mia Italiana*, (2), 437–461 [In Italian].

Pugliese, E. 2002 and 2006. *Italy Between International Migration and Internal Migration*. Bologna: Il Mulino [In Italian].

Pugliese, E. 2011. Internal migration in the Italian migration scene: Innovation, persistences, common places. In *Sociologia del Lavoro*, 121, 19–29. [In Italian].

Rees, P., Todisco, E., Terra Abrami, V., Durham, H. and Kupiszewski, M. 1998. Internal migration and regional population dynamics in Italy. *Essays* 3. Rome: Istat.

Reher, D.S. 1998. Family ties in Western Europe: Persistent contrasts. *Population and Development Review*, 24(2), 203–234.

Rogers, A. 1975. *Introduction to Multiregional Mathematical Demography.* New York: John Wiley.

SVIMEZ. 2015. *Annual Report of SVIMEZ on the Economy of the Mezzogiorno 2015.* Bologna: Il Mulino [In Italian].

Treves, A. 1976. *Internal Migration in Fascist Italy. Politics and Demographic Reality.* Turin: Einaudi [In Italian].

United Nations. 1998. *Recommendations on Statistics of International Migration. Revision 1.* ST/ESA/STAT/SER.M/58/Rev.1. New York: United Nations.

Viesti, G. 2005. New migrations: the transfer of young and qualified workforce from the South to the North. *Il Mulino*, LIV(4), 678–688 [In Italian].

# Part III

# Commentary and synthesis

# 12 Internal migration

## What does the future hold?

*William H. Frey*

The work presented in this volume is path breaking and, in many ways, attempts to pull off the impossible with its central aim of making an international assessment of recent internal migration trends during a period of dramatic global economic and demographic change. In particular, it has sought to discern whether or not the decline in domestic migration rates witnessed in the USA over the past three decades is symptomatic of worldwide changes—a tall order indeed. Fortunately, the editors have assembled a group of scholars who are more than up to the task. They are not only experienced observers of the multi-layered drivers of population movement within their respective countries, but are also expert at piecing together statistics for this most difficult-to-measure of all demographic phenomena.

The comprehensive analysis of Bell, Charles-Edwards, Bernard and Ueffing in Chapter 4 provides the suggestion that there is, indeed, a broad trend toward internal migration decline. This is not only the case in the so-called 'frontier nations' of the USA and Australia, but is also occurring in many other countries, whether their overall level of migration intensity is high (e.g. Taiwan) or low (e.g. Egypt). Nevertheless, migration decline is not found to be universal and especially it is not the case for many countries of the more developed world, including most of those in Europe. As a sociologist demographer who has worked largely on the USA, it is eye-opening for me to now be able to observe both the similarities and the sharp differences that exist across countries in the forces that are impacting internal migration. Yet the key question is: what do they imply about internal migration in the future?

### New generations and age-related migration

Being a demographer, I am especially interested in how changing demographic composition impacts a nation's migration shifts. Clearly, population ageing is at the forefront of these compositional shifts. Given increased life expectancy and lowered fertility in most countries, an ageing population

suggests that there will be proportionately fewer people in what have traditionally been peak migration ages—young adults in their 20s and early 30s
(Rogers and Castro, 1981). As such, the changing age composition should
point to lower overall migration.

Yet there is an important modification of this ageing effect, particularly
in the developed world. This involves some 'flattening' of the long-standing
age/migration relationship because of the reduced migration being observed for today's young adults. While the recent US migration decline
has been pervasive among all demographic groups, it has been especially
pronounced among the group termed the 'millennial generation', born between 1980 and 1995, who is currently ensconced in the peak migration
ages. As young adults, millennials have taken perhaps the greatest economic hit, of any age group, from the Great Recession of 2007–2009 and
its aftermath. As a result, employment opportunities for its members have
dried up, their capacity to buy homes has diminished and, as a generation,
they have put off standard life-course phenomena such as getting married,
bearing children and making the moves that are generally associated with
those events.

Two questions arise in generalizing this US millennial experience to the
future and to other countries. First, is their behaviour simply a 'period effect' which will impact this generation alone? If this is the case, then as the
economy improves, jobs become more available again and housing gets
back within their financial reach, the next generation of young adults will
follow more traditional life course patterns of marriage and childbearing.
Secondly, does this reflect the experience of the USA alone? Even if this is
a period effect, it could be more evident in America than in other nations.
While the Great Recession has made its mark on other countries around the
world, it is possible that its impact is greater in the USA, because migration
is more strongly related to the young adult's sequences of employment and
life course events there.

There is much to suggest that the recent US millennial experience is neither a period effect nor a uniquely American experience. Rather, it may
reflect the fact that future generations in America and elsewhere may show
different patterns than in the past and from each other—trends which will
modify the standard age/migration relationship and, thus, migration levels
in the future. While the Great Recession certainly had an impact on US migration, there are signs that the millennial low-migration phenomenon will
not be short-lived, but reflects a permanent change in living tastes toward
inner city rented housing, a substitution of longer commuting for residential
migration and improved accessibility to a more internet-connected labour
market which could change the age relationship as well as the intensities of
long-distance migration. Just as previous social changes such as increased
women's labour force participation led to migration-changing household
patterns (such as less mobile dual-earner couples), so too can new lifestyles
choices (such as those occurring in a more flexible labour market with

different transport and communication options) alter those for future generations. Such choices may not necessarily lower future migration rates, but they may make residential mobility less dependent on traditional life-cycle patterns.

Of course, upcoming generations will adapt differently depending on their national contexts. While there is evidence of similar urban-based millennial migration patterns in the UK as related by Lomax and Stillwell in Chapter 6, Japan, Italy and Australia provide different social and economic contexts leading to lower young-adult migration and the flattening of the age migration curve. In Chapter 8 on Japan, Fielding tells how many young people, discouraged by their employment and home-buying prospects, remain dependent on their parents for housing and other support into their 30s and sometimes their 40s (though others go off for education and jobs). In Chapter 11 on Italy, Bonifazi, Heins and Tucci emphasise the strong supportive role of the family in delaying migration of young adults in a country with already low migration rates along with rigid housing and labour markets. In Chapter 7 on Australia, Bell, Wilson, Charles-Edwards and Ueffing show a continued right-ward shift in the age migration schedule due to delays in marriage, child-bearing and rising tertiary education, all consistent with reduced overall migration levels.

Not all countries project a reduction in migration because of a changed age/migration relationship. In Chapter 9 Shuttleworth, Osth and Niedomysl indicate a continued rise in Swedish internal migration due to increased migration rates within each age group. They attribute this to rising education- and employment-related movement, especially among young people who are willing to relocate long distances for both. Of course, each country holds a different economic context for this movement, with Sweden making a more swift transition from a manufacturing-based to a post-industrial economy than some other countries. There are other features of a nation's demographic structure which can impact age-related movement such as the dominance of the large baby boom generation in America's labour and housing markets which, as Cooke demonstrates in Chapter 5, has contributed to lower young-adult migration for the past several decades.

Nonetheless, there is plenty of evidence to suggest that members of future generations in most of the countries studied will not neatly adhere to past age-related markers such as post-secondary school enrolment, employment attainment and household formation—which have been closely tied to changes in residence. Lifestyle changes, occupational shifts, more flexible employment opportunities and the substitutability of commuting and tele-commuting for changes of residence may make the age/migration patterns of the future more difficult to predict. Thus, although most populations will be ageing, it is now not as clear as in the past what this means for overall migration levels.

## Spatial restructuring and internal migration

Demographic structure provides only one prism through which to view changing migration levels. The spatial structure of the nation's population can also affect its internal migration. In their comparative national analysis, Bell *et al.* (2015) show a strong positive relationship between a country's urbanisation level and its internal migration intensity.

Of course, internal settlement structures differ across countries and shift over time in ways that may or may not have an impact on national migration levels. In Chapter 8, Fielding chronicles shifts in Japan's spatial economic evolution from the strong rural-to-big-city migration in the 1950s and 1960s to more complex patterns including city-to-city moves and some return migration. This kind of evolution is apparent in some form for many of the developed countries focused on in this book, where the importance of the major urban core areas as migrant destinations has changed as the spatial system has evolved. Irrespective of these national developments, however, broad shifts in the nature of production and commerce can foster migration patterns that may pervade many areas, as evidenced by the widespread counter-urbanisation of the 1970s (Champion, 1989; Frey, 1988).

It is also the case that episodic economic forces can impinge on spatial economies in ways that impact internal migration. Past economic cycles, and especially the recent Great Recession, have shifted migration flows between peripheral and core regions in many of the countries reviewed here, including between the North and South of the UK, between the *Mezzogiorno* and North-Central regions of Italy, the movement to greater Tokyo in Japan and between Northeast and Midwest 'snowbelt' regions and South and West 'sunbelt' regions in the USA. In most cases, broad economic downturns reduced the flows across regions leading to more modest migration gains or even losses for previously growing areas. In the USA, two longstanding rapidly growing sunbelt-region states, Florida and Nevada, registered unprecedented net losses of internal migrants for one or more years in the wake of the Great Recession.

While it is clear that spatial restructuring and episodic economic events can impact the *direction* of migration flows, the question still remains: what does this imply for the nation's overall migration intensity? The US experience during the Great Recession suggests that reductions in net migration flows go hand-in-hand with reductions in overall gross migration intensities. That is, at about the same time as the snowbelt-to-sunbelt migration began to wane (around 2007), so too did internal migration rates at all geographical levels—inter-state, inter-county, and within county. As of 2015, each of these migration levels is near the lowest of any year since annual statistics were first compiled for 1947–1948. The question remains as to whether there will be a pick-up in internal migration, nationwide, as the economy revives and flows to the sunbelt begin to rise.

A more compelling case that shifts in migration direction correlate with shifts in national migration intensity is found in Australia. In Chapter 7,

Bell and colleagues show the evolution of migration flows across a spectrum of places from 'metro remote' to 'metro core' between 1976–1981 and 2006–2011. Shifts, such as those associated with changes in regional housing costs, altered location tastes for young adults and retirees, and remote-area resource booms, changed a distinctly up-the-urban hierarchy flow structure to one that has become far more stable. This has not only led to a lower efficiency of migration flows but also to a downturn in national migration intensity.

At the same time, one should not assume that greater economic and social homogeneity across regions necessarily lowers national migration intensity. Sweden provides a counterexample where there is a relatively even distribution of services, healthcare and education (if not higher education) along with low inequality levels among municipalities. Yet it shows a rising internal migration rate.

Thus, one cannot necessarily generalise about the relationship between spatial restructuring, the direction of dominant migration flows and a nation's internal migration intensity. Again taking into account the generational perspective discussed earlier, the evolving nature of work, lifestyle tastes and telecommuting among new waves of young adults may call these relationships into question.

### The impact of rising international migration

Beyond these restructuring shifts, and new generational preferences, there is another exogenous factor which can affect internal migration levels. This is the rise of foreign immigration which is occurring in many developed countries. While the USA has long prided itself as a nation of immigrants, it faced a long immigration lull from the 1920s through to the 1960s, but then—and especially since the 1980s—foreign migration from Latin America and Asia has made its presence felt. Since then, too, many European countries have shown greater movement across national borders resulting from the Schengen agreement which took effect in 1995, as well as, more recently, greater immigration from countries outside the European Union. More broadly, the increased international migration from developing to developed countries should impact internal movement within both those sets of countries.

Just how can immigration affect national internal migration intensities? To the extent that immigrants compete with native-born migrants, one might anticipate a reduction in the latter's internal migration. But there are two reasons why new immigrants can increase the overall rate of internal movement. In the first place, given that immigrants from other nations tend to cluster, at least initially, in gateway cities that house same-nationality residents, this can lead to increased outflows of longer-term residents who move elsewhere to find housing and employment. This was the case as new waves of immigrants first clustered in US gateway areas such as New York and Los Angeles, provoking a domestic migrant outflow to other parts of the country (Frey, 1996). Now it is also the case for London (Champion, 2016) as well as other global cities.

The second reason is that over time, as immigrants become permanent residents or citizens, they bolster the size of the labour force, particularly in the younger ages. Moreover, because immigrant families already have previous experience of moving, they will be more amenable to moving internally than the native-born. In the USA, due to the high volume of recent immigration, the foreign-born population and children born to them subsequent to their arrival represent almost a quarter of the population, potentially bolstering both the size and the youthfulness of future internal migration. Immigrants and their descendants have now begun to disperse across all parts of the country (Frey, 2015).

Of course, as international migration swells the labour forces of many countries, other countries—in both more and less economically advanced parts of the world—will experience out-migration. In their comparative analysis, Bell *et al.* (2015) find a positive correlation between a nation's internal migration and its international migration rates, but a negative correlation for its remittance rate. This suggests that nations with substantial international out-migration will profit from remittances sent back from those out-migrants, reducing the need for movement within those countries in search of more income.

## Looking ahead

What can be said about internal migration in the future? Quite clearly most countries will be ageing, not least those in the developed world. In Europe, one third of the population will be older than age 60 in the year 2050 (UN, 2015). Thus, a straight-line projection would suggest that internal migration will decline precipitously due to age-composition effects. While this is probably broadly true, there are unknown factors that will impact patterns of population movement within nations.

One of these will be the migration preferences of future generations in light of the changing nature of work, lifestyles and modes of information. It could mean a substitution of circular migration, commuting and even telecommuting for internal migration, thereby making traditional migration rates even lower than projected. The relationships between the spatial restructuring of settlement systems and directional migration flows within nations are also unknown, as networked global knowledge-based industries become more pervasive, making employment more flexible and perhaps leading to less permanence in residential relocations.

Yet as international migration becomes more dominant, especially as ageing industrialised nations face shrinking workforces, many of those nations may well see greater internal movement of the native-born, along with new waves of foreign-born and second-generation residents as well as part-time, part-year additions to their labour forces.

So, while traditional internal migration—as now measured in censuses, population registers and surveys—may decline in the future, it could also become less meaningful as a measure of overall population mobility, given the increased importance of other varieties of movement which will differ both geographically and temporally. This will make the spatial mobility of future national populations even more difficult to monitor than it is today, providing ever greater challenges for demographers.

## References

Bell, M., Charles-Edwards, E., Kupiszewska, D., Kupiszewski, M., Stillwell, J. and Zhu, Y. 2015. Internal migration and development: Comparing migration intensities around the world. *Population and Development Review*, 41(1), 33–58.

Champion, A.G. (ed) 1989. *Counterurbanisation: The Changing Pace and Nature of Population Deconcentration*. London: Edward Arnold.

Champion, T. 2016. Internal migration and the spatial distribution of population. In Champion, A.G. and Falkingham, J. (eds) *Population Change in the United Kingdom*. London: Rowman & Littlefield, 125–142.

Frey, W.H. 1988. Migration and metropolitan decline in developed countries: A comparative study. *Population and Development Review*, 14(4), 595–628.

Frey, W.H. 1996. Immigration, domestic migration and demographic balkanisation in America: New evidence for the 1990s. *Population and Development Review*, 22(4), 741–763.

Frey, W.H. 2015. *Diversity Explosion: How New Racial Demographics are Remaking America*. Washington DC: Brookings Institution Press.

Rogers, A. and Castro, L. 1981. *Model Migration Schedules*. Laxenburg: International Institute for Applied Systems Analysis.

UN. 2015. *World Population Prospects: The 2015 Revision: Key Findings and Advance Tables*.Working Paper No. ESA/P/WP.241. New York: Department of Economic and Social Affairs, United Nations Population Division. https://esa.un.org/unpd/wpp/Publications/Files/Key_Findings_WPP_2015.pdf

# 13 Sedentary no longer seems apposite

## Internal migration in an era of mobilities

*Keith Halfacree*

As foregrounded in Chapter 1 of this volume, the world appears to be getting progressively more mobile and has been doing so for some considerable time. Indeed, as early as 1971, Wilbur Zelinsky asserted in his celebrated 'mobility transition' hypothesis that, 'the most advanced and affluent societies have now achieved a state in which the term "sedentary" no longer seems apposite for their members. Almost constant change and movement have truly become a way of life' (Zelinsky, 1971, p. 247). People are 'in almost non-stop daily, weekly or seasonal oscillation across and within spatial and social zones, indulge in a vast range of irregular temporary excursions, and frequently migrate, in the sense of formal change of residence' (*idem*). More than four decades later, according to Gössling and Stavrinidi (2016, p. 723), the situation has now been reached where '[m]obility, in contemporary society, is not only an option, but also an obligation'.

This is a societal development with huge ramifications, not least in terms of widespread reactions against migration, which were successfully harnessed in the political campaigns that led in 2016 to the 'Brexit' vote in favour of the UK leaving the European Union and to the election of Donald Trump as US President. Concerns have ranged from outright racism to more diffuse worries about access to jobs, health services, housing and so on for existing residents. People, it seems, have been left feeling no longer at home in (local) community but, through external forces of both homogenisation and differentiation, becoming existentially scattered to the four winds. This feeling is not limited to international migration, but has parallels in the antipathy often shown towards city migration to suburbs and rural areas, which try to protect themselves through exclusionary zoning and similar NIMBY (Not In My Back Yard) ploys.

To achieve a better understanding of internal migration today, it is vital to place it within the concept of an 'era of mobilities'. To this end, this assessment of the messages coming through in the previous chapters begins by introducing this era and its relationship to migration and by reflecting critically on how scholarship has traditionally, often implicitly, presented and understood migration. It then focuses in on the present status of internal migration and its links with mobilities more generally.

# Migration in an era of mobilities

## *Movements, liquidities, flows: a zeitgeist?*

In Urry's (2002, p. 161) words: 'there are countless mobilities, physical, imaginative and virtual, voluntary and coerced. There are increasing similarities between behaviours that are "home" and "away"'. The arguments propounded by influential books such as Castles and Miller's (1993) *Age of Migration* and Bauman's (2000) *Liquid Modernity* have strongly converged with more sociological expressions such as Urry's (2007) *Mobilities* and geographical contributions such as Cresswell's (2006) *On the Move*. Together, they suggest how '[a]ll the world seems to be on the move' (Sheller and Urry, 2006, p. 207; see also Larsen *et al.*, 2006; Adey, 2010; Sheller, 2011). Of course, people have moved residentially from place to place throughout humanity's existence (Brettell, 2013), with migration in the early 'modern' period tending to be under-appreciated (Pooley and Turnbull, 1998), but the magnitude and complexity of population flows today is unparalleled. This is the case for more than just people, however, as flux—both experiential and metaphorical—has increasingly displaced fixity and come to predominate within daily life and consciousness. Its central consequence has been what Gale (2009) terms 'de-differentiation', including the transgressing of the categories of 'home' and 'away' that Urry suggested. Thus, much of humanity has seemingly entered—exactly when can be debated elsewhere—an era of mobilities (Halfacree, 2012).

What has caused the epochal shift to an era of mobilities remains a moot point and one which also cannot be engaged with here. However, as implied throughout this chapter, a condition of mobility can be strongly allied to the evolution of capitalism into its present-day, dominant, more 'flexible' or neo-liberal forms. This is not to assert economic determinism but does re-emphasise how daily life is not (re)formed either independently or at any great distance from the underlying economic domain. Clearly, any 'mobilities paradigm' (Sheller and Urry, 2006) for examining the present day must also recognise the place of neo-liberalism and other related conditions such as globalisation as central within its explanatory framework.

Returning to migration, this clearly has a very central place within the era of mobilities, as is consistently noted by its leading scholars (e.g. Cresswell and Merriman, 2011). However, by starting with mobilities—both material and, as noted shortly, immaterial—rather than with the third component of Population Geography's classic demographic triumvirate of births, deaths and migrations (Barcus and Halfacree, 2017), embodied human migration becomes just one element of interest to mobilities scholars. Thus, we may well note the current salience in daily news bulletins of the subjects of the *Age of Migration* (Castles *et al.* 2014), but the scope and impact of mobilities certainly does not stop there. In fact, in part perhaps because migration is now so very prominently studied across numerous sub-disciplines (Brettell and Hollifield, 2008), it can seem relatively rather neglected or at least taken for granted within the mobilities canon, where more novel expressions of movement have grabbed most scholarly attention.

Through a mobilities lens, migration needs to be emplaced first within what Pooley *et al.* (2005, p. 2) termed a 'mobility continuum' (Figure 13.1), whose time-space mapping Bell and Ward (2000) had pioneered. Table 13.1 introduces five broad families of mobilities, with migration again seeming to be rather 'lost' within the first category, not least with its inclusion of quotidian movements. This is a status, however, that is revisited below. Second, as Table 13.1 makes clear and Figure 13.1 hints in its final category, migration needs to be emplaced in the context of diverse material and immaterial mobilities, perhaps the most prominent of the latter being the vast volumes of data flowing through the internet and other telecommunications media. It is movement or travel which can be virtual, imaginative and/or communicative (Gale, 2009). The importance of this material-immaterial engagement was pioneered by Urry's (2002) concept of the 'post-tourist', whereby an element of the perceived de-differentiation (Gale, 2009) between tourism and the daily life was the enhanced immersion in mediated images of tourist places. Subsequent physical travel may still be central to most tourist experiences but is now moulded via the immaterial as never before, not least through social media (e.g. Cohen *et al.*, 2013; Gössling and Stavrinidi, 2016).

Being mobile in diverse ways is, moreover, not only an empirical state within the era of mobilities but also typically portrayed as a normative state

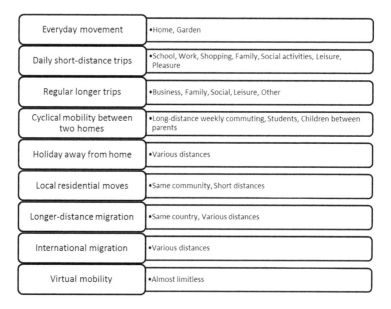

| Everyday movement | •Home, Garden |
| Daily short-distance trips | •School, Work, Shopping, Family, Social activities, Leisure, Pleasure |
| Regular longer trips | •Business, Family, Social, Leisure, Other |
| Cyclical mobility between two homes | •Long-distance weekly commuting, Students, Children between parents |
| Holiday away from home | •Various distances |
| Local residential moves | •Same community, Short distances |
| Longer-distance migration | •Same country, Various distances |
| International migration | •Various distances |
| Virtual mobility | •Almost limitity |

*Figure 13.1* The mobility continuum.
Source: Substantially adapted from Pooley *et al.* (2005, Figure 1.1).

*Table 13.1* Five families of contemporary mobility

| Type | Subject | Examples |
|------|---------|----------|
| Material | People | Work, leisure, family, safety: from quotidian mobility through internal migration to international migration |
| Material | Commodities | Raw and finished goods to producers, retailers, consumers |
| Immaterial | Imaginations | Other places via written word, photographs, film and TV, memories, conversations, dreams |
| Immaterial | Virtual worlds | Internet exploration of places |
| Material and/or Immaterial | Communications | Letters, cards, phone calls, emails, texts, online conversations (e.g. Skype) |

Source: Substantially adapted from Gale (2009, p. 133) and Larsen *et al.* (2006, p. 4).

of being. Living mobile lives (Elliott and Urry, 2010) is a condition widely promoted and even glamorised through diverse means, from the blatant seductions of advertising to a more entrenched sense that 'contemporary societies assign high social value to the consumption of distance' (Cohen and Gössling, 2015, p. 1663). From this perspective, to be immobile is thus to be a 'problem', a source of shame, embarrassment and inadequacy.

One consequence of recognising an era of mobilities in all its scope and dimensions, therefore, is the need for all migration scholars to give greater attention to people's residential relocations and their relational connections to other forms of mobility. From this it immediately follows, as Chapter 1 of this volume sensibly acknowledges, that mobilities as a concept cannot be dismissed—or supported—simply from one element alone. As already explained in Chapter 1, observations of rising mobility and declining internal migration rates are not necessarily incompatible. However, from within this 'mobile' perspective, understanding of 'migration', too, needs further critical attention.

### Re-specifying migration

The era of mobilities may raise the existential significance of migration from occasional life course disruption to a more regular part of a commonplace cultural texture of normative flux, but this perspective does not leave the concept of migration untouched—even repositioned within the continuum of mobilities—as conventionally delineated. Our concept of migration also feels the force of the de-differentiating wave that articulates mobilities' 'liquefaction of *social forms*' (Gale, 2009, p. 132), since these forms include socially constructed cognitive objects (Halfacree, 2001).

As defined in a recent textbook, migration is 'the movement of people to live in a different place' (Holdsworth *et al.*, 2013, p. 96). It is a 'permanent change in residence', as deemed in most censuses to have occurred when one's 'usual address' is different from that of one or five years ago. Nonetheless, as Holdsworth *et al.* (2013, p. 98) also noted, careful consideration of these definitions immediately raises a host of questions over the precise meaning of terms such as 'different place', 'live in', 'permanent' and 'usual address'. Such terms do not escape the attention of the mobilities critique.

In particular, mobility scholarship's rejection of an assumption of a sedentarist norm also causes a questioning of the component terms of migration as conventionally understood. Now widely recognised (e.g. Cresswell, 2006; Gustafson, 2014) and reinforced by philosophical reflections on human dwelling (e.g. Heidegger, 1971), 'sedentarism' expresses the idea that being still, bounded and 'authentic' through being-in-place is a foundational feature of human life. In consequence, mobility is regarded with suspicion. It is at root 'inauthentic', even potentially deviant, inherently disruptive of normal settled states of affairs (Cresswell, 2006). Thus, people 'live' 'permanently' at a single 'usual address' unless residential relocation impels them to a 'different place' where a (re-)building of sedentarist roots automatically begins again.

Rejection of an *assumption* of sedentarism—as opposed to recognizing it as an achievement to be worked at—also rejects the certainty that one can always recognise a single 'usual address'. It likewise throws into the air the notion that migrations are 'permanent'. Indeed, as most people move residence many times in their lives, how can anyone ever declare a move to be permanent? Furthermore, with this implied fuzzier sense of both place and time, the notion of 'living in' somewhere and the boundedness of 'different places' become equally uncertain. In short, the mobilities paradigm works to undermine the predominant significance of the empirical fact of residential relocation from Point A to Point B.

This critique and reappraisal can be taken even further to challenge the taxonomic delineations that surround and regulate migration scholarship (Halfacree, 2001) in three ways. Firstly, attention can be paid to the enduring distinction (as in the present book) between internal migration and international migration (King, 2002). Whilst the act of crossing a national boundary is certainly very likely to be highly significant for a migration and its experience, its core and primary importance is not necessarily true *a priori*. Favell (2008a, p. 270, emphasis added) has argued how the 'defining' role of state boundaries can be over-emphasised since 'the world is not *only* one of nation-state units'. For example, for lifestyle migrants (discussed below), whether internal or international, their urban-to-rural relocation may well be more significant to their daily lives than the international scale of the relocations of Favell's (2008b) economically elite *Eurostars in Eurocities*.

Secondly, within just internal migration, what is seen as migration is often itself separated from short-distance residential mobility. This, too, can be problematic if also assumed to be a hard divide (Coulter *et al.*, 2016). As noted in Chapter 3 of this book, research has suggested that there is often 'no obvious or easily defined cut off between local and long distance migration', whilst in Chapter 4 Bell and colleagues found on their IMAGE project how declines or rises in internal migration were frequently apparent at all spatial scales.

Thirdly, even the distinction between (internal) migration and more quotidian mobility (such as commuting for work or travelling to shops) is not to be assumed as paramount. Indeed, it can sometimes be more useful for researchers to examine migration according to themes cross-cutting both the internal/international and the migration/mobility divides, such as lifestyle-prompted movement (Cohen *et al.*, 2013; see Barcus and Halfacree, 2017), than simply to work within pre-ordained categories. In the era of mobilities, these social forms are more liquid than we have tended to acknowledge.

## Internal migration within an era of mobilities

### *Every day, not everyday*

From the preceding discussion it might be concluded that internal migration acts as a relatively minor member of the mobilities cast, certainly not its star player. This needs qualifying. While internal migration represents an everyday component, occurring ubiquitously, within the general cacophony of mobile rhythms of lives that writes and reproduces the era of mobilities, such migrations are rarely every day for those involved (Schillmeier, 2011). On the one hand, internal migration is an everyday mobility, in that almost everyone migrates during their life, most of us many times. As a form of mobility, as argued above, it is thus not so clearly distinguishable or as unique as Population Geographers have tended to imply. Yet, on the other hand, internal migration is not an everyday quotidian mobility, mundane and largely taken-for-granted, in that the significance of a residential relocation is likely to be wildly considered and the action itself can have profound and long-term life course consequences (Fielding, 1992; Halfacree and Rivera, 2012). In other words, whilst arguing for internal migration to be understood relationally within its mobilities context, it must still generally be acknowledged as somewhat more existentially significant than strolling to the shops for a newspaper.

### *Neo-liberal expressions*

In the twenty-first century, neo-liberal capitalism's central demand, simply put, is predominantly for a fluid or flexible workforce, where 'flexibility' is understood in at least three ways. It is needed in terms of what tasks can be performed, with workers developing a portfolio of skills and experiences rather than pursuing one career, whether defined by job or occupation. It must be spatial, with workers willing to move almost at the drop of a hat to access the latest work opportunity and over any distance from the local to the international. Thirdly, flexibility is demanded of the life course priorities of the workers, such that existential needs for ties to people, places or practices should not impede the other two areas of flexibility.

In this context, one might immediately expect internal migration to be enhanced in these neo-liberal times, as indeed is implied by the frequent association made between neo-liberalism and mobilities. However, this interpretation is wholly inadequate, as it presents workers in the kind of atomised ways that bedevil early neo-classical migration theories (Barcus and Halfacree, 2017). In short, whilst neo-liberal capitalism might desire and prompt enhanced internal migration with one hand, with many of its other hands it can hold back such migration. There are many ways that (internal) migration can be suppressed by, or displaced by and thus dispersed among, other categories within the mobilities continuum. Three examples will suffice.

One brake on internal migration is that the process of moving house is not as straightforward as the estate agency and removal businesses would have us believe. Leaflets posted through the door may promise a no-hassle house sale but they cover only part of the relocation story (Halfacree, 2012). Moving home is widely recognised as stressful (e.g. Mann and Seager, 2007), disruptive (Fielding, 1992) and usually very costly in terms of time and money. There is also the prominent barrier of the geographical unevenness of housing costs, which are widely noted with preventing movement from less to more expensive places (Cameron and Muellbauer, 1998; Rabe and Taylor, 2010; see also Coulter, 2013). Instead of internal migration, therefore, other forms of mobility may be adopted to compensate. The most obvious example is the rise of long-distance commuting (Green *et al.*, 1999), facilitated by developments in transport mobilities, but others include such unstable and temporary living arrangements as 'sofa surfing' (Schwartz, 2013).

Secondly, fluidity in the character of jobs undertaken—flexible jobs, zero-hours contracts, employment precarity—can undermine the incentive to migrate if the job in question is consequently seen as insecure or unrewarding. Whilst the idea of economic calculation within the migration decision-making process has been widely critiqued (Barcus and Halfacree, 2017), a perhaps more qualitative sense of 'is it worth it?' undoubtedly informs this process. When a job is certainly not 'for life', then what is the (rational) point of making a 'permanent' move? How potential migrants engage with risk and uncertainty (Williams and Baláž, 2012) is clearly of central significance here.

A third example is provided by the breakdown of any norm of a family having one predominant income earner or 'breadwinner'. The growth of the 'dual-career household' (Green, 1997) means that finding suitable jobs accessible through commuting from a single 'usual address' can be extremely challenging. Hence, the rise of 'dual-location households' (e.g. Green *et al.*, 1999), 'commuter marriages' (e.g. van der Klis and Mulder, 2008) and the 'living-apart-together' (LAT) relationship (e.g. Levin, 2004), all displacing a potential internal (or international) migration. The existence of such households is, as with long-distance commuting, facilitated by the development of transport mobilities. Furthermore, they demonstrate the de-differentiating force of mobilities; for example, with LAT expressing 'neither a new family form... nor... a simple reaction to constraints' (Duncan *et al.*, 2013, p. 337) but new flexible articulations of inter-personal relationships.

### Resisting neo-liberalism and de-differentiation

In the era of mobilities, as just seen, internal migration is normatively promoted in response to demands for economic flexibility, but then in practice undermined by other aspects of neo-liberalism and either blocked or displaced into other mobilities. In addition, negative existential consequences of the contemporary mobilities experience (Cohen and Gössling, 2015) can prompt both migration and non-migration as critical and resistant rather than compliant and acceding responses. Indeed, focusing simply on movement forms only part of the scope of the mobilities paradigm. In the words of Sheller (2011, p. 1), mobilities research 'emphasises the relation of such mobilities to associated immobilities and moorings, including their ethical dimension; and it encompasses both the embodied practice of movement and the representations, ideologies and meanings attached to both movement and stillness'.

Resisting de-differentiating mobilities can—at first sight, rather paradoxically—prompt other forms of internal migration, thereby somewhat ironically re-inscribing a positive association between mobilities and internal migration. In particular, whilst a strong individualism as well as (allied) neo-liberalism may stimulate the mobilities condition (Bauman, 2007), this can also promote more social- or community-seeking responses, as with many of Duncan *et al.*'s (2013) LATs envisaging future co-habitation. At least two further responses merit fuller discussion.

One mobilities expression of resistance to de-differentiation stems from the potential for social worlds to be ever more geographically scattered, not least due to numerous mobilities developments that range from improved transportation, allowing longer distances to be travelled regularly, to social media and the internet prompting ever-distant social links. Consequently, the most significant and valued social links to immediate family and close friends are often no longer tied down to local, regional or even national scales (Larsen *et al.*, 2007). Social media and other communications mobilities appear insufficient to keep these links flourishing. Embodied propinquity is still needed, expressed by enhanced Visiting Friends and Relatives (VFR) travel. Janta *et al.* (2015) associates this type of mobility with five types of practice. As well as maintaining social relationships, it also about care provision (e.g. to elderly parents), affirming or even discovering place-based roots and identities, asserting territorial rights (e.g. for voting) and pursuing leisure and tourism activities. All but the last of these express a critical response to mobilities' de-differentiating liquefactions and neo-liberalism's abstraction of the individual.

Secondly, reduced ability to entangle oneself in a place-based community—or, as put by Cohen and Gössling (2015, p. 1672), 'decreasing time for co-present social life at home and locally'—is a further existential experience consequent from enhanced mobilities. This emerges not only from the need for the VFR mobilities just mentioned, but also from the extensive time-space demands of hyper-mobile business travel and from a more general flexible precarious economic existence. The resulting social or communitarian cost of mobilities can be manifested critically in many ways, including through a rising 'desire for connectedness' (Gössling and Stavrinidi, 2016, p. 724) and a 'rootedness' that, as Cooke (2011, 2013; see also Chapter 5 of this book) observes for the USA, is not just driven by material priorities.

Such desire for (re-)connection is expressed particularly strongly through the imaginative geographical lure of 'a place in the country' (Halfacree, 2008). This refutes liquid modernity's treatment of space as 'ceas[ing] to count for much at all' (Gale, 2009, p. 132) by (re-)emphasizing rurality's status as a source for articulating a critical form of place consumption. Specifically, as providing metaphorical 'bolt-holes', 'castles' or 'life-rafts', consuming rural places through residence can express 'critical responses to mainstream everyday life' (Halfacree 2010, p. 250). The mobilities associated with such forms of rural consumption range from those linked to accessing rural leisure and living within second homes (e.g. Halfacree, 2012) to more permanent counter-urban relocations in search of an assumed more sedentarist rural *gemeinschaft* existence (e.g. Halfacree, 2008; Halfacree and Rivera, 2012). Thus, Cognard (2014, p. 216) could depict even relatively poor urban residents relocating to rural upland areas of France as being motivated in part by the lure of 'a place that is reassuring in its permanence in these uncertain times'. The burgeoning lifestyle migration literature (e.g. Benson and Osbaldiston, 2014) illustrates these pro-rural quests extremely well. Even the amenity migration literature is now recognising how the appeal of many rural places is often their supposed promise of the 'slow life' and 'stillness' as much as their more active recreational offer (Moss, 2014).

Taken together, these two responses illustrate how many forms of what Cohen *et al.* (2015) term 'lifestyle mobilities' may be facilitated by Table 13.1's five families of contemporary mobilities but nonetheless express a critical narrative on the overall 'liquid' condition. It is a narrative with which internal migration is deeply enmeshed. However, critique may also be expressed through non-migration, through efforts to try to stay put and dwell within relatively established and secure locally emplaced moorings. In other words, the presence of non-migration must not be seen solely in terms of constraints preventing relocation—although these are very widespread,

*Table 13.2* Internal migration and neo-liberal mobilities

|  | Increase in internal migration | Decrease in internal migration |
|---|---|---|
| Neo-liberal consequences: mobilities | Flexible work and workers—precarity<br>Normative 'nomadic' identities | Displacement to other mobilities<br>Generalised conditions—no point to moving<br>Monetary costs, geographically highly variable |
| Neo-liberal resistances: community | Enhanced importance of visiting friends and relations<br>Second home consumption<br>Pro-rural lifestyles | Staying put—building place-based communities<br><br>Dropping-out—Brexit? |

Source: Compiled by the author.

as noted above (e.g. Cooke, 2013; Coulter, 2013)—but as an expression of asserting more 'rooted' social forms of dwelling. Hence Italy's more 'familial' society promoting varied forms of commuting more than internal migration, documented by Bonifazi and colleagues in Chapter 11 of this book.

In summary, there is no clear or singular relationship between our present mobile times and moving house. Table 13.2 therefore attempts to bring together the diversity of internal migration experiences illustrated in this book in the context of their associations with neo-liberal mobilities. It shows how positive association between internal migration and mobilities is but one box from four. In addition, mobilities can be expressed in reduced migration due to displacement, lack of necessity or costs. Furthermore, both heightened and reduced internal migration can resist neo-liberal mobilities through seeking to access what is presumed lost or simply opting out of migration practices, respectively. It is therefore clear that, whilst mobility today may well be an 'obligation', how individuals and families fulfil it is very variable indeed.

## Conclusion: beyond mobility saturation

It is clear that internal migration has a central place within any present-day era of mobilities and will continue to do so into the foreseeable future. However, this chapter has argued that it is simply too one-dimensional to expect to see any clear positive relationship between the two. The whole basis of any mobilities era or *zeitgeist* is that the whole is greater than the sum of its individual parts: 'mobilities' is much more than internal migration or even migration in total. Indeed, the chapter identified an ambiguous relationship, with the former sometimes encouraging and facilitating the latter but at other times discouraging and preventing it. At the same time, both migration and non-migration may be seen as attempts to resist and counter the fluid logics of the era of mobilities.

   In conclusion, nearly half a century ago Zelinsky was extremely perceptive with his assertion that the idea of humans being sedentary was no longer satisfactory. Sedentarism is always in dialogue with nomadism (Deleuze and Guattari, 1987) and how we dwell today will thus implicate countless forms of mobility, of which internal migration remains a major player. Yet the present era of mobilities simultaneously highlights the limits of the mobile nomadic life, as was also hinted at by Zelinsky:

> it is more difficult to fix an effective upper limit to human mobility, even if the phenomenon is obviously finite. Is there a point beyond which mobility becomes counterproductive economically and socially or even psychologically and physiologically? ... When and how will mobility saturation be reached?
>
> (Zelinsky, 1971, pp. 247–248)

It is important to recognise that mobility is prevented for many and is also resisted for its consequences. Indeed, to dwell in an existentially satisfying manner in the twenty-first century requires much more effort in terms of producing both settlement and mobility practices, including those of internal migration, than has been realised to date.

## References

Adey, P. 2010. *Mobility*. London: Sage.

Barcus, H. and Halfacree, K. 2017. *An Introduction to Population Geographies: Lives Across Space*. London: Routledge.

Bauman, Z. 2000. *Liquid Modernity*. Cambridge: Polity.

Bauman, Z. 2007. *Liquid Times*. Cambridge: Polity.

Bell, M. and Ward, G. 2000. Comparing permanent migration with temporary mobility. *Tourism Geographies*, 2, 97–107.

Benson, M. and Osbaldiston, N. (eds) 2014. *Understanding Lifestyle Migration*. Basingstoke: Palgrave Macmillan.

Brettell, C. 2013. Anthropology of migration. In Naess, I. (ed) *The Encyclopedia of Global Human Migration*. New Jersey: Wiley-Blackwell. http://onlinelibrary.wiley.com/doi/10.1002/9781444351071.wbeghm031/abstract, accessed 7 Aug 2016.

Brettell, C. and Hollifield, J. (eds) 2008. *Migration Theory: Talking Across Disciplines*. 2nd edition. London: Routledge.

Cameron, G. and Muellbauer, J. 1998. The housing market and regional commuting and migration choices. *Scottish Journal of Political Economy*, 45(4), 420–446.

Castles, S. and Miller, M. 1993. *The Age of Migration*. Basingstoke: Macmillan.

Castles, S., de Haas, H. and Miller, M. 2014. *The Age of Migration*. 5th edition. Basingstoke: Palgrave Macmillan.

Cognard, F. 2014. Forgotten faces of amenity migration: Poor migrants moving to the uplands of France. In Moss, L. and Glorioso, R. (eds) *Global Amenity Migration*.

British Columbia, and Port Townsend, Washington State: New Ecology Press, 203–218.

Cohen, S. and Gössling, S. 2015. A darker side of hypermobility. *Environment and Planning A*, 47(8), 1661–1679.

Cohen, S., Duncan, T. and Thulemark, M. 2013. Introducing lifestyle mobilities. In Duncan, T., Cohen, S. and Thulemark, M. (eds) *Lifestyle Mobilities: Intersections of Travel, Leisure and Migration*. Farnham: Ashgate, 1–18.

Cooke, T. 2011. It is not just the economy: Declining migration and the rise of secular rootedness. *Population, Space and Place*, 17(3), 193–203.

Cooke, T. 2013. Internal migration in decline. *Professional Geographer*, 65(4), 664–675.

Coulter, R. 2013. Wishful thinking and the abandonment of moving desires over the life course. *Environment and Planning A*, 45(8), 1944–1962.

Coulter, R., Van Ham, M. and Findlay, A. 2016. Re-thinking residential mobility. Linking lives through time and space. *Progress in Human Geography*, 40(3), 352–374.

Cresswell, T. 2006. *On the Move: Mobility in the Modern Western World*. London: Routledge.

Cresswell, T. and Merriman, P. (eds) 2011. *Geographies of Mobilities*. Farnham: Ashgate.

Deleuze, G. and Guattari, F. 1987. *A Thousand Plateaus*. Minneapolis, MN: University of Minnesota Press.

Duncan, S., Carter, J., Phillips, M., Roseneil, S. and Stoilova, M. 2013. Why do people live apart together? *Families, Relationships and Societies*, 2(3), 323–338.

Elliott, A. and Urry, J. 2010. *Mobile Lives*. Abingdon: Routledge.

Favell, A. 2008a. Migration theory rebooted. Asymmetric challenges in a global agenda. In Brettell, C. and Hollifield, J. (eds) *Migration Theory. Talking Across Disciplines*. 2nd edition. London: Routledge, 318–328.

Favell, A. 2008b. *Eurostars in Eurocities. Free Movement and Mobility in an Integrating Europe*. Oxford: Blackwell.

Fielding, A. 1992. Migration and culture. In Champion, T. and Fielding, T. (eds) *Migration Processes and Patterns. Volume 1. Research Progress and Prospects*. London: Belhaven Press, 201–212.

Gale, T. 2009. Urban beaches, virtual worlds and 'the end of tourism'. *Mobilities*, 4(1), 119–138.

Gössling, S. and Stavrinidi, I. 2016. Social networking, mobilities, and the rise of liquid identities. *Mobilities*, 11(5), 723–743.

Green, A. 1997. A question of compromise? Case study evidence on the location and mobility strategies of dual-career households. *Regional Studies*, 31, 641–657.

Green, A., Hogarth, T. and Shackleton, R. 1999. *Long Distance Living. Dual Location Households*. Bristol: Policy Press.

Gustafson, P. 2014. Place attachment in an age of mobility. In Manzo, L. and Devine-Wright, P. (eds) Place Attachment: Advances in Theory, Methods and Applications. London: Routledge, 37–48.

Halfacree, K. 2001. Constructing the object: Taxonomic practices, counter urbanisation and positioning marginal rural settlement. *International Journal of Population Geography*, 7, 395–411.

Halfacree, K. 2008. To revitalise counter urbanisation research? Recognising an international and fuller picture. *Population, Space and Place*, 14, 479–495.

Halfacree, K. 2010. Reading rural consumption practices for difference: Bolt-holes, castles and life-rafts. *Culture Unbound*, 2, 241–263.

Halfacree, K. 2012. Heterolocal identities? Counter-urbanisation, second homes and rural consumption in the era of mobilities. *Population, Space and Place*, 18, 209–224.

Halfacree, K. and Rivera, M.J. 2012. Moving to the countryside... and staying: Lives beyond representation. *Sociologia Ruralis*, 52, 92–114.

Heidegger, M. 1971. Building dwelling thinking. In Heidegger, M. (ed) *Poetry, Language, Thought*. Trans. A. Hofstadter. London: Harper and Row, 145–161.

Holdsworth, C., Finney, N., Marshall, A. and Norman, P. 2013. *Population and Society*. London: Sage.

Janta, H., Cohen, S. and Williams, A. 2015. Rethinking visiting friends and relatives mobilities. *Population, Space and Place*, 21(7), 585–598.

King, R. 2002. Towards a new map of European migration. *International Journal of Population Geography*, 8, 89–106.

Larsen, J., Urry, J. and Axhausen, K. 2006. *Mobilities, Networks, Geographies*. Aldershot: Ashgate.

Larsen, J., Urry, J. and Axhausen, K. 2007. Networks and tourism: Mobile social life. *Annals of Tourism Research*, 34(1), 244–262.

Levin, I. 2004. Living apart together: A new family form. *Current Sociology*, 52(2), 223–240.

Mann, S. and Seager, P. 2007. *Upping Sticks: How to Move House and Stay Sane*. Devon: White Ladder Press.

Moss, L. 2014. The rural change agent amenity migration: Some further explorations. In Moss, L. and Glorioso, R. (eds) *Global Amenity Migration*. Kaslo, British Columbia, and Port Townsend, Washington State: New Ecology Press, 11–30.

Pooley, C. and Turnbull, J. 1998. *Migration and Mobility in Britain since the Eighteenth Century*. London: UCL Press.

Pooley, C., Turnbull, J. and Adams, M. 2005. *A Mobile Century?* Aldershot: Ashgate.

Rabe, B. and Taylor, M. 2010. Differences in opportunities? Wage, unemployment and house-price effects on migration. Institute for Social and Economic Research, University of Essex, Working Paper No. 2010–05. www.iser.essex.ac.uk/research/publications/working-papers/iser/2010-05.pdf, accessed 7 Aug 2016.

Schillmeier, M. 2011. Unbuttoning normalcy—on cosmopolitical events. *Sociological Review*, 59, 514–534.

Schwartz, M. 2013. Opportunity costs: The true price of internships. *Dissent*, 601, 41–45.

Sheller, M. 2011. Mobility. *Sociopedia.isa*. www.sagepub.net/isa/resources/pdf/mobility.pdf, accessed 7 Aug 2016.

Sheller, M. and Urry, J. 2006. The new mobilities paradigm. *Environment and Planning A*, 38, 207–226.

Urry, J. 2002. *The Tourist Gaze*. 2nd edition. London: Sage.

Urry, J. 2007. *Mobilities*. Cambridge: Polity Press.

Van der Klis, M. and Mulder, C. 2008. Beyond the trailing spouse: The commuter partnership as an alternative to family migration. *Journal of Housing and the Built Environment*, 23(1), 1–19.

Williams, A. and Baláž, V. 2012. Migration, risk, and uncertainty: Theoretical perspectives. *Population, Space and Place*, 18(2), 167–180.

Zelinsky, W. 1971. The hypothesis of the mobility transition. *Geographical Review*, 61, 219–249.

# 14 Conclusions and reflections

*Tony Champion, Ian Shuttleworth and Thomas Cooke*

The central question that the book set out to answer (see Chapter 1) was whether the USA was alone amongst high-income countries in experiencing a decline in internal migration rates or whether it was an example of a generalised trend. It seemed unlikely that the USA was a unique case, given that the factors put forward to explain this decline are many and varied, as illustrated by the list of 15 explanations presented in Table 1.1. Even that array was not exhaustive, as in Chapter 5 Cooke demonstrated the importance of a further dimension, namely cohort effects relating to changes in the relative size of younger and older working age groups resulting from previous baby booms and busts. While some of these explanations for the slowing of migration might be considered particular to the United States (such as state differentials with regard to health insurance and professional registration), the vast majority are forces that can be observed to be in play quite widely across the developed world. The USA has not been alone in experiencing such societal developments as de-industrialisation, population ageing and the increasing use of advanced ICT and so, to the extent that other countries have been through the same processes, they might also be expected to share similar patterns of internal migration decline. If the same migration trends seen in the USA are observed across all the case-study countries, it is interesting. However, it is also interesting (for different reasons) if there are differences in trends across the selection. Either way, the findings will reveal something about how internal migration is related to stage of development and also address those theoretical contributions that assume that migration will increase through time as national income increases.

These considerations therefore suggested to us the need for comparative international research and provided the motive for examining internal migration levels and trends in depth in the seven case study countries included in this book. There are some caveats. If we were to have found migration declines in all these seven countries, this would not necessarily imply that internal migration decline is being experienced across all advanced high-income economies. After all, information is not available in the same depth for all high-income countries. Likewise, if one or more case-study countries have not seen migration rate decreases, this does not disprove that something more general is happening: it might be that general downward pressures on migration are being counteracted there or do not work with the same vigour as in other countries. With these limitations in mind, what does the collective evidence of this book show, and what is the best judgement that can be made about the role of general forces operating across national boundaries as opposed to that of possibly widely divergent national experiences? This concluding chapter summarises the book's findings on the trends in internal migration rates observed in the seven countries before reflecting on their significance for theory and for a future research agenda.

## Observed trends in internal migration rates

The first point to make is that the country studies seem to have met with rather variable success in tracking down data with which they felt confident about piecing together long-term trends in internal migration intensities. This patchiness was foreshadowed in the IMAGE-based work, though—as reported in Chapter 4—through its sterling efforts this project has now succeeded in getting some data for nine-tenths of the 193 UN member states and data for at least three of the four census rounds 1980–2010 for around a quarter of them. Even so, some of the case studies have been able to set the experience of recent decades into a much longer time frame, with an annual time series for Sweden back to 1900 (Figure 9.1), Italy back to 1929 (Figure 11.1), USA back to 1947 (Figure 5.2) and Japan back to 1959 (Figure 8.1). The chapter on Australia picks up the story from 1971 (effectively 1966 as the data refer to five-year migration, see Table 7.1) and the UK's primarily from 1975 (Figure 6.1, though there is one-year migration data back to the 1961 Census), while data on Germany as currently constituted obviously runs only from reunification (though from 1974 for West Germany, see Figure 10.1).

A second type of patchiness arises from the fact that the migration data for these countries are rarely measuring the same phenomenon, as was made clear in Chapter 3, in its discussion of the challenges facing international comparative study. Among our seven countries, the most common type of time-series data is that derived from administrative records that count all address-changes that cross a boundary between places (normally municipalities and equivalent local government areas), with the problem that these places can vary greatly in size, thus affecting the chances of an address-change crossing such a boundary. By contrast, data from censuses and other surveys that are derived from comparing current usual address with one a fixed time ago normally cover local residential mobility as well as between-area migration but miss any multiple or return moves made within the reference period, with this latter problem increasing in severity as the period lengthens. The soundest basis for comparing migration intensities between countries is for all distances of move over the shortest possible reference period, such as the one-year Crude Aggregate Migration Intensities shown in Figure 4.1. Happily, either by direct measurement or by estimation using the Courgeau method (see Chapter 3), this basis allowed all our seven countries to be ranked on their latest overall address-changing rate, putting the USA at the top, closely followed by Australia and Sweden and then with progressively lower rates for the UK, Germany, Japan and Italy (see also Table 1.2).

Our key interest in this book, however, has been in how migration intensities have changed significantly over time and in whether any such changes have been in a similar direction for all types of migration as defined in terms of distance of move (normally proxied by the granularity of the reporting geographies with fewer areas meaning a greater share of longer-distance moves). Table 14.1 attempts to answer these two questions by using the information provided by the seven country case studies (Chapters 5–11) to build on the relevant results from the IMAGE project presented in Chapter 4. With the latter being shown in bold type, it is clear that the case studies have provided a wealth of extra detail, extending the picture back before the 1980s and also including a wider range of migration types. The latter is partly based on additional calculations made by us, notably in terms of pinning down specific 'distances' of move, such as the 'within county' and 'between county within state' types in the US case. Nevertheless, both the variation between countries in the number of migration types listed and the numerous blank cells in the table are testament to the fact that, despite the best efforts of our researchers, considerable patchiness still remains—a point that we will return to later when we set out our thoughts on priorities for further research.

The most robust and readily interpretable approach to comparing the seven countries is, as just mentioned, on the basis of all address-changes: it is therefore a shame that sufficiently reliable data on this measure were not available for the three countries currently with the lowest overall rate, except in the 2000s for Italy and Japan. As regards this latest decade, this measure indicates that the USA was by that time not alone in experiencing a fall in total migration rate, with significant reductions (i.e. by at least 5%) being found for Australia and Japan too and also being likely for Germany, judging by the significant fall in its between-municipality rate which, given that most migration is short-distance, is the nearest equivalent to the overall rate. By contrast, Italy saw a rise of more than 5% in its total migration rate between the start and end of the 2000s (according to Table 11.3), while Sweden and the UK are classified as stable as their rates for 2010 were almost identical to those for 2000 (see Table 4.3). The overall impression gained from this information is that there was a tendency towards convergence during this decade, with the falls for the two highest-rate countries, stability in the middle order and the rise for lowest-rate Italy. Moreover, the actual rates (not presented in the table for simplicity's sake) show that it is the falls for the USA and Australia that are leading this convergence, as these were much larger in percentage point terms than the rise for Italy. The

*Table 14.1* Trends in migration rates, by migration type, for the seven case study
countries (arranged in order of current rate of all address-changing)

| Country, and migration type (N areas) | 1960s | 1970s | 1980s | 1990s | 2000s |
|---|---|---|---|---|---|
| **USA** | | | | | |
| All address-changes | Fall | Fall | **Stable** | **Fall** | **Fall** |
| Between state (51) | Rise | Fall | **Stable** | **Fall** | **Fall** |
| Between county (3,144) | Rise | Rise | Stable | Fall | **Fall** |
| Between county within state | Stable | Rise | Stable | Fall | Fall |
| Within county | Fall | Fall | Stable | Fall | Fall |
| **Australia** | | | | | |
| All address-changes | | Stable | **Stable** | **Stable** | **Fall** |
| Between state (8) | | | **Stable** | **Fall** | **Fall** |
| Between TSD (69) | | | **Stable** | **Stable** | **Fall** |
| Between TSD within state | | | Fall | Fall | Fall |
| Within TSD | | | Stable | Rise | Fall |
| **Sweden** | | | | | |
| All address-changes | | Fall | Fall | Stable | **Stable** |
| Between county (21) | | Fall | Stable | Rise | **Stable** |
| Between municipality (290) | | Fall | Stable | Rise | Stable |
| Between parish (2,512) | Rise | Fall | Stable | Rise | **Stable** |
| Between municipality within county | | Fall | Stable | Rise | Rise |
| Between parish within municipality | | Fall | Stable | Rise | Stable |
| Within parish | | Fall | Fall | Stable | Stable |
| **UK** | | | | | |
| All address-changes | Rise | Fall | Stable | Rise | **Stable** |
| Between region (E&W, 9) | | | Rise | Rise | Fall |
| Between CHA (E&W, 80) | | | Stable | Rise | Fall |
| Between CHA within region (E&W) | | | Stable | Rise | Stable |
| <10 km (E&W) | | | Fall | Fall | Fall |
| **Germany** | | | | | |
| All address-changes | | | | | |
| Between state (16) | | | | | **Fall** |
| Between county (412) | | Fall | Stable | Rise | **Stable** |

| Country, and migration type (N areas) | 1960s | 1970s | 1980s | 1990s | 2000s |
|---|---|---|---|---|---|
| Between municipality (12,227) | | | | | **Fall** |
| *Japan* | | | | | |
| All address-changes | | | | | **Fall** |
| Between prefecture (47) | Rise | Fall | **Fall** | **Fall** | **Fall** |
| *Italy* | | | | | |
| All address-changes | | | | | Rise |
| Between province (107) | | | | **Rise** | **Fall** |
| Between municipality (8,100) | Stable | Fall | Fall | **Rise** | **Stable** |
| 50 km and over | | Fall | Fall | Stable | Stable |
| under 50 km | | Fall | Fall | Rise | Rise |
| Within municipality | | | | | Rise |

Sources: In **bold**, IMAGE-based results as reported in Table 4.3 or authors' calculations from Tables 4.6, 4.7, 4.9 and 4.10. Not in bold, based on authors' reading of Chapters 1 and 5–11.

Notes: Stable refers to a change in rate of less than 5% over the period. Blank cells indicate lack of reliable data. $N$ refers to number of statistical areas at the specified administrative level, using latest number where this has changed over time; TSD Temporal Statistical Division; CHA Consistent Health Area; E&W England and Wales only. See also text, and notes to Tables 4.3 and 4.5.

further reductions in already rather low-migration Germany and Japan, however, do not fit neatly within this interpretation.

At the same time, in taking the record back to the 1960s, Table 14.1 provides only partial support for the idea of convergence in address-changing rates across the seven countries. The picture is perhaps clearest at the bottom end of the mobility scale, where the suite of rates shown for Italy suggests that the reductions seen there in the 1970s and 1980s may have resulted in unsustainably low rates, leading to a subsequent rebound. For the USA at the other end of the scale, we have already noted (see Figure 1.1) its overall level of address-changing dropping since the early 1960s, with just a temporary lull in the 1980s, but it is evident from Table 14.1 that this was primarily driven by a slowdown in the more localised 'within-county' moves. Nevertheless, the USA's longer-distance moves seem to have reached a plateau in the 1980s and then also begun the convergence process.

Table 14.1 also makes it possible to summarise the broad pattern of changes over time across all seven countries, though caution needs to be exercised because of the coverage and number of cases varying between decades. Though only 11 observations are available for the 1960s, the main impression is of rising rather than falling migration rates, with some evidence of rises in four of the five countries with data. In the 1970s, by contrast, the vast majority of the observations indicate declining migration rates, which is perhaps a reflection of the end of the post-war economic boom in many of our countries then. The predominant trend in the 1980s is one of stability, which for some countries appears to the prelude to falling rates in the 1990s but in others seems to represent a bottoming-out process that leads to subsequent increase in rates. Finally, in the first decade of the twenty-first century, the modal trend is one of falling rates, making up 18 of the 31 observations or 58%, followed by stability (a change of less than 5% either way) and with only four cases of rising rates. How far these falls in rates between 2000 and 2010 result from economic recession, as in the 1970s, is a key question. Looking back over the five decades, however, there has been a general tendency for migration to decline across the case-study countries, although this is by no means universal either by country or distance of move.

In sum, two of the case-study countries, Australia and the USA, have seen falls in internal migration across all spatial scales. As settler countries, they have been historically associated with footloose and migratory populations, but since the late twentieth century they have seen declining internal migration. As just suggested, this pattern might perhaps be interpreted as a convergence towards the 'norm' of other countries, as their two economies and populations have matured from the rapid growth and turbulence of the nineteenth and early and middle twentieth centuries. However, these two large continental-sized countries, with their common receptiveness to free-market varieties of Capitalism, differ from the other case-study countries. Although the UK embraced free-market reform with considerable vigour and thus falls into the same camp as Australia and the USA, its migration response was not the same. Long-distance moves remained constant from the 1970s even though there is evidence that short-distance moves, over ranges of 10 km or less, have fallen. In contrast, Sweden has always been a relatively high-migration country and internal migration rates at multiple spatial scales have recently shown no signs of decrease and, if anything, the opposite is the case. For Germany, internal migration rates have remained fairly steady in recent years, albeit at a relatively low level, as flows into some areas for some age groups have increased while others have fallen. Italy's migration rates seem to have bottomed out in the 1980s at a much lower level than currently recorded by the other six countries, but Japan is continuing its long-term downward trajectory, much more akin to recent US and Australian experience despite being a much less migratory society in the recent past than those two nations.

## Interpretation and discussion

The overall migration picture revealed through examination of the case-study countries is therefore one of complexity—there appears to be no one story of migration decline that can be told in common across all the countries, and there seems to be no neat account that can be read off from abstract concepts about the transformations of high-income countries and their impact upon internal migration. However, whilst grand explanations in terms of stage of development are superficially attractive, it would be naïve to expect them to apply unproblematically in the 'real world', because the latter is just too geographically variegated for general social, economic and historical processes to have the same outcomes everywhere. Perhaps the neatest encapsulation of this notion is by Massey (1995) who noted that the historic trajectory of nations and regions through the capitalist economy—and through successive spatial divisions of labour—generates palimpsests where legacies shape current and future trends. A window into this can be opened if the case-study countries are considered in depth.

In many ways, if we look back to Table 1.2, there are a lot of commonalities between the seven case-study countries. They all are highly urbanised, all have high incomes and high human development index scores, and all have ageing populations. They might all reasonably be placed in Phase IV or Phase V of Zelinsky's schema: clearly, they share more similarities between themselves than they do with the low- or middle-income countries that predominate in Africa, Asia and Latin America. Yet, despite these commonalities when considering them as a subset of all countries, in more detail the seven countries also exhibit considerable divergences which might have implications for an understanding of the variations between them in their levels and trends in address-changing rates. These obviously include geographic size where the US and Australia dwarf the European countries and Japan, but also go beyond this aspect. The USA, for instance, has high-income inequality, whilst Sweden is one of the most even in the world. Although all seven are undergoing population ageing, some, notably Italy and Japan, have gone further along that path than others. Additionally, there are differences in political and social culture, which are harder to capture but some of which are at least partially reflected in terms of type of capitalism and the strength of the safety net for the unemployed. There are also features that are even less observable and quantifiable such as urban structure and hierarchy, political traditions and regional development patterns. European countries are diverse in their social, political and economic settlements, too, with the case study chapters suggesting that migration is facilitated in Sweden by the welfare state minimising risk, whilst in Italy family oriented culture and poorer welfare provision make internal migration, especially for the young, a much riskier proposition and hence less attractive. Context matters but the question raised in this case is how and in what ways.

In terms of Massey's palimpsest, there are good reasons why each country is different. In a geographically uneven world, each country has begun from a different starting point and has therefore followed a different path to reach its current high-income position. The UK, as the first industrial nation, followed a different route to later comers like Japan and Italy; and the experience of the European nations diverges from that of the settler societies of the

USA and Australia. These differences—and the above is not an exhaustive list—make a comparative analysis challenging, because the focus is not just on one or two easily measurable indicators but instead concerns a whole host of observed and unobserved factors. This is both one of the strengths and also one of the weaknesses of a comparative approach, especially when it is impossible to conduct a controlled experiment. The interpretation of the case-study evidence therefore relies on a sizeable addition of judgement and is more akin to a qualitative assessment than a quantitative analysis that formally tests hypotheses. Bearing this caveat in mind, what can be drawn from the country-based evidence?

Firstly, concentrating just on countries with migration decline, it is probable that either there are different drivers at work in these countries or that they are more susceptible than those others with stable or increasing migration rates to the forces pushing migration down. For example, in the USA and Australia, declines were noted across all distance bands whereas in the UK this was found to be just for short distances. This has important implications for explanation, since the correlates (and the assumed causes) of changing address vary by distance (Niedomysl, 2011). Typically, shorter-distance address-changes are prompted by housing, family and environmental factors whereas longer-distance migratory moves are more often linked to tertiary education and employment. The latter factors are often more important for younger people, those associated with short-distance moves for older people. This implies in the USA and Australia that either a number of labour and housing market forces are working in the same way to reduce migration rates across all spatial scales and ages or else there is something else—perhaps technological change (e.g. Cooke, 2011, 2013)—which is leading towards 'secular rootedness' across the board. In contrast, in the UK it is possible that change in the housing market constitutes the most important driver because of the main decline being noted for short-distances moves, with the stability of inter-regional rates since the 1970s perhaps reflecting the relative absence of fundamental change in regional economic structures and labour market opportunities compared to the USA and Australia. The evidence that the book presents is sufficient to say that the UK is not like the USA or Australia, and it permits an informed opinion as to a possible line of investigation even though it cannot fully explain the differences between these three countries. Even though the UK, the USA and Australia are exemplars of the Anglo-American model of capitalism, there are major differences in migration outcomes between them—and this indicates the limited utility of broad brush classifications in terms of exploring national differentials.

The cases of Japan and Italy differ from the other countries in that they lie towards the lower end of the internal migration intensity rankings. Indeed, both have had historically low rates of migration and the evidence presented in the book suggests that their rates have fallen further since the 1960s, though apparently bottoming out in Italy more recently. This indicates that the migration decline has not been concentrated in formerly high-migration countries that are now moving to convergence but is also happening in nations with traditions of relatively low migration. These countries are also interesting in that they show that declining migration intensity is not just a phenomenon of the Anglosphere of English-speaking societies that have largely adopted a neo-liberal economic model, but that it can also be found in countries with different social and economic systems and which have followed different routes to membership of the club of developed high-income states.

The examples of Germany and Sweden provide further contrast. In the case of Germany, it is difficult to chart internal migration over the long term because what is 'internal' has altered with changes in the borders of the country, the most recent being reunification in 1990. Nevertheless, the available evidence suggests that it has been a relatively low-migration country since the end of World War II except during the Fordist economic boom of the 1950s and 1960s. Moreover, migration rates have tended to follow the business cycle without, unlike the USA and Australia, a long-term downward trend. Finally, with the benefit of a relatively long time series, Sweden is perhaps the most distinctive of the seven countries. While it has historically been a relatively high-migration country, there has been little sign of migration decline over the last quarter of a century. Indeed, there are some signs of uplift in its rates of between-area migration and also of much more stability in its short-distance address-changing than in most of the other countries.

Despite the caveats mentioned at the outset of this section, the evidence presented in the book seems sufficient to falsify the hypothesis that internal migration decline is a feature of all advanced economies in the late twentieth and early twenty-first centuries and therefore related to stage of development. It is therefore tempting to retreat from overarching explanations couched in terms of late Capitalism, phase of Zelinsky mobility transition model or similar abstract concepts. However, this is perhaps an overhasty verdict because it might be that general forces are at work but are being experienced differently in different national contexts or that national migration and population systems are responding differently to the same forces according to country-specific factors.

It is also overhasty to throw away approaches that emphasise commonalities between countries because of the case-study findings themselves. Although there are national differences in migration trends, when the chapters are surveyed and synthesised, there are many similarities in the possible drivers of migration that are identified. These add to the literature but also raise other questions. In Italy, Australia and Japan, for instance, there is a common emphasis on the delaying of life events such as leaving the parental home and starting a family as an explanation for low or decreasing migration rates. Looking across the case studies, another theme which emerges from the UK, but is also seen in Italy, Australia and Germany, concerns rigidities in the housing market. This is interesting, not least because the housing markets in these countries are different in terms of composition and policy environment but, despite this, it seems to depress migration rates. In the Japanese and Italian case studies, changes in the regional geography of economic activity are also argued to lower migration rates, especially with the decline of manufacturing, but in Sweden recent changes in regional opportunity structures have perhaps driven internal migration upwards. The role of higher education is key in influencing the migration of the youngest (and most mobile) adult age groups and in the UK and in Sweden it has been argued that the expansion of this sector in the 1990s drove up or maintained migration rates for young people in these countries.

Considering these factors in the round, there are fertile grounds for comparative research aimed at a better understanding of how they each operate in different contexts. They also raise important conceptual questions, for while it is feasible to consider these as low-level explanations 'as they are', it may also be argued that, taken together, they are perhaps symptomatic of high-income, developed countries and so there might be a return to more abstract cross-national explanations/classifications via this route. They also raise questions for further, more targeted, empirical investigation. It would seem useful, for instance, to explore the relationship between delayed life transitions and migration events for young adults in a selection of national contexts. Likewise, the comparative operations of the housing market seem a possible further avenue. These, and other ideas, are explored in more detail in the final section.

## Looking ahead

Some future elements of a comparative research agenda can be traced in the discussion above. The provision of better data is vital as a necessary foundation. As noted in Chapter 3, there are now robust ways to make cross-national comparisons of internal migration that minimise the problems of the modifiable areal unit problem. To build on this, better time-series data are needed for a wider variety of countries. All seven of the case-study countries have time series going back at least to the 1990s, and the majority have information running from the 1960s, with Sweden and Italy having information on migration flows from the early twentieth century. Similar efforts as in this book might be made to collate and review migration data through time for other countries, starting with the other high-income countries like Canada and New Zealand and the other member states of the European Union, but then extending outwards to other world areas where records of the various types of internal migration may currently be less comprehensive and reliable.

There is also a good case to be made for using other sources such as historical census and administrative data records to try to understand internal migration in a very long-term perspective. After all, the types of questions raised in the book, especially with regard to abstract models of social change such as that of Zelinsky, realistically require data runs of considerable temporal depth in order to make full assessments. Information on migration flows between places would work well as it would fit within the techniques developed by IMAGE. However, linked individual data from censuses, administrative data sources and population registers would be the Holy Grail. They would not only permit analyses with a wide range of individual and household characteristics, but they would also allow the application of statistical techniques to explore age, cohort and period effects on change over time in migration intensities, dealing not just with descriptions and correlates but also moving towards considering causes.

Whilst this would be the 'gold standard' for research, it would not be trivial to achieve. The simplest approach would be to make use of aggregate data on migration and other demographic characteristics such as age and cohort size to extend the approach introduced in Chapter 5 by Cooke on the United States. Naturally, there would be foreseen (and unforeseen) pitfalls in data harmonisation, but concentrating on relatively easy-to-define explanatory variables such as age, cohort size, and unemployment (if defined according to international standards) could be one way to develop a more controlled quantitative comparative analysis across two or more countries.

A second approach would be to seek to use census and/or administrative data across two or more countries. This would permit techniques that require individual-level data to be used and thereby improve the prospects for rigorous comparative analysis. The experience of the case-study chapters indicates that this would be very challenging given nationally specific data sources and definitions. It should also be noted that, judging by the experience of Champion and Shuttleworth (2016b), performing a consistent migration analysis even within one country using one data source (in this case the Censuses of England and Wales) is difficult because of the shifting sands of variable definitions and scope.

Thus, although these first two options are worth exploring further, it is a third avenue that seems to us the most promising for taking forward comparative international research. This is to tap into existing cross-national surveys and also, if necessary, to seek to develop bespoke cross-national survey data. International survey resources have been an established way to tackle social science questions and there are already-existing examples of Europe-wide surveys such as the European Social Survey or the European Labour Force Survey. A thorough scoping exercise might yield some returns but, since such sources tend to put greater emphasis on international migration at the expense of internal migration, it is highly probable that dedicated new surveys will be required to answer the questions raised in the book. This position is not unique: Willekens *et al.* (2016) have suggested a number of ways of improving the robustness of international migration data, including a World Migration Survey, and it is perhaps only a small step from this to add measures of internal migration, drawing on the experience of studies such as those included in this book.

The earlier mention of age, cohort and period effects raises wider questions. One application of an analysis in these terms might be in examining the migratory fortunes of different generations within advanced economies. However, there is a much broader application which is suggested by the findings in Chapter 4 that some lower-income countries are experiencing migration declines just like most of the advanced countries that formed the case studies. This raises the possibility that the migration declines observed in the developed world are not an 'age effect' (in the sense of being driven by progression through stages of development or Zelinsky's Phases) but instead might be a general long-term period effect—a 'sign of the times'—that is operating across the world to depress internal migration rates, such as tendencies to globalisation or technological change. This supposition, of course, begs the question of what this period effect (or effects) might be. One (short-term) candidate might be global economic cycles where, for example, the 2008 Great Recession has had an impact on all national economies to a greater or lesser extent but is normally a short-term period effect that operates on the scale of a few years or a decade at the most. There may, however, be other longer period effects which are far more structural and far deeper in their impact. One instance of this is the diffusion of transport and ICT technologies which is pervasive throughout the world regardless of a country's income level. This might, for instance, consider how ICT technologies influence migration intentions and behaviour across the global development/income continuum. These operate on multi-decade temporal scales and are akin to the historical processes underlying deep-rooted economic shifts, for example to and from industrial society. This is one way to incorporate a wider political economy perspective into migration analysis but at the abstract level of long-term social, economic and political shifts. In Chapter 8 on Japan, Fielding provides a good example of how to achieve this particularly in terms of understanding how national space economies have evolved, but—as with the limitations noted above of our comparative analysis in terms of stage of development (or Zelinsky phase)—it is easy to be too abstract.

If, then, this course of action is rejected as too ambitious in a cross-national framework, there are still many research questions that need to be addressed in better understanding migration trends and their drivers in advanced economies. A number of issues have been identified in this book. These include delving deeper into the impact on migration decisions of such developments as delays in leaving the parental home and subsequent life events and transitions, immobility associated with housing markets, changing space economies and labour markets, the effect of increased participation in higher education and changing attitudes towards families and expectations for caring. There is a good case for in-depth study and analysis of each of these domains. These naturally allow a more grounded (and thus more easily handled) political-economy approach to be introduced. There are concrete differences between different national contexts in social and political regulation of domains such as education, housing and the labour market, not also to say welfare provision. One message of the book is that national differences and contexts seem to matter, but we leave hanging questions of how and why. There is thus surely scope for cross-national qualitative and quantitative analyses of internal migration that take up some of these themes in more depth.

At the same time, calling upon the contribution of Chapter 13, there are sound arguments for adopting an integrative and holistic approach to the gamut of mobility types. This approach involves viewing internal migration as relational rather than treating it as a separate domain that is isolated from other forms of mobility at different temporal and spatial scales and from the inter-related decisions that are made about these mobilities. A sensible central focus for this type of perspective might be on the family and household, especially the way that internal migration—as defined as change of usual home address—fits into life-course trajectories and how these vary between different national contexts. This type of analysis might consider how delays in life-course events and transitions are influenced by developments in housing and labour markets and how internal migration relates to these circumstances. In the same way, internal migration can be viewed as one type of possible spatial mobility, with whether it is chosen or not depending on how it relates to these other options and wider social, economic, and demographic conditions.

These considerations lead towards qualitative work and an investigation of migration intentions, community satisfaction, social capital and individual embeddedness in place. As suggested by Champion and Shuttleworth (2016a), one major unanswered question in explaining migration declines is whether this is by default or design: are there are now more 'frustrated stayers' who want to change address but cannot, or more 'happy stayers' who do not move home because they do not want to do so? A biographical approach, with the obvious caveat of inaccuracies in recall of past events and feelings, might be one avenue, but another more robust approach might be in comparing and contrasting citizens of high- and low-migration countries in their attitudes towards moving house and their degrees of place satisfaction, community satisfaction and local embeddedness.

It might reasonably be asked why there is any further need for research into the levels and trends of internal migration in different countries. The answer to this question is three-fold. Firstly, internal migration through a permanent change of address has historically been one of the main ways by which labour supply and demand have been matched at regional and sub-regional scales. Its decline may therefore have significant knock-on effects that impact negatively on labour markets and economic performance, everything else being equal. Of course, everything else might not be equal, and that might be of interest in itself. Other forms of spatial mobility might have substituted for address-changing in the labour market or else patterns of labour demand may have changed. However, whatever has been happening, there is a strong argument that migration declines should be of interest to economic policymakers.

Secondly, in an era that has seen rising concerns about decreasing social mobility across a range of countries, there is a need to take seriously its geographical aspects. It is well known in the literature that has explored the concept of the 'escalator region' (e.g. Fielding, 1992; Gordon *et al.*, 2015) that geographical mobility at the regional scale is associated with more rapid career progression, implying that migration declines reduce the chances of upward social mobility. There is also abundant literature on neighbourhood context and life chances at smaller spatial scales, and the ways in which people are sifted between places by address-changes which take them up, or down, the social deprivation hierarchy (Bailey, 2012; Bailey and Livingstone, 2008; Connolly *et al.*, 2007; Hedman *et al.*, 2011). The outcomes of this are complex; more people staying in place might mean more now remain stuck in social disadvantage than would have been the case in a more mobile past. Alternatively, more immobility might lead to stronger social capital in communities and some areas remaining more stable when better-off people do not now leave.

Finally, politically, less address-changing at regional and smaller spatial scales might encourage people to live more in their 'silos' as social segregation patterns become resistant to change, thereby assisting in the development of separate and parallel lives (Phillips, 2006). The possibility that this is happening is worth more analysis. All in all, it is therefore hoped that the book has answered some questions but, more importantly, that it has also raised other questions that will be the focus for a developed future research agenda.

## References

Bailey, N. 2012. How spatial segregation changes overtime: Sorting out the sorting processes. *Environment and Planning A*, 44, 705–722.

Bailey, N. and Livingstone, M. 2008. Selective migration and neighbourhood deprivation: Evidence from 2001 Census migration data for England and Scotland. *Urban Studies*, 45, 943–961.

Champion, A.G. and Shuttleworth, I. 2016a. Is longer-distance migration slowing? An analysis of the annual record for England and Wales. *Population, Space and Place,* published online in Wiley Online Library. doi:10.1002/psp.2024

Champion, A.G. and Shuttleworth, I. 2016b. Are people moving address less? An analysis of migration within England and Wales, 1971–2011, by distance of move. *Population, Space and Place,* published online in Wiley Online Library. doi:10.1002/psp.2026

Connolly, S., O'Reilly, D. and Rosato, M. 2007. Increasing inequalities in health: Is it an artefact caused by the selective movement of people? *Social Science & Medicine*, 64(10), 2008–2015.

Cooke, T.J. 2011. It is not just the economy: Declining migration and the rise of secular rootedness. *Population, Space and Place*, 17(3), 193–203.

Cooke, T.J. 2013. Internal migration in decline. *The Professional Geographer*, 65(4), 664–675.

Fielding, A.J. 1992. Migration and social mobility: South East England as an escalator region. *Regional Studies*, 26(1), 1–15.

Gordon, I., Champion, T. and Coombes, M. 2015. Urban escalators and interregional elevators: The difference that location, mobility, and sectoral specialisation make to occupational progression. *Environment and Planning A*, 47(3), 588–606.

Hedman, L., Van Ham, M. and Manley, D. 2011. Neighbourhood choice and neighbourhood reproduction. *Environment and Planning A*, 43(6), 1381–1399.

Massey, D. 1995. *Spatial Divisions of Labour: Social Structures and the Geography of Production*. Basingstoke: Macmillan.

Niedomysl, T. 2011. How migration motivations change over migration distance: Evidence on variations across socioeconomic and demographic groups. *Regional Studies*, 45(6), 843–855.

Phillips, D. 2006. Parallel lives? Challenging discourses of British Muslim self-segregation. *Environment and Planning D: Society and Space,* 24(1), 25–40.

Willekens, F., Massey, D., Raymer, J. and Beauchemin, C. 2016. International migration under the microscope: Fragmented research and limited data must be addressed. *Science*, 352(6288), 897–899.

# Index

Bold page numbers indicate tables, *italic* numbers indicate figures.

For Product Safety Concerns and Information please contact our EU
representative GPSR@taylorandfrancis.com
Taylor & Francis Verlag GmbH, Kaufingerstraße 24, 80331 München, Germany

www.ingramcontent.com/pod-product-compliance
Ingram Content Group UK Ltd.
Pitfield, Milton Keynes, MK11 3LW, UK
UKHW021017180425
457613UK00020B/965